Lecture Notes in Physics

Springer
Berlin
Heidelberg
New York
Barcelona
Hong Kong
London
Milan
Paris
Singapore
Tokyo

The Editorial Policy for Proceedings

The series Lecture Notes in Physics reports new developments in physical research and teaching – quickly, informally, and at a high level. The proceedings to be considered for publication in this series should be limited to only a few areas of research, and these should be closely related to each other. The contributions should be of a high standard and should avoid lengthy redraftings of papers already published or about to be published elsewhere. As a whole, the proceedings should aim for a balanced presentation of the theme of the conference including a description of the techniques used and enough motivation for a broad readership. It should not be assumed that the published proceedings must reflect the conference in its entirety. (A listing or abstracts of papers presented at the meeting but not included in the proceedings could be added as an appendix.)
When applying for publication in the series Lecture Notes in Physics the volume's editor(s) should submit sufficient material to enable the series editors and their referees to make a fairly accurate evaluation (e.g. a complete list of speakers and titles of papers to be presented and abstracts). If, based on this information, the proceedings are (tentatively) accepted, the volume's editor(s), whose name(s) will appear on the title pages, should select the papers suitable for publication and have them refereed (as for a journal) when appropriate. As a rule discussions will not be accepted. The series editors and Springer-Verlag will normally not interfere with the detailed editing except in fairly obvious cases or on technical matters.
Final acceptance is expressed by the series editor in charge, in consultation with Springer-Verlag only after receiving the complete manuscript. It might help to send a copy of the authors' manuscripts in advance to the editor in charge to discuss possible revisions with him. As a general rule, the series editor will confirm his tentative acceptance if the final manuscript corresponds to the original concept discussed, if the quality of the contribution meets the requirements of the series, and if the final size of the manuscript does not greatly exceed the number of pages originally agreed upon. The manuscript should be forwarded to Springer-Verlag shortly after the meeting. In cases of extreme delay (more than six months after the conference) the series editors will check once more the timeliness of the papers. Therefore, the volume's editor(s) should establish strict deadlines, or collect the articles during the conference and have them revised on the spot. If a delay is unavoidable, one should encourage the authors to update their contributions if appropriate. The editors of proceedings are strongly advised to inform contributors about these points at an early stage.
The final manuscript should contain a table of contents and an informative introduction accessible also to readers not particularly familiar with the topic of the conference. The contributions should be in English. The volume's editor(s) should check the contributions for the correct use of language. At Springer-Verlag only the prefaces will be checked by a copy-editor for language and style. Grave linguistic or technical shortcomings may lead to the rejection of contributions by the series editors. A conference report should not exceed a total of 500 pages. Keeping the size within this bound should be achieved by a stricter selection of articles and not by imposing an upper limit to the length of the individual papers. Editors receive jointly 30 complimentary copies of their book. They are entitled to purchase further copies of their book at a reduced rate. As a rule no reprints of individual contributions can be supplied. No royalty is paid on Lecture Notes in Physics volumes. Commitment to publish is made by letter of interest rather than by signing a formal contract. Springer-Verlag secures the copyright for each volume.

The Production Process

The books are hardbound, and the publisher will select quality paper appropriate to the needs of the author(s). Publication time is about ten weeks. More than twenty years of experience guarantee authors the best possible service. To reach the goal of rapid publication at a low price the technique of photographic reproduction from a camera-ready manuscript was chosen. This process shifts the main responsibility for the technical quality considerably from the publisher to the authors. We therefore urge all authors and editors of proceedings to observe very carefully the essentials for the preparation of camera-ready manuscripts, which we will supply on request. This applies especially to the quality of figures and halftones submitted for publication. In addition, it might be useful to look at some of the volumes already published. As a special service, we offer free of charge LaTeX and TeX macro packages to format the text according to Springer-Verlag's quality requirements. We strongly recommend that you make use of this offer, since the result will be a book of considerably improved technical quality. To avoid mistakes and time-consuming correspondence during the production period the conference editors should request special instructions from the publisher well before the beginning of the conference. Manuscripts not meeting the technical standard of the series will have to be returned for improvement.

For further information please contact Springer-Verlag, Physics Editorial Department II Tiergartenstrasse 17, D-69121 Heidelberg, Germany

Ingo Peschel Xiaoqun Wang
Matthias Kaulke Karen Hallberg (Eds.)

Density-Matrix Renormalization

A New Numerical Method in Physics

Lectures of a Seminar and Workshop Held at the
Max-Planck-Institut für Physik komplexer Systeme
Dresden, Germany, August 24th to September 18th, 1998

Springer

Editors

Prof. Dr. Ingo Peschel
Dr. Matthias Kaulke
FB Physik
Freie Universität Berlin
Arnimallee 14, D-14195 Berlin, Germany

Dr. Xiaoqun Wang
Institut Romand de Recherche Numérique en Physique des Matériaux
PPH 333-EPFL, CH-1015 Lausanne, Switzerland

Dr. Karen Hallberg
Centro Atomico Bariloche
8400 Bariloche, Argentina

Library of Congress Cataloging-in-Publication Data.

Die Deutsche Bibliothek - CIP-Einheitsaufnahme

Density matrix renormalization : a new numerical method in physics ; lectures of a seminar and workshop, held at the Max-Planck-Institut für Physik Komplexer Systeme, Dresden, Germany, August 24th to September 18th, 1998 / Ingo Peschel ... (ed.). - Berlin ; Heidelberg ; New York ; Barcelona ; Hong Kong ; London ; Milan ; Paris ; Singapore ; Tokyo : Springer, 1999
(Lecture notes in physics ; Vol. 528)
ISBN 3-540-66129-8

ISSN 0075-8450
ISBN 3-540-66129-8 Springer-Verlag Berlin Heidelberg New York

Printed in Germany

Typesetting: Camera-ready by the authors/editors
Cover design: *design & production*, Heidelberg

SPIN: 10731988 57/3144 - 5 4 3 2 1 0 – Printed on acid-free paper

Preface

This book deals with a method which is still not widely known, although it caused a revolution in the numerical treatment of low-dimensional quantum systems. How this started about ten years ago, is described in detail in a separate contribution. The reason for the impact it had, on the other hand, is easy to give. Density-matrix renormalization has two features which normally do not come together. With it, one can treat large systems containing, for example, several hundred spins and, at the same time, reach spectacular accuracies such as ten decimal places in ground-state energies. Thus the method combines the advantages of Monte Carlo calculations and of exact diagonalisations and it is not surprising that it was applied soon after its invention to a variety of problems, both old and new, and mostly connected with quantum chains. Parallel to that, it was extended in other directions, for example to finite temperatures, which increased the area of applications further. The number of groups using it grew also and in 1997 about sixty articles appeared in which it played an essential role.

This seemed a good time to bring the experts in the field together. On the one hand to summarize the status and to discuss open questions as well as future directions as, for example, the extension of the method to higher dimensions. On the other hand to share their experience with younger researchers interested in learning and applying the technique, including practical work on the computer. This was the concept of a four-week seminar and workshop at the Max-Planck-Institut für Physik komplexer Systeme in Dresden in the fall of 1998. It was the first meeting on density-matrix renormalization, and among its about 50 participants from all over the world were the leaders in the field. The workshop was viewed as very successful and the proposal to collect the contributions in the form of lecture notes was strongly supported from all sides, even though this had not been planned initially. This has led to the book which is presented here.

The volume is divided into two parts of roughly equal size which are closely related. The first one contains a series of seven longer lectures which present the method from various angles. This serves as a general introduction on the level of an advanced textbook. A central chapter is the second one which explains the basic principles and also contains, in line with the aims of the workshop, a computer program for a simple example. Other chapters deal with more advanced aspects, like the application to time-dependent problems or to finite temperatures. Two-dimensional systems, both classsical and quantum-mechanical, are also discussed. Finally, to put things into a broader perspective, there are contributions on Wilson's original renormalization method as well as on variational approaches.

The second part deals with the application of the technique to various physical problems, arranged in five different groups. The main themes of the first part appear again, but also some additional aspects. However, the contributions are shorter and the focus is somewhat different. The physical questions and properties stand in the center and the method is in general only briefly discussed, except in those cases where new aspects come in. In this way, the second part is also a review of actual one-dimensional problems, interesting questions and their present status. Of course, it reflects the research topics of the participants, but it covers a large portion of the applications and thus gives a very good overview even though the additional posters are not included.

Throughout the book considerable effort has been made to obtain a clear, structured and self-contained presentation of the material, but also as much coherence as possible. Due to the structure, a certain amount of duplication in the references was inevitable, but we think that it can be accepted. We thank all authors for their, sometimes enthusiastic, cooperation in the project. We are particularly glad that Steven White agreed to write down his personal recollections on how it all began.

Of course, the whole undertaking would have been impossible without the Max-Planck-Institut in Dresden and its outstanding facilities. We thank its director, Peter Fulde, and its staff for their constant support, and also Springer-Verlag which agreed to publish the book in this well-known series. We hope that, besides being a source for everyone interested in the topic, it will also stimulate new research and thereby contribute to the further development of this field.

Berlin, April 1999

Ingo Peschel
Xiaoqun Wang
Matthias Kaulke
Karen Hallberg

Contents

How It All Began: A Personal Account

Steven R. White

Department of Physics and Astronomy, University of California
Irvine, CA 92697-4575, USA

In this chapter I'd like to take the opportunity to relate some of the history leading up to the development of DMRG. I think it is always interesting to look back, from a perspective where we know what works and what doesn't, and see the sometimes tortured path that got us here. The history I present here is a personal one: the work I was involved in, and the development of my ideas, which eventually led me to develop DMRG. I am not attempting to give proper credit to others who may have had similar ideas earlier, which an ordinary review should do – for that I refer to other chapters in this book. I will include many incidental details which one would not find in a review. Some of the events that happened along the way I found quite amusing, and I hope you will too.

Much of the development of DMRG traces back to my years as a graduate student at Cornell, from 1982 to 1987. I had first wanted to work for Michael Fisher, who worked on statistical mechanics and phase transitions, but during the first year that I would be working on my thesis, he was going to be on sabbatical at Caltech and so he wasn't taking any new students.

During the fall of 1983, the start of my second year at Cornell, Ken Wilson announced a set of lectures on "Chemistry". In these lectures, Wilson described an approach to solving the electronic structure of atoms and molecules using renormalization group and other ideas borrowed from particle physics and statistical mechanics. None of Wilson's approach had been carried out beyond the planning stages, and it is a testament to Wilson's prestige at the time that he could announce a set of lectures on what he hoped to do on a very difficult problem, and have everyone come. Wilson was strongly influenced by his father, E. Bright Wilson, a noted chemist at Harvard, who, among other things, wrote one of the first quantum mechanics books with a chemistry orientation in collaboration with Linus Pauling. This influence led Wilson to have as a long term goal the development of new electronic structure techniques, and Wilson's only publication during his first several years as an assistant professor at Cornell was on "Model Atomic Hamiltonians" [1], in which a lattice model was introduced for first and second row atoms.

I had worked for a chemist for three years as an undergraduate at UC San Diego, doing molecular dynamics simulations. (Coincidentally, this chemist's name is Kent Wilson. Kent, like Ken, is an unsual and remarkable person.) I thought Wilson's plan sounded like a good thing to work on, and I approached him about being his student. I remember one of the first things he said: "You know there will be some computer work involved?" I assured him that that was right up my alley. However, he told me that he was only going to have shared students, not students of his own, so I asked John Wilkins, who worked on a variety of topics in solid state theory, if he would be my other thesis advisor. He agreed, and so I was a shared student with Wilson and Wilkins, and the eventual title of my thesis was "New methods for electronic structure calculations."

The main ideas of Wilson's plan were: 1) One can use RG language to describe very weak bound states of model "atoms". 2) With Grassmann algebra, one can integrate out fermions in an imaginary time path integral, and thus apply renormalization group ideas to electronic structure. This involved setting up a fine real-space lattice around a molecule, and then applying block-spin RG methods, using Grassmann integration, to "decimate" the lattice until the lattice spacing was comparable to an atomic radius. At that point, the problem would become nonperturbative, and decimation would stop. 3) The final part of the problem would be to apply the Langevin equation as a simulation tool to finish the problem. The Langevin approach had been developed by Klauder and others as a sort of early quantum Monte Carlo method. The effective action from the blocking procedure would be used as input to the Langevin equation.

When I started, I really didn't understand the big picture of what we were doing, but I was willing to dive in right away and learn as I went along. I spent a year working on a toy model which tested the block-spin approach with Grassmann variables, and then I spent a year working on the Langevin approach. This second year was actually spent at UC Santa Barbara – Wilkins had taken six students with him to spend a year there. At the end of this time, however, we realized the entire project was a failure. We ran into two serious, seemingly insurmountable problems: first, we weren't able to devise an effective method for integrating out core electrons in the Grassmann RG approach; and second, we found that the Langevin approach became numerically unstable and also gave the wrong answer, due to something similar to the fermion minus sign problem in quantum Monte Carlo.

I spent my last year at Cornell trying out some other approaches to electronic structure. Wilson didn't seem especially interested in these, and so I mostly talked to Wilkins about them. One of these was a variation on the configuration interaction method used in quantum chemistry. This approach is interesting in retrospect because a little of the flavor of DMRG is there: the wavefunction was expanded in an optimal way using the largest eigenvalues and eigenvectors of a matrix. I also worked on a variation of quantum Monte Carlo in the canonical ensemble and in applying the finite element method to electronic structure. Wilson continued to be interested in electronic structure methods in general, and he attended a regular group meeting of mostly Wilkins people in which this was discussed.

In the fall of my last year at Cornell, Wilson gave an interesting talk on the application of his numerical RG approach, which he developed for the Kondo impurity problem, to 1D lattice systems. This approach used a Hamiltonian, as opposed to an Lagrangian, and there was no extra time dimension. Another student of Wilkins at that time was Barbara Jones, who was applying this approach to two Kondo impurities. I had talked with Barbara about this method and thought it was very interesting. After Wilson had applied it to impurity problems, others applied it to 1D lattice systems, where it was called "real space RG". Wilson did not think very highly of real space RG, and in this talk Wilson used a toy model – a discretized particle in a 1D box – to analyze why it failed. Wilson showed how the truncation of the interactions at the edge of a block was unjustified, and led to nonsensical results. In pictorial terms, ignoring the interactions between blocks means that the retained states vanish at the edge of the blocks, and any subsequent states at later iterations always have kinks where the two blocks were joined. How would one fix this problem? Wilson suggested correcting for the neglected terms using a complicated form of perturbation theory. However, he concluded that this would be very difficult to carry out even in the 1D noninteracting case.

Perhaps Wilson was hoping someone would try out his ideas for fixing real space RG, but it was to be five years before this happened.

I had gotten to know Doug Scalapino, Bob Sugar, and their students quite well during our year at UCSB, and in September of 1987 I went back to UCSB for a post doc with Scalapino. There I focused on quantum Monte Carlo (QMC), Hubbard models, and in general the field of strongly correlated electrons and high temperature superconductivity, which I didn't know very much about at the start. We made good progress in improving the Monte Carlo methods, and produced some influential papers on the Hubbard model. For example, we found some of the first indications that d-wave pairing, rather than s-wave pairing, was prevalent. After two years as a post doc, I went to UC Irvine as an Assistant Professor.

While still at UCSB and for my first year or two at UCI I worked on methods for extracting dynamics from Monte Carlo. Some of our early results using a method I developed prompted Richard Silver and Jim Gubernatis to apply the maximum entropy method, which turned out to be superior. I switched to using maximum entropy coupled with QMC, and studied some of the dynamical properties of the Hubbard model. However, I also wanted to become more familiar with exact diagonalization methods, and a student and I developed a Lanczos program to study dynamics on 4 × 4 Hubbard systems.

During this period, I would occasionally try out some new idea to solve the fermion minus sign problem, the main bottleneck in the QMC method for fermions. Each of these attempts failed. (I would also occasionally hear about attempts by others to fix the minus sign problem, with at best only partial success.) For example, my student Marco Vekić and I tried a variation on the world-line QMC method in which the basic unit was a block of sites rather than a single site. We found that this actually made the minus sign problem much worse – Marco found that there was a new type of "internal" minus sign which isn't there when you work with single sites! I began to conclude that the minus sign problem was really quite fundamental, and not likely to allow any easy solutions. I was also getting discouraged by the limitations of QMC on strongly-coupled systems, such as very slow evolution through phase space, large statistical errors, and poor scaling with system size, even in cases when there was no minus sign problem. This led me to once again think about numerical RG methods. (Since then, there has been more progress in dealing with the minus sign problem, but it is still a fundamental difficulty.)

I was not tempted to return to the Grassmann imaginary time formulation I had used as a graduate student. My work with quantum Monte Carlo had led me to conclude that it was very hard, and not very natural, to reach zero temperature by making the time direction of the lattice very long. Furthermore, I was very impressed with how the Lanczos method in exact diagonalization could pick out the ground state even with a very small gap to the first excited state. So, I began thinking back to Wilson's Hamiltonian-based numerical RG. Besides working at zero temperature directly, this approach allowed one to deal directly with composite particles or states: for example, suppose all of the electrons in a particular system (say, one with an attractive on-site interaction) were tightly bound in pairs. In the Hamiltonian approach, most of your states describing the Hilbert space would consist of pairs, with little trace of the original electrons. In the Grassmann approach, you would have to work with the original electrons, which seemed to me to be severe drawback.

However, because of the problems with real space RG, I thought that it might be promising to try to solve a finite Hubbard lattice using the numerical RG method in momentum space. In other words, one sets up a "lattice" where each site represents a particular allowed momentum, and the states of the site indicate whether a particle with that momentum is present. Another reason for trying out the momentum space RG was previous experience: my Lanczos work with a student on Hubbard systems had been in momentum space, in order to implement momentum conservation in a simple way. I carried out this approach on a 4×4 Hubbard system, and wrote a paper on it. I concluded that the results were pretty reasonable for energy differences, although total energies were poor. I wanted to know what Wilson thought of this work, and I sent him some email and a copy of the paper [2].

Wilson did not seem overly impressed. He stated emphatically that the particle-in-a-box problem of real space RG needed to be solved first before trying out many particle systems. He also suggested the perturbation approach to fix it, which he had mentioned in his talk at Cornell. He said I should talk to Wilkins and one of his post docs, who had also asked him recently about the particle-in-a-box problem again.

I took Wilson's opinions very seriously. I contacted Wilkins, who referred me to his post doc Kevin Ingersent. I proposed to Kevin, who had some additional notes on Wilson's perturbation approach, that we try out Wilson's ideas together.

Kevin and I tried Wilson's perturbation theory and many other variations, as many as we could think of. This was during the early spring of 1991. All of these approaches failed – in fact, the results were almost always worse than not doing the perturbative corrections at all. We ran out of ideas, concluding it was hopeless, and gave up. In retrospect, I know that the starting point for the perturbation theory was just too poor.

When the summer came, I couldn't resist thinking again about the particle-in-a-box problem, and thought more about boundary conditions(BCs). I knew that simply changing the BCs didn't fix the problem, although it changed what the errors looked like: for example, with fixed BCs, you had problems with slopes at the edges of a block, while with free BCs you got discontinuities. Trying a single BC which was in between fixed and free was possible, but it didn't work any better. Finally I got the idea of trying two different BCs at the same time – getting some states from one BC, some from another. I was still thinking in terms of fixed and free BCs, and it seemed that based on what the wavefunctions for each boundary condition looked like, this might work much better–you could combine the different states to get rid of the various discontinuities. I immediately tried out the "fixed-free" approach numerically – I always try things out numerically very soon after I get an idea, since I enjoy programming and trying out the idea more than trying to work out the details on paper. The test consisted of comparing energies of low lying states after ten RG iterations with exact results.

When I saw the results from the first fixed-free test, I was astonished! Previously, the results from all my numerical tests were garbage, orders of magnitude off from the exact solution. Not only had the fixed-free method worked, the results now seemed exact! The results agreed to at least ten digits, keeping only eight states, and I didn't know whether the miniscule differences were simply the results of round off error. (Some time later, I carried out the calculations again in quadruple precision, and found that the results were good only to ten digits, no more.)

I played around with this for just a short while more, and then I returned to thinking about interacting systems. I felt that this had indeed given me the key to understanding the many particle problem as well as the single particle one. I first tried out the fixed free technique on Heisenberg spin chains. I ended up trying several variations of this approach, including the antiperiodic/periodic combination of boundary conditions, without success. However, this led me to the conclusion that I needed extra boundary sites, which were not part of the block, in order to induce the right boundary condtions. This led to the idea of solving a "superblock" system, and then projecting the superblock state onto the block. I devised a projection method, which was conceptually similar to the optimal configuration interaction approach from my thesis. The status of this work, as of September 1991, was written up in a paper for a conference in Osaka [3]. Already I was becoming confident that this was going to be a very promising approach, and I said so at the conference.

In September, 1991 Reinhard Noack came to work for me as a post doc. Reinhard had been a student of Scalapino, where he worked on quantum Monte Carlo calculations of a two chain Hubbard model. Reinhard had had a lot of trouble in studying the two chain system because of the minus sign problem. I suggested to him that he take over the particle-in-a-box work and turn it into a paper. He got going very quickly on it and the paper was written by December [4]. Subsequently, Reinhard worked on applying these single particle RG methods to problems in Anderson localization. (Eventually, Reinhard returned to studying the two chain Hubbard model, this time using DMRG and with much more success.)

In the late fall or early winter of 1992 I returned to the interacting problem and realized that my projection idea of the superblock onto a block was actually equivalent to a singular value decomposition:

$$\psi_{ij} = U_{i\alpha} D_{\alpha\alpha} V_{\alpha j} \tag{1}$$

where i denotes states of the block, and j denotes states of the remainder of the superblock. I had taught graduate statistical mechanics a few times since coming to UCI, for which I had read parts of Feynman's Lectures on Statistical Mechanics [5]. One of my favorite parts of the book was his introduction to density matrices. I soon realized that performing the singular value decomposition was equivalent to diagonalizing a density matrix:

$$\rho_{ii'} = \psi_{ij}\psi_{i'j} = U_{i\alpha} D^2_{\alpha\alpha} U_{i'\alpha}. \tag{2}$$

This made me very happy: now the technique was becoming quite elegant.

I then began numerical tests and more extensive program development. One early test was to diagonalize a finite Hubbard chain, split it in half, and look at the density matrix eigenvalues for the two sides. This behaved quite nicely. I then began writing a program to do the infinite-system DMRG method for the Heisenberg spin 1/2 chain, for which there is the exact Bethe ansatz solution. I was getting odd-even bond dimerization, so it didn't look like it was working well. However, I remember checking the average energy over two adjacent bonds and being very surprised at how close that was to the exact result – about 5 or 6 digits were correct! It was at that point that I realized things could work very well with this method even with interacting systems.

At that point I had a good understanding of all the pieces of the puzzle. I figured out how the finite system method should work. When I got that program working, and found it worked extremely well, I remember being most concerned about how I would explain this method to others – it was completely unlike any other numerical method I had heard of. I felt the finite system method would look very strange and ad hoc, and I dreaded trying to write a paper to explain the algorithm and explain why, for example, there were two sites in the middle in the standard superblock configuration. It was not that I couldn't explain why, it was just that the explanation would be complicated, as would the description of the algorithm in general. I tried out periodic boundary conditions, found that they didn't work as as open BCs, and also developed an intuitive understanding of why that was.

I wanted a good problem to demonstrate the technique on and I settled on the spin-1 Heisenberg chain, where I got some nice results. I wanted to publicize the method a little. I had a family trip arranged to the East coast in May, and so I contacted friends at Bell Labs and IBM Yorktown Heights, telling them I had some results from an exciting new method and I wanted to give a talk. (I remember Andy Millis at Bell Labs introducing me, saying something like "This is Steve White who has some new results that *he* is very excited about...") The reception I got at Bell was very enthusiastic, and I started a collaboration with David Huse to study the spin-1 Heisenberg chain in more detail [6]. (At IBM, however, no one seemed particularly interested – they weren't working on those kinds of problems.)

I'd like to finish this chapter with some stories about the first DMRG-related papers and some interactions with the referees. For the particle-in-a-box paper [4], with Reinhard Noack, we felt that the editors would have a hard time finding someone to referee the paper, since this was not an active field. We suggested in our submittal letter to Physical Review Letters several people as referees, including Ken Wilson and Patrick Lee, who had worked in numerical RGs in the past. After some time we got a very sensible referee report back, pointing out that while this was very interesting, it was still just for a single particle, and might not generalize to many particle systems. Therefore, the referee recommended that it appear in Physical Review B, Rapid Communications. I thought this was completely reasonable and we told the editors to put it in Rapid Communications.

Very soon after I got a very surprising letter from the editors of PRL. It said that because of a late referee's report, they were now going to accept it for PRL. (We could also still send it to Rapid Communications if we wanted.) This didn't make any sense to me – if you get one good and one bad report, you always have to resubmit, and then maybe you get the paper in because the third referee likes it. So what happened is very unusual – I call it *Spontaneous Unrejection.* Then I read the new referee's report, and I realized why. The report was incredibly insightful. After saying that we had solved an important (toy) problem, and saying that the paper should be accepted, it went on to speculate on the prospects for applying the methods to many particle systems. It said that the fixed-free method probably wouldn't work for interacting systems, since the boundary conditions would apply to all particles identically, whereas it would need to apply in different ways to different particles. It said that the superblock method might work, but that putting three or more blocks together would be very hard to diagonalize. Now, I had already come to these conclusions (or very similar ones) after months of further work. I realized that this report had to be from Wilson – only he had this level of insight about this problem – and that this explained why the editor would now accept the paper: Wilson was the final word in RG methods.

A few months after this exchange I submitted to PRL the original DMRG paper [7]. I got back one positive and one negative report. Both reports were critical of the presentation, which I was sympathetic with – I found it very hard to explain the method in a modest number of words. I rewrote it, taking out a description of the finite system method, which was too brief to make sense, and resubmitted. I got back a third report, which said, "I agree with Referee B. This paper is not appropriate for publication in Physical Review Letters. The reason is as follows. This paper is making a calculation on the Heisenberg linear chain problem. In my opinion, the method is not very new and the above problem has been widely worked in connection with Haldane problem. Hence this paper is inappropriate for publication in PRL..."

I have never appealed a PRL decision before or since, but in this case I did. I asked it be sent to a divisional editor, and I believe they sent it to Peter Young. The divisional editor accepted it at once.

It took me quite a while to write the longer paper describing the algorithms in detail [8]. I also had a hard time getting started because it seemed like too much work and I enjoyed applying DMRG to problems more than I liked writing. However, the long paper did finally get written, and shortly thereafter DMRG began to be used by a number of other groups.

I have been absolutely delighted by the marvelous work done by many other groups using DMRG since then. One of the biggest compliments to a scientist is to have his work used by others, and seeing all the uses, improvements, and extensions to DMRG, as discussed in the rest of this book, is a source of great pleasure to me.

References

1. K.R. Piech and K.G. Wilson, J. Chem. Phys. **48**, 3740 (1968)
2. S.R. White, Phys. Rev. B **45**, 5752 (1992).
3. S.R. White, in Computational Approaches in Condensed-Matter Physics, edited by S. Miyashita, M. Imada, and H. Takayama, Springer, p 97 (1992)
4. S.R. White and R.M. Noack, Phys. Rev. Lett. **68**, 3487 (1992)
5. R.P. Feynman, Statistical Mechanics: A Set of Lectures, Benjamin (1972)
6. S.R. White and D.A. Huse, Phys. Rev. B**48**, 3844 (1993)
7. S.R. White, Phys. Rev. Lett. **69**, 2863 (1992)
8. S.R. White, Phys. Rev. B**48**, 10345 (1993)

Part I

Introductory Lectures

1 Wilson's Numerical Renormalization Group

Theo Costi

Theoretische Physik III, Universität Augsburg
D-86135 Augsburg, Germany

The idea of the numerical renormalization group (NRG) for a quantum-mechanical system with Hamiltonian H is to obtain the many-body eigenstates and eigenvalues on all energy scales $\omega_1 > \omega_2 > \ldots$ in a sequence of steps, with each step corresponding to a distinct energy or length scale [1]. This is achieved by a formal procedure of tracing out high energy states to give effective Hamiltonians $H_N, N = 0, 1, \ldots$ describing the physics on successively lower energy scales $\omega_N, N = 0, 1, \ldots$. The renormalization group (RG) transformation R relates effective Hamiltonians describing the physics on successive energy scales : $H_{N+1} = R[H_N]$. Although the idea is straightforward, its practical implementation turns out to be a difficult task, especially if one wants the RG transformation to give the eigenstates and eigenvalues of H accurately down to the lowest energies.

A complete implementation of the method for a quantum mechanical problem was given by Wilson, who succeeded in constructing a non-perturbative RG transformation for the Kondo model valid on all energy scales [1]. This gave the first convincing calculation of the crossover from the high temperature weak coupling regime of the Kondo model to the low temperature strong coupling regime. The same transformation has since been applied to a range of similar quantum impurity models with equal success [2]. On the other hand, it has proven more difficult to develop similar RG transformations for quantum lattice models. For example, calculations using an approximate RG transformation for the one-dimensional Hubbard model [3] showed that the spectra became inaccurate after four RG iterations, corresponding to a maximum system size of 16 lattice sites, clearly not sufficient to access the lowest energies of interest. These difficulties with the real space approaches provided the motivation for the development of the density matrix renormalization group (DMRG) [4,5] described in this book.

In this chapter we describe a specific implementation of the NRG idea, Wilson's non-perturbative NRG method for the Kondo model [1]. This may serve as a useful background for the later chapters on the DMRG. A detailed description of the formalism of the NRG, the analysis of fixed points and full details on the calculation of thermodynamic quantities can be found in [1,2]. We also describe the calculation of dynamic and transport properties of quantum impurity models via the NRG. The outline is as follows: Sect. 1 introduces the Anderson and Kondo models [6], which are used in subsequent sections to illustrate various aspects of the NRG. The basic ideas of the RG, such as fixed points, crossover scales, RG trajectories and associated universal scaling functions are also discussed there. Wilson's method is described in Sect. 2 and its application to dynamic and transport properties of quantum impurity models in Sect. 3. Finally, Sect. 4 concludes with some remarks on possible future directions using the NRG.

1 Quantum Impurity Models

Examples of quantum impurity models include single and multi-channel Kondo models, the Anderson impurity model and the dissipative two-state system. The simplest versions of these models will be introduced in detail below. They describe a large number of physical systems of current experimental and theoretical interest. Thus, transport through single-electron devices, such as quantum dots, and certain local properties of heavy fermion compounds are modeled by Kondo or Anderson impurity like models. Quantum tunneling between macroscopic fluxoid states in a superconducting quantum interference device [7], tunneling between edge states in the $\nu = 1/3$ fractional Quantum Hall effect [8], two-level atoms coupled to the electromagnetic field in optical fibers [9] and tunneling of defects in solids [10] are examples which can be modeled by the dissipative two-state system [11,12]. The latter is also equivalent to the anisotropic Kondo model [13,14].

The Anderson model

Let us start with the Hamiltonian of the Anderson impurity model, the prototype model of strongly correlated impurity systems. This was introduced in [16] as a microscopic model for local moment formation in non-magnetic metals. Its Hamiltonian is

$$H_{AM} = \varepsilon_d n_d + U n_{d\uparrow} n_{d\downarrow} + \sum_{\mathbf{k},\mu=\uparrow,\downarrow} (V_{\mathbf{k}d} c^+_{\mathbf{k}\mu} d_\mu + H.c) + \sum_{\mathbf{k},\mu=\uparrow,\downarrow} \varepsilon_{\mathbf{k}} c^+_{\mathbf{k}\mu} c_{\mathbf{k}\mu}$$

The first two terms describe the impurity, which, for simplicity, is represented here by a non-degenerate s-level of energy ε_d. Electrons in the local level are subject to a Coulomb repulsion U which acts between spin-up and spin-down electrons. The local level hybridizes with the Bloch states of a non-interacting s-wave conduction band, the last term in H_{AM}, with amplitude $V_{\mathbf{k}d}$. The properties of the model are determined by the hybridization function $\Delta(\omega) = \pi \sum_{\mathbf{k}} |V_{\mathbf{k}d}|^2 \delta(\omega - \varepsilon_{\mathbf{k}})$, which, like the conduction density of states $\rho(\omega) = \sum_{\mathbf{k}} \delta(\omega - \varepsilon_{\mathbf{k}})$, will in general be a complicated function of energy. For most magnetic impurity systems, where the main interest is in the low energy behaviour, it is a good approximation to set $\Delta(\omega) \approx \Delta(\varepsilon_F) \equiv \Delta$. It can also be shown that a *weak* energy dependence of $\Delta(\omega)$ does not affect the low energy behaviour of the model [2].

For a numerical treatment, it is useful to reformulate the Anderson model in the form of a linear chain model. This allows the model to be iteratively diagonalized by a procedure to be described in Sect. 2. The first step in this procedure will be carried out explictly below. This simple calculation also shows how the Kondo model arises as the low energy effective model of the Anderson model in the strong correlation limit. We consider only the case of orbital symmetry, $V_{\mathbf{k}d} = V_{kd}$ and $\varepsilon_{\mathbf{k}} = \varepsilon_k$. Then, only s-wave conduction states couple to the impurity (for details see Appendix A of [2]), and we can replace $c_{\mathbf{k},\mu}$ by $c_{k,l=m=0,\mu}$ and abreviate the latter to $c_{k,\mu}$. We first notice that the impurity state in the Anderson model hybridizes with a local Wannier state $|0,\mu\rangle = f^+_{0,\mu}|\text{vac}\rangle$, with $|\text{vac}\rangle$ the vacuum state, and $f_{0,\mu}$ given by

$$V f_{0,\mu} = \sum_k V_{kd} c_{k,\mu}. \tag{1}$$

The value of V follows from the normalization $\{f_{0,\mu}, f^+_{0,\mu}\} = 1$

$$V = (\sum_k |V_{kd}|^2)^{1/2}. \tag{2}$$

Using the above local state one can apply the Lanczos procedure for tridiagonalizing a hermitian operator, such as H_c, to obtain

$$H_c = \sum_{k,\mu} \varepsilon_k c^+_{k,\mu} c_{k,\mu} \rightarrow \sum_{\mu,n=0}^{\infty} \epsilon_n f^+_{n,\mu} f_{n,\mu} + \lambda_n (f^+_{n,\mu} f_{n+1,\mu} + H.c.) \tag{3}$$

with site energies, ϵ_n, and hoppings, λ_n, depending only on the dispersion ε_k and hybridization matrix elements V_{kd} through the hybridization function $\Delta(\omega)$ [2]. The Anderson model then takes the linear chain form

$$\begin{aligned} H_{AM} &= \varepsilon_d n_d + U n_{d\uparrow} n_{d\downarrow} + V \sum_\mu (f^+_{0,\mu} d_\mu + d^+_\mu f_{0,\mu}) \\ &+ \sum_{\mu,n=0}^{\infty} \epsilon_n f^+_{n,\mu} f_{n,\mu} + \lambda_n (f^+_{n,\mu} f_{n+1,\mu} + f^+_{n+1,\mu} f_{n,\mu}) \end{aligned}$$

depicted in Fig. 1. Although, formally, this model looks like the one-dimensional

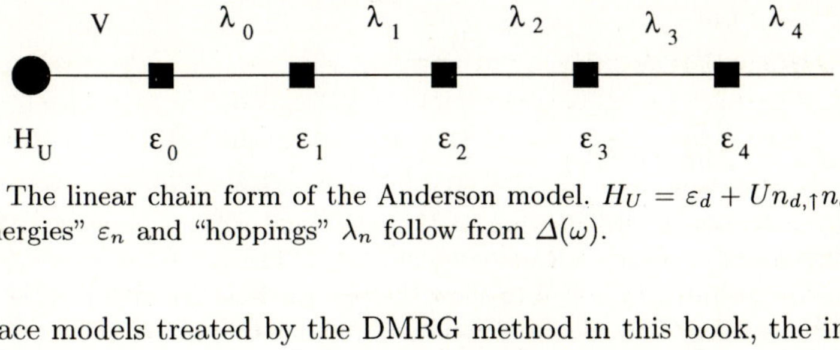

Fig. 1. The linear chain form of the Anderson model. $H_U = \varepsilon_d + U n_{d,\uparrow} n_{d,\downarrow}$. The "site energies" ε_n and "hoppings" λ_n follow from $\Delta(\omega)$.

real-space models treated by the DMRG method in this book, the interpretation here is not in terms of electrons hopping on a one-dimensional lattice in real-space. Instead, as will become clearer in Sect. 2, each successive site added along the chain corresponds to adding lower energy degrees of freedom, measured relative to the Fermi level. By considering longer chains one can then access lower energies.

A zeroth-order (high-energy) approximation to the spectrum of the Anderson model can be obtained by considering just the coupling of the $n = 0$ Wannier state to the impurity and neglecting all others,

$$H_{AM} \approx H_0 = \varepsilon_d n_d + U n_{d\uparrow} n_{d\downarrow} + V \sum_{\mu} (f^{+}_{0,\mu} d_{\mu} + d^{+}_{\mu} f_{0,\mu}) \tag{4}$$

There are 16 many-electron states $|n_d, n_0\rangle$, which can be classified by the conserved quantum numbers of total electron number N_{el}, total z-component of spin S_z^{tot} and total spin $\boldsymbol{S}$. Using these symmetries we can diagonalize the block matrices $H^0_{N_e,S,S_z}$ to obtain the many-body eigenstates $|N_{el}, S, S_z, r\rangle$ and the corresponding eigenvalues. They are shown schematically in Fig. 2 for the symmetric case $\varepsilon_d = -U/2$ in the strong correlation limit $U \gg V^2$. The spectrum separates into two groups of states, one group of low energy states lying close to the groundstate with spacings $\mathcal{O}(V^2/U)$ and one group of high energy states lying at energies $\mathcal{O}(U/2)$ higher and also split by $\mathcal{O}(V^2/U)$.

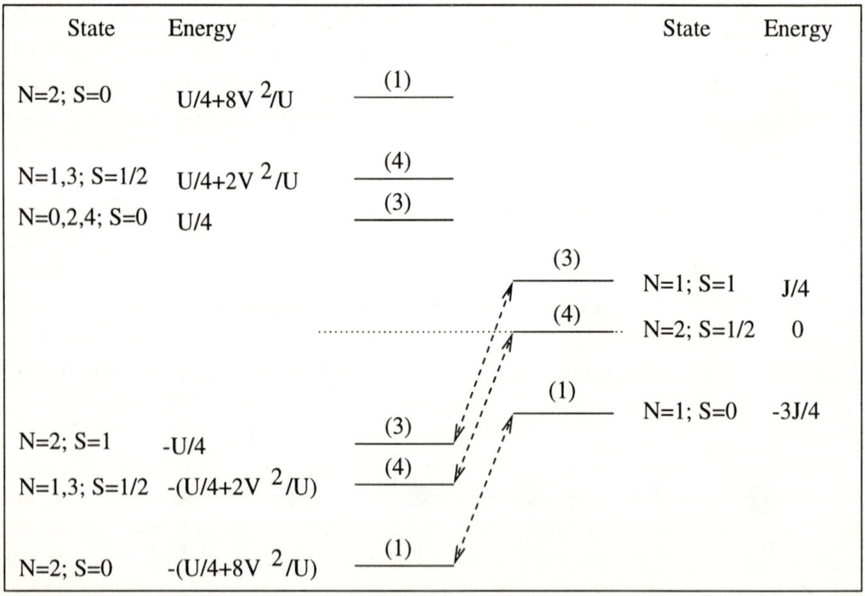

Fig. 2. Zeroth-order approximation to the spectrum of the symmetric Anderson model (left) and the isotropic Kondo model (right). The spectrum of the Anderson model has been shifted by $-U/4$ to show the particle-hole symmetry of the model. The states are labeled by electron number N and total spin S. The numbers in brackets are the degeneracies of the energy levels.

The Kondo model

Consider, now, the (orbitally) isotropic Kondo model,

$$H_{KM} = \sum_{k,\mu} \varepsilon_{k,\mu} c^{+}_{k,\mu} c_{k,\mu} + J\boldsymbol{S}.\boldsymbol{s}_0 \tag{5}$$

where $\boldsymbol{S}$ is the impurity spin and

$$\boldsymbol{s}_0 = f^{+}_{0,\mu} \boldsymbol{\sigma}_{\mu\nu} f_{0,\nu}$$

is the conduction electron spin-density at the impurity site with $f_{0,\mu} = \sum_k c_{k,\mu}$ the local Wannier state. This describes a $S = 1/2$ local moment interacting antiferromagnetically ($J > 0$) with the conduction electron spin-density at the impurity. As for the Anderson model, a zeroth-order (high-energy) approximation to the spectrum of the Kondo model is obtained by considering only the coupling of the impurity spin to the $n = 0$ Wannier state,

$$H_{KM} \approx H_0 = J\boldsymbol{S}.\boldsymbol{s}_0 = \frac{J}{2}[(\boldsymbol{S} + \boldsymbol{s}_0)^2 - \boldsymbol{S}^2 - {\boldsymbol{s}_0}^2]. \tag{6}$$

Its spectrum of 8 eigenstates is also shown in Fig. 2. It consists of a single group of states close to the singlet groundstate with spacings $\mathcal{O}(J)$. This is the same group of states found above for the low energy part of the zeroth approximation to the Anderson model. So at energies, $\omega \ll U/2$, we have that $H_{AM} \approx H_{KM}$, i.e. the low energy behaviour of the Anderson model is governed by the Kondo model with a weak antiferromagnetic exchange interaction J. Identifying the splittings in the spectra for the two models (e.g. that between the lowest two states $E_2^{AM} - E_1^{AM} = 6V^2/U = 3J/4 = E_2^{KM} - E_1^{KM}$) gives the relation between the bare couplings $J = 8V^2/U$ (in agreement with the value found from the Schrieffer-Wolff transformation [17]). Within this zeroth order approximation the spin ($E_3^{KM} - E_1^{KM}$) and single-particle ($E_2^{KM} - E_1^{KM}$) excitations are not strongly renormalized (being of order the bare scale J). The key ingredient of Wilson's NRG, to be discussed in Sect. 2, is a controlled procedure for adding the remaining states $n = 1, 2, \ldots$ neglected in the above approximation. As we shall see in the calculation of dynamical quantities below this leads to a drastic renormalization of the spin and single-particle excitations, such that the relevant excitations of the Kondo model (those at $\omega \ll J$) are on a scale of $k_B T_K \ll J$ (e.g., see Figs. 7-8 in Sect. 3). The bare scale J is renormalized down to the Kondo scale $k_B T_K = D(\rho J)^{1/2} \exp(-1/\rho J)$, where D is the band width and $\rho = 1/2D$ the density of conduction states (a physical interpretation of this large renormalization $J \to k_B T_K$ is given below in terms of tunneling and dissipation).

The Anisotropic Kondo model

The anisotropic Kondo model (AKM) played a key role in the development of the RG approach to quantum impurity models. It was introduced by Anderson as a device for studying the RG flow of the isotropic Kondo model [18].

The AKM also turns out to be of interest in its own right. It is closely related to one-dimensional models with long-range interactions, such as the inverse-square Ising model and the Coulomb gas model [18], which, in contrast to one-dimensional models with short-range interactions, exhibit *finite*-temperature phase transitions (of the Kosterlitz-Thouless type). The connection of the AKM to the dissipative two-state system with Ohmic dissipation (the Ohmic two-state system) [11,12], discussed below, makes the model relevant to the dynamics and thermodynamics of a large class of physical systems. We shall see below that it gives a new perspective on the Kondo effect in terms of the familiar notions of tunneling and dissipation.

The AKM results from the Kondo model by breaking the isotropy of the exchange interaction into a transverse ($J_\perp$) and a longitudinal ($J_\parallel$) part:

$$H = \sum_{k,\sigma} \varepsilon_k c^\dagger_{k\sigma} c_{k\sigma} + \frac{J_\perp}{2} \sum_{\mathbf{k},\mathbf{k}'} (c^\dagger_{k\uparrow} c_{k'\downarrow} S^- + c^\dagger_{k\downarrow} c_{k'\uparrow} S^+)$$
$$+ \frac{J_\parallel}{2} \sum_{\mathbf{k},\mathbf{k}'} (c^\dagger_{k\uparrow} c_{k'\uparrow} - c^\dagger_{k\downarrow} c_{k'\downarrow}) S^z + g\mu_B h S_z. \quad (7)$$

For later purposes we have introduced, in the last term, a local magnetic field, h, coupling only to the impurity spin. The significance we attach to this model is the following: it describes quantum mechanical tunneling between two impurity spin states denoted by $\uparrow$ and $\downarrow$ with a level asymmetry $\varepsilon = g\mu_B h$ and connected by a bare tunneling matrix element $J_\perp$. The longitudinal term $J_\parallel$ acts as a dissipative coupling of the impurity spin to the electronic environment. Large $J_\parallel$ favours spin-flips (tunneling) whereas small $J_\parallel$ reduces spin-flips (tunneling). The former case therefore corresponds to weak dissipation (indeed the case $J_\parallel = \infty$ reduces to a two-state system decoupled from the electronic environment, i.e. zero dissipation) and the latter case to strong dissipation.

The Ohmic two-state system

It can be shown in several ways, including bosonization, that the AKM reduces to the Ohmic two-state system [11,13,14], whose Hamiltonian is given by [11]

$$H_{SB} = -\frac{1}{2}\hbar\Delta_0\sigma_x + \frac{1}{2}\varepsilon\sigma_z + \sum_i \omega_i (a^\dagger_i a_i + \frac{1}{2}) + \frac{1}{2}\sigma_z \sum_i \lambda_i (a_i + a^\dagger_i). \quad (8)$$

The first two terms describe a two-level system with a bare tunneling matrix element Δ_0 and a level asymmetry ε. The third term represents the environment. It consists of an infinite set of harmonic oscillators with frequency spectrum ω_i. In the mapping from the AKM these arise from the particle-hole excitations of the electron sea. The interaction between the two-level system and the environment is described by the last term. The form of the spectral function $J(\omega) = \frac{\pi}{2}\sum_i \lambda_i^2 \delta(\omega - \omega_i)$ characterizes the model. The explicit derivation of the model from the AKM leads to a linear frequency dependence $J(\omega) = 2\pi\alpha\omega$, which is termed *Ohmic*, and which holds for $\omega \ll \omega_c$, where $\omega_c = 2D$ is a high energy cut-off. The dimensionless parameter, α, characterizes the strength of the dissipation. The precise correspondence between the models is given by $\varepsilon = g\mu_B h$, $\frac{\Delta_0}{\omega_c} = \rho J_\perp$, and $\alpha = (1 + \frac{2\delta}{\pi})^2$ where $\tan\delta = -\frac{\pi\rho J_\parallel}{4}$ [11,14].

Scaling analysis

The scaling idea of Anderson for the AKM consists of eliminating high energy conduction electron states lying in a thin shell of width $|dD|$ close to the band edges $\pm D$ to obtain a new effective Hamiltonian with reduced band edges $\pm(D - |dD|)$ and renormalized couplings. The renormalizations of the dimensionless couplings $\rho J_\perp, \rho J_\parallel$ with decreasing D are described by differential equations, known as the "Poor Man's" scaling equations when the derivation is perturbative in both $\rho J_\perp$ and $\rho J_\parallel$ [19]. The "Anderson-Yuval" scaling equations are perturbative only in $\rho J_\perp$. By using the correspondence between the AKM and the Ohmic two-state system given above, we can translate the well-known Anderson-Yuval scaling equations for the AKM [18] into the following scaling equations for the Ohmic two-state system [11,14]

$$\frac{d\alpha}{d\ln\omega_c} = \alpha(\frac{\Delta}{\omega_c})^2 + \mathcal{O}(\frac{\Delta}{\omega_c})^4, \qquad \alpha(\omega_c = \omega_0) = \alpha_0 \tag{9}$$

$$\frac{d(\Delta/\omega_c)}{d\ln\omega_c} = -(1-\alpha)(\frac{\Delta}{\omega_c}) + \mathcal{O}(\frac{\Delta}{\omega_c})^3, \quad \Delta(\omega_c = \omega_0) = \Delta_0. \tag{10}$$

In these equations α and Δ are variables depending on the running cut-off, ω_c. For the purposes of this section, we denote by α_0 and ω_0 the parameters appearing in the bare Hamiltonian (where they formerly appeared as α and ω_c). The infinitesimal RG transformation (9–10) can be restated in terms of a RG transformation R relating effective Hamiltonians H, H' with different cutoffs ω_c, $\omega_c + d\omega_c$, $(d\omega_c < 0)$, $H = H(\omega_c, \alpha, \Delta/\omega_c) \rightarrow H' \equiv R[H]$. In addition to renormalizing the bare interactions, such transformations will also generate new interactions. These are difficult to treat analytically, so usually only the few most relevant interactions are included. In the numerical approach, described in Sect. 2, one works with general effective Hamiltonians, which allow a large number of new interactions to be included.

From (9) and (10) we see that there is a line of *fixed points* at $\Delta_0/\omega_0 = 0$ for $\alpha_0 \geq 0$ (at which the couplings remain invariant on reducing ω_c). Their stability to a finite Δ_0/ω_0 follows from (10), which states that Δ/ω_c is *relevant, marginal* or *irrelevant* depending on whether the dissipation strength α_0 is less than, equal to or larger than 1. Hence, the fixed points at $\Delta_0/\omega_0 = 0$ and $\alpha_0 > 1$ are stable low-energy fixed points (the ferromagnetic fixed points of the AKM or the "self-trapped" fixed points of the two-state system), whereas the line of fixed points at $\Delta_0/\omega_0 = 0$ for $\alpha_0 \leq 1$ are unstable high-energy fixed points. The *scaling trajectories* (shown in Fig. 3) are calculated by dividing the two equations (9–10) and integrating the resulting equation from ω_0 down to ω_c:

$$\frac{1}{2}\left[(\frac{\Delta}{\omega_c})^2 - (\frac{\Delta_0}{\omega_0})^2\right] = -((\ln\alpha - \alpha) - (\ln\alpha_0 - \alpha_0)) \tag{11}$$

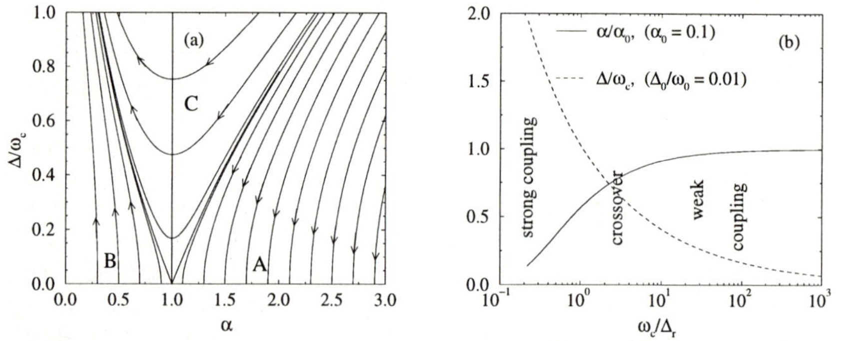

Fig. 3. (**a**) The scaling trajectories of the Ohmic two-state system [14]. The left and right separatrices at $\alpha = 1, \Delta/\omega_c = 0$ define the regions labeled A, B and C and the arrows indicate the direction of decreasing ω_c. In (**b**), the flow of the running couplings in the tunneling regime B is shown.

In regime A of Fig. 3 ($\alpha_0 > 1$), the dimensionless tunneling amplitude Δ/ω_c is irrelevant and the scaling equations remain valid as $\omega_c \to 0$. Integrating (10) down to ω_c gives a renormalized tunneling amplitude $\Delta_r \equiv \Delta(\omega_c) = \Delta_0(\omega_c/\omega_0)^{\alpha_0}$ which vanishes at $T = 0$ at low energies. Quantum mechanical tunneling is absent for $\alpha_0 > 1$ at $T = 0$. This region corresponds to the localized regime of the Ohmic two-state system (or the ferromagnetic sector of the AKM).

In regime B ($\alpha < 1$), which corresponds to the antiferromagnetic sector of the AKM ($J_{\parallel} > 0$), Δ/ω_c is relevant and the flow is towards the strong-coupling fixed point at $\alpha = 0$ and $\Delta/\omega_c = \infty$. This is suggested by the numerical solution of (9) and (10), which is shown in Fig. 3. The scaling analysis, which is perturbative in $\Delta(\omega_c)/\omega_c$, of course, breaks down when $\Delta/\omega_c = \mathcal{O}(1)$. The NRG confirms (Fig. 6 in Sect. 2) that the low energy fixed point is at $\Delta/\omega_c = \infty$ and $\alpha = 0$ (corresponding to a flow of the AKM towards the isotropic strong-coupling fixed point $\rho J_{\perp} = \rho J_{\parallel} = \infty$). In this regime, $\Delta(\omega_c)$ tends to a finite value, the renormalized tunneling amplitude Δ_r, as $\omega_c \to 0$. In the Kondo model this low-energy scale is the Kondo scale, generalized to the anisotropic case. Thus the renormalization of the bare tunneling amplitude $\Delta \rightarrow \Delta_r$ by frictional effects of the environment corresponds to the renormalization of the bare transverse exchange interaction in the AKM from $J_\perp \rightarrow T_K$. This gives a physical significance to the Kondo scale as a *renormalized tunneling amplitude.* It can be estimated as the *crossover scale* separating weak ($\Delta/\omega_c \ll 1$) and strong-coupling ($\Delta/\omega_c \gg 1$) regimes of the model. Define $\Delta_r = \Delta(\tilde{\omega}_c)$ where $\tilde{\omega}_c$ is the crossover scale such that $\Delta(\tilde{\omega}_c)/\tilde{\omega}_c = 1$. Integrating (10) down to this crossover scale and approximating α by α_0 in the range $[\tilde{\omega}_c, \omega_0]$ gives $\Delta_r/\omega_0 = (\Delta_0/\omega_0)^{1/(1-\alpha_0)}$, which, up to prefactors depending on α_0, is the correct low energy scale for the Ohmic two-state system [11,12]. The isotropic weak-coupling Kondo model corresponds to $\alpha_0 \rightarrow 1^-$. The frictional effects are largest in this case and give rise to the expected exponential dependence of $\Delta_r = T_K$ on $-1/\rho J$.

The RG flow described above consists of a one parameter family of scaling trajectories labeled by a parameter $\gamma = \gamma(\alpha, \Delta)$ which takes a constant value along each trajectory. This *scaling invariant* can be found from (11), to within an additive constant. We briefly mention the idea of *universality*, that Hamiltonians with different initial couplings $(\alpha_0, \Delta_0/\omega_0)$ and $(\alpha_0', \Delta_0'/\omega_0')$ but with the same scaling invariant $\gamma = \gamma(\alpha_0, \Delta_0/\omega_0) = \gamma(\alpha_0', \Delta_0'/\omega_0')$ flow along the same RG trajectory specified by γ and that therefore their physical properties are characterized by the same *universal scaling functions.* The scaling functions for the isotropic Kondo model are those on the RG trajectory labeled by initial $\alpha = \alpha_0 = 1^-$. They have also been calculated for general α (arbitrary anisotropy) in [14] by using the Bethe-Ansatz solution of the AKM [15], and in [42] via the NRG.

2 Wilson's Numerical Approach

Wilson's formulation of the RG for the Kondo model is similar in spirit to Anderson's scaling method. The main difference lies in the non-perturbative construction of the RG transformation. The scaling approach uses perturbation theory in the initially small dimensionless coupling ($\rho J_\perp$) to construct such a transformation, but since $\rho J_\perp$ increases with decreasing energy scale this approach eventually becomes inaccurate. In the Wilson approach the RG transformation is perturbative only via a small parameter $\Lambda^{-1/2} < 1$ which is related to the momentum rescaling factor $\Lambda > 1$. The accuracy of the transformation is the same at each step and is independent of the size of the running couplings. For this reason it gave the first correct description of the crossover from the weak coupling to the strong coupling regime of the Kondo model.

Separation of scales

In the Kondo problem, as in other quantum impurity problems, the behaviour of the system changes qualitatively over many energy scales as it passes through a crossover between fixed points (e.g. from behaviour characteristic of a well defined magnetic moment at high temperature to behaviour characteristic of a Fermi liquid at temperatures below the crossover scale). In order to describe this crossover the idea is to separate out the many energy scales in the problem and to set up a procedure for treating each scale in turn. A separation of energy scales can be achieved by discretizing the conduction band $[-1 < \varepsilon_k/D < 1]$ into positive and negative energy intervals, $D_n^+ = [\Lambda^{-(n+1)}, \Lambda^{-n}]$ and $D_n^- = [-\Lambda^{-n}, -\Lambda^{-(n+1)}]$, $n = 0, 1, \ldots$, about the Fermi level $\varepsilon_F = 0$ as shown in Fig. 4 Assuming orbital isotropy, only s-

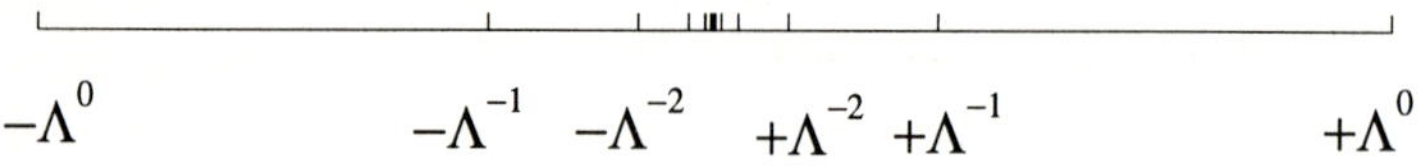

Fig. 4. Logarithmic discretization of the conduction band

wave conduction electrons interact with the impurity, and we can rewrite the continuum version of the Kondo Hamiltonian in the following dimensionless form (henceforth, summation over repeated spin indices is implied)

$$H = \int_{-1}^{+1} d\varepsilon\, \varepsilon\, c^{+}_{\varepsilon,\mu} c_{\varepsilon,\mu} + \underbrace{J\rho \int_{-1}^{+1} d\varepsilon \int_{-1}^{+1} d\varepsilon'\, c^{+}_{\varepsilon,\mu} \boldsymbol{\sigma}_{\mu,\nu}\, c_{\varepsilon',\nu}.\boldsymbol{S}}_{2J\rho f^{+}_{0,\mu} \boldsymbol{\sigma}_{\mu\nu} f_{0,\nu}.\boldsymbol{S}}, \tag{12}$$

A constant density of states, ρ, has been assumed and, as in Sect. 1,

$$f_{0\mu} = \frac{1}{\sqrt{2}} \int_{-1}^{+1} d\varepsilon\, c_{\varepsilon,\mu} \tag{13}$$

is the Wannier state at the impurity. The logarithmic discretization approximation now consists of choosing from each interval $D_n^{\pm}$ just one state, the average electron state

$$c_{-n,\mu} \sim \int_{-\Lambda^{-n}}^{-\Lambda^{-(n+1)}} d\varepsilon\, c_{\varepsilon,\mu}$$

and the average hole state

$$c_{+n,\mu} \sim \int_{+\Lambda^{-(n+1)}}^{+\Lambda^{-n}} d\varepsilon\, c_{\varepsilon,\mu}$$

These states have energies

$$\varepsilon_{\pm n} = \pm\frac{1}{2}(\Lambda^{-n} + \Lambda^{-(n+1)}) = \pm\frac{1}{2}\Lambda^{-n}(1 + \Lambda^{-1}) \tag{14}$$

Of all the states one can construct in each interval $D^{\pm}_{-n}$, these are the states which are most localized near the impurity [2]. The infinite number of states $p = 1, 2, \ldots$ neglected in each interval $D^{\pm}_{n}$ are required to be orthogonal to the states defined above. This suggests that the states neglected $p = 1, 2, \ldots$ will be centred at sites away from the impurity. A more precise argument shows that they are centred at distances $r \sim \Lambda^{p}$ from the impurity and that they only couple indirectly to the impurity [2]. Consequently they can be neglected for the calculation of impurity properties. We therefore arrive at the discretized Kondo Hamiltonian

$$H \approx \sum_{n=0}^{\infty}(\varepsilon_{-n}c^{+}_{-n,\mu}c_{-n,\mu} + \varepsilon_{+n}c^{+}_{+n,\mu}c_{+n,\mu}) + 2J\rho f^{+}_{0,\mu}\boldsymbol{\sigma}_{\mu\nu}f_{0,\nu}.\boldsymbol{S}$$

which as in (3) can be put into the linear chain form

$$H = \frac{1}{2}(1+\Lambda^{-1})\sum_{n=0}^{\infty}\Lambda^{-n/2}(f^{+}_{n,\mu}f_{n,\mu}+f^{+}_{n+1,\mu}f_{n,\mu})+2J\rho f^{+}_{0,\mu}\boldsymbol{\sigma}_{\mu\nu}f_{0,\nu}.\boldsymbol{S}. \quad (15)$$

Here, we have used the explicit form of the Lanczos coefficients ϵ_n, λ_n appearing in (3) which were calculated analytically in [1] for a logarithmically discretized conduction band: $\epsilon_n = 0$ and $\lambda_n \approx \frac{1}{2}(1+\Lambda^{-1})\Lambda^{-n/2}, n >> 1$. This form of the Hamiltonian provides a clear separation of the energy scales $\frac{1}{2}(1+\Lambda^{-1})\Lambda^{-n/2}, n = 1, 2, \ldots$ in H and allows the diagonalization of the Hamiltonian in a sequence of controlled steps, each step corresponding to adding an orbital $f_{n,\mu}$ which is a relative perturbation of strength $\Lambda^{-1/2} < 1$.

RG transformation

A RG transformation relating effective Hamiltonians on successive energy scales $\Lambda^{-n/2}$ and $\Lambda^{-(n+1)/2}$ can be set up as follows. First, H in (15) is truncated to N orbitals to give H_N, whose lowest scale is $D_N = \frac{1}{2}(1+\Lambda^{-1})\Lambda^{-(N-1)/2}$. In order to look for fixed points we define rescaled Hamiltonians $\bar{H}_N \equiv H_N/D_N$:

$$\bar{H}_N = \Lambda^{(N-1)/2}[\sum_{n=0}^{N-1}\Lambda^{-n/2}(f^{+}_{n,\mu}f_{n,\mu} + f^{+}_{n+1,\mu}f_{n,\mu}) + \tilde{J}f^{+}_{0,\mu}\boldsymbol{\sigma}_{\mu\nu}f_{0,\nu}.\boldsymbol{S}],$$

$$\tilde{J} = \frac{2J\rho}{\frac{1}{2}(1+\Lambda^{-1})},$$

from which we can recover H as

$$H = \lim_{N\to\infty}\frac{1}{2}(1+\Lambda^{-1})\Lambda^{-(N-1)/2}\bar{H}_N. \quad (16)$$

The sequence of rescaled Hamiltonians $\bar{H}_N$ satisfies the recursion relation

$$\bar{H}_{N+1} = \Lambda^{1/2}\bar{H}_N + (f^{+}_{N,\mu}f_{N+1,\mu} + f^{+}_{N+1,\mu}f_{N,\mu}), \quad (17)$$

and allows a RG transformation R to be defined:

$$\bar{H}_{N+1} = R[\bar{H}_N] \equiv \Lambda^{1/2}\bar{H}_N + (f^+_{N,\mu} f_{N+1,\mu} + f^+_{N+1,\mu} f_{N,\mu}) - \bar{E}_{G,N+1} \quad (18)$$

with $\bar{E}_{G,N+1}$ the ground-state energy of $\bar{H}_{N+1}$.

Iterative scheme

The transformation R relates effective Hamiltonians $H_N = D_N \bar{H}_N$ and $H_{N+1} = D_{N+1}\bar{H}_{N+1}$ on decreasing scales $D_N > D_{N+1}$. It can be used to iteratively diagonalize the Kondo Hamiltonian by the following sequence of steps:

1. the local part

$$\bar{H}_0 = \Lambda^{-1/2}\, \tilde{J}\, f^+_{0,\mu}\sigma_{\mu\nu} f_{0,\nu}.\boldsymbol{S}, \quad (19)$$

   which contains the many-body interactions, is diagonalized (the "zeroth"-order step described in Sect. 1),
2. assuming that $\bar{H}_N$ has been diagonalized,

$$\bar{H}_N = \sum_\lambda \bar{E}^N_\lambda |\lambda\rangle\langle\lambda| \quad (20)$$

   we add a "site" and use (18) to set up the matrix for $\bar{H}_{N+1}$ within a product basis $|\lambda, i\rangle = |\lambda\rangle_N |i\rangle_{N+1}$ consisting of the eigenstates $|\lambda\rangle_N$ of $\bar{H}_N$ and the 4 states $|i\rangle_{N+1}$ of the next orbital along the chain (i.e. $|i\rangle_{N+1} = |0\rangle, |\uparrow\rangle, |\downarrow\rangle, |\uparrow\downarrow\rangle$). The resulting matrix

$$\begin{aligned}\langle\lambda, i|\bar{H}_{N+1}|\lambda', i'\rangle = {} & \Lambda^{1/2}\delta_{i,i'}\delta_{\lambda,\lambda'}\bar{E}^N_\lambda \\ & + (-1)^{N_{e,\lambda'}}\langle\lambda|f^+_{N,\mu}|\lambda'\rangle\langle i|f_{N+1,\mu}|i'\rangle \\ & + (-1)^{N_{e,\lambda}}\langle i|f^+_{N+1,\mu}|i'\rangle\langle\lambda|f_{N,\mu}|\lambda'\rangle,\end{aligned}$$

   with $N_{e,\lambda}, N_{e,\lambda'}$ the number of electrons in $|\lambda\rangle, |\lambda'\rangle$ respectively, is diagonalized and the procedure is repeated for the next energy shell as depicted in Fig. 5. Since $\bar{H}_N$ is already diagonalized, the off-diagonal matrix elements, involving ${}_N\langle\lambda|f_{N,\mu}|\lambda'\rangle_N$, can be expressed in terms of the known eigenstates of $\bar{H}_N$ (see [2] for explicit expressions).

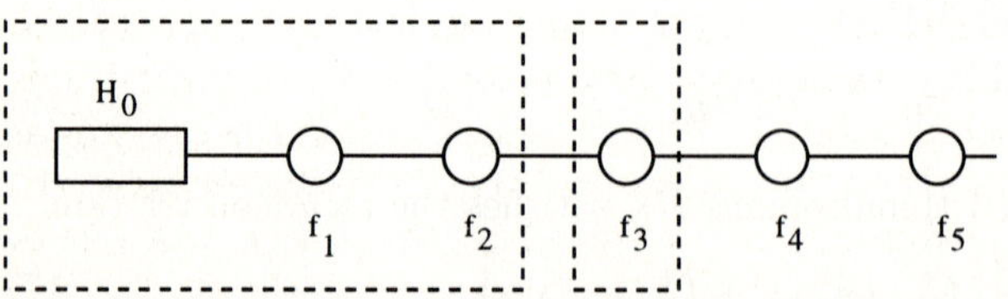

Fig. 5. Iterative diagonalization scheme for H, starting with H_0 and then adding successive orbitals $f_1, f_2, \ldots$.

In practice since the number of many-body states in $\bar{H}_N$ grows as 4^N it is not possible to retain all states after about $N = 5$. For $N > 5$ only the lowest 1000 or so states of $\bar{H}_N$ are retained. The truncation of the spectrum of $\bar{H}_N$ restricts the range of eigenvalues in $H_N = D_N \bar{H}_N$ to be such that $0 \le E_\lambda^N \le KD_N$ where $K = K(\Lambda)$ depends on Λ and the number of states retained. For 1000 states and $\Lambda = 3$, $K(\Lambda) \approx 10$. However, eigenvalues below D_N are only approximate eigenvalues of the infinite system H, since states with energies below D_N are calculated more accurately in subsequent iterations $N+1, N+2, \ldots$. Therefore the part of the spectrum of H_N which is close to the spectrum of H is restricted to $D_N \le E_\lambda^N \le K(\Lambda)D_N$. This allows the whole spectrum of H to be recovered by considering the spectra of the sequence of Hamiltonians H_N, $N = 0, 1, \ldots$. In this way the many–body eigenvalues and eigenstates are obtained on all energy scales.

There is no guarantee that the neglect of the higher energy states in the procedure described above will not make inaccurate the low-energy part of the spectrum calculated in subsequent iterations. However, in the Kondo model, as well as in other quantum impurity models, the addition of an energy shell is a perturbation of relative strength $\Lambda^{-1/2} < 1$ and the neglect of the high-energy states turns out not to spoil the low-energy spectrum. Herein lies the difference to the naive real-space NRG approaches to one-dimensional lattice problems. For the latter, the addition of a site is never a small perturbation since the hopping matrix elements on a one-dimensional lattice are constant. Consequently real-space RG methods which neglect states on the basis of energy are uncontrolled approximations. Another factor which makes the Kondo calculation accurate is the relatively large fraction of states which can be retained at each iteration. This is 1/4 for the Kondo model and can be made slightly larger by using symmetries. For models where this fraction is smaller (e.g. 1/6 for the local Cu – O model [51], 1/16 for the two-channel Kondo model) explicit calculations show that the neglect of high-energy states remains a valid approximation.

The smallness of the perturbation when adding successive energy shells appears to be the main reason for maintaining the accuracy of the low-energy spectrum for each N and therefore for each energy scale. How accurate is the spectrum obtained from the iterative procedure? For the Kondo model, the fixed point spectrum (that of $\bar{H}_N$ for $N \to \infty$) corresponds to setting $\tilde{J} = \infty$ in H and can be calculated exactly. Typically, the relative error in the lowest eigenvalues calculated from the iterative procedure is found to be of $\mathcal{O}(10^{-7})$[1]. This holds even for $N \sim 100$, which for $\Lambda = 2$ corresponds to eigenvalues with absolute value $\Lambda^{-(N-1)/2} \sim 10^{-15}$. That the eigenstates are also obtained accurately follows indirectly from the fullfilment of exact Fermi-liquid relations for dynamical properties (e.g. the Friedel sum rule of Sect. 4).

Comparison with DMRG

The DMRG method, described in the next chapter, differs from the usual NRG approach (and that used in the Kondo calculation) in several ways. The most important, and the reason for its success as applied to one-dimensional lattice models, is the criterion for choosing the basis states of the subsystems (blocks) used to extend the size of the system (superblock). These are chosen according to their weight in a reduced density matrix built from a few eigenstates of the larger system (superblock). That is, the states retained in the subsystems (similar to the lowest states retained in $\bar{H}_N$ in the Kondo calculation) are in this case not necessarily the lowest energy states, but they are the states which couple most strongly, in the sense of having large eigenvalues in the reduced density matrix describing the subsystem, to the ones of interest, the target states of the larger system (in the Kondo calculation these might be taken to be the lowest few eigenstates of $\bar{H}_{N+1}$). The procedure gives highly accurate results for these target states, something which is not possible within the conventional real-space NRG. A second difference of the DMRG approach to the usual NRG approach, is precisely the focus of the DMRG on obtaining just a few low-lying states accurately, instead of a group of several hundred states on each energy scale which one is in principle trying to obtain within the usual NRG approach. However, as explained above, the latter aim is currently only realized for quantum impurity models. Conventional real-space approaches fail in this respect, and the DMRG is currently the most accurate method for obtaining low-lying eigenstates of one-dimensional lattice models.

3 Applications

Applications of the NRG to quantum impurity models fall into three areas: analysis of fixed points, calculation of thermodynamics and calculation of dynamic and transport properties.

Following Wilson's formulation of the NRG for the Kondo problem, the fixed points and thermodynamics of a number of other quantum impurity models were investigated, including the Anderson impurity model [2], the resonant level model [20], the two-impurity [21,22] and two-channel Kondo models [23,24], Anderson models with further screening [25,26] or conduction channels [24,27,28,32] and Kondo impurities in superconductors [29] and pseudogap systems [30,31,33].

A further development was the calculation of dynamical quantities, starting with the $T = 0$ absorption and photoemission spectra of the X-ray Hamiltonain [34], followed by the calculation of $T = 0$ spectral densities and response functions of the non-degenerate and degenerate Anderson impurity models [35–40] and related models [41–44]. Finite-temperature dynamic and transport properties of the Anderson impurity model have also been calculated [40,45–47]. The linear conductance of a quantum dot, modeled by an

Anderson impurity model, has been calculated as in [40] and used to interpret recent experiments [48] (see [49,50] for further applications to mesoscopic systems). The calculation of non-equilibrium transport through correlated impurity systems has been discussed in [51].

Another recent direction for applications of the NRG has been to the study of Anderson and Kondo impurity models with energy-dependent hybridization functions, $\Delta(\omega)$, or momentum-dependent exchange interactions. These are of interest in the context of pseudogap systems, Kondo insulators and in the dynamical mean field theory of strongly correlated lattice systems [53]. The possibility of using the NRG for the latter was noted in [45,52]; applications to the metal-insulator transition in the Hubbard model include [54,55].

Fixed points and thermodynamics

A fixed point H^* of R is defined by

$$R[H^*] = H^*. \tag{21}$$

In practice, proximity to a fixed point is identified by ranges of N, $N_1 \leq N \leq N_2$, where the energy levels $\bar{E}_p^N$ of $\bar{H}_N$ are approximately independent of N: $\bar{E}_p^N \approx \bar{E}_p$ for $N_1 \leq N \leq N_2$. A typical energy level flow diagram showing regions of N where the energy levels are approximately constant is shown in Fig. 6 for the AKM [42]. There is an unstable high-energy fixed point (small N) and a stable low-energy fixed point (large N). The low-energy spectrum is identical to that of the isotropic Kondo model at the strong-coupling fixed

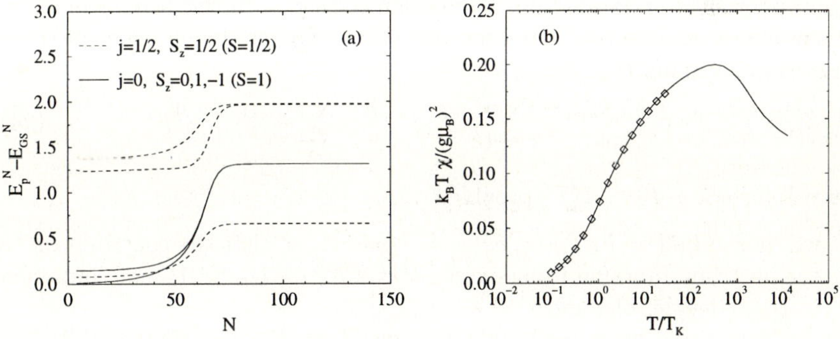

Fig. 6. (**a**) The lowest rescaled energy levels of the AKM for $J_\parallel = 0.443$ and $J_\perp = 0.01$. The states are labeled by conserved pseudospin j and total S_z [42]. In (**b**) the static susceptibility of the Anderson impurity model for $U/\pi\Delta = 6, \varepsilon_d/\Delta = -5$ (full curve) is shown. The symbols are from the universal susceptibility curve for the isotropic Kondo model (taken from Table V of [2]), which agrees with the low temperature susceptibility of the Anderson model.

point $J = \infty$ in [1] (e.g. the lowest single-particle excitations in Fig. 6, $\eta_1 = 0.6555$, $\eta_2 = 1.976$ agree with the $\Lambda = 2$ results of the isotropic model in [1]). The crossover from the high-energy to the low-energy fixed point is associated with a scale $T_K(J_\perp, J_\parallel)$ (or Δ_r in the equivalent Ohmic two-state system of Sect. 1). Spin-rotational invariance, broken at high energies, is restored below this scale (e.g. the $j = 0$ states with $S_z = 0$ and $S_z = \pm 1$ become degenerate below T_K and can be classified by the same total spin S as indicated in Fig. 6). Analytic calculations can be carried out in the vicinity of fixed points by setting up effective Hamiltonians $H_{eff} = H^* + \sum_\lambda \omega_\lambda O_\lambda$, where the leading deviations O_λ about H^* can be obtained from general symmetry arguments. This allows, for example, thermodynamic properties to be calculated in a restricted range of temperatures, corresponding to the restricted range of N where $\bar{H}_N$ can be described by a simple effective Hamiltonian H_{eff}. We refer the reader to the detailed description of such calculations in [1,2], and we turn now to the numerical procedure for calculating thermodynamics, which can give results at all temperatures, including the crossover regions.

Suppose we have diagonalized exactly the Hamiltonian for a quantum impurity model such as the Kondo model and that we have all the many-body eigenvalues E_λ and eigenstates $|\lambda\rangle$:

$$H = \sum_\lambda E_\lambda |\lambda\rangle\langle\lambda| \equiv \sum_\lambda E_\lambda X_{\lambda\lambda}. \tag{22}$$

We can then calculate the partition function

$$Z(T) \equiv Tr\ e^{-H/k_B T} = \sum_\lambda e^{-E_\lambda/k_B T}, \tag{23}$$

and hence the thermodynamics via the impurity contribution to the free energy $F_{imp}(T) = -k_B T \ln Z/Z_c$, where $Z_c = Tr\ e^{-H_c/k_B T}$ is the the partition function for the non-interacting conduction electrons. In the NRG procedure we can only calculate the "partition functions" Z_N for the sequence of truncated Hamiltonians H_N:

$$Z_N(T) \equiv Tr\ e^{-H_N/k_B T} = \sum_\lambda e^{-E_\lambda^N/k_B T} = \sum_\lambda e^{-D_N \bar{E}_\lambda^N/k_B T} \tag{24}$$

We will have $Z_N(T) \approx Z(T)$ provided

1. we choose $k_B T = k_B T_N \ll E_{max}^N = D_N K(\Lambda)$ so that the contribution to the partition function from excited states $E_\lambda^N > D_N K(\Lambda)$, not contained in Z_N, is negligible, and
2. the truncation error made in replacing H by H_N in equating (23) and (24) is small. This error has been estimated in [2] to be approximately $\Lambda^{-1} D_N/k_B T_N$.

Combining these two conditions requires that

$$\frac{1}{\Lambda} \ll \frac{k_B T_N}{D_N} \ll K(\Lambda). \tag{25}$$

The choice $k_BT = k_BT_N \approx D_N$ is reasonable and allows the thermodynamics to be calculated at a sequence of decreasing temperatures $k_BT_N \sim D_N, N = 0, 1, \ldots$ from the truncated partition functions Z_N. The procedure is illustrated in Fig. 6 for the impurity static susceptibility of the Anderson impurity model

$$\chi_{imp}(T) = \frac{(g\mu_B)^2}{k_BT}\left[\frac{1}{Z}Tr\ (S_z^{tot})^2 e^{-H/k_BT} - \frac{1}{Z_c}Tr\ (S_{z,c}^{tot})^2 e^{-H_c/k_BT}\right].$$

Dynamic properties

We consider now the application of the NRG method to the calculation of dynamic properties of quantum impurity models. For definiteness we consider the Anderson impurity model and illustrate the procedure for the impurity spectral density $\rho_{d,\mu}(\omega, T) = -\frac{1}{\pi}\text{Im}G_{d,\mu}(\omega, T)$, with $G_{d,\mu}(\omega, T)$ the impurity Green's function. Response functions can be calculated in a similar way.

Suppose we have all the many-body eigenstates $|\lambda\rangle$ and eigenvalues E_λ of the Anderson impurity Hamiltonian H. Then, the impurity Green's function can be written in the Lehmann representation as

$$G_{d,\mu}(\omega, T) = \frac{1}{Z(T)}\sum_{\lambda,\lambda'}|\langle\lambda|d_\mu|\lambda'\rangle|^2 \frac{e^{-E_\lambda/k_BT} + e^{-E_{\lambda'}/k_BT}}{\omega - (E_{\lambda'} - E_\lambda)} \tag{26}$$

and the corresponding impurity spectral density $\rho_{d,\mu}$ as

$$\rho_{d,\mu}(\omega, T) = \frac{1}{Z(T)}\sum_{\lambda,\lambda'}|M_{\lambda,\lambda'}|^2(e^{-E_\lambda/k_BT} + e^{-E_{\lambda'}/k_BT})\delta(\omega - (E_{\lambda'} - E_\lambda))$$

with $M_{\lambda,\lambda'} = \langle\lambda|d_\mu|\lambda'\rangle$.

Consider first the $T = 0$ case ($T > 0$ is described in the next section), then

$$\rho_{d,\mu}(\omega, T = 0) = \frac{1}{Z(0)}\sum_{\lambda}|M_{\lambda,0}|^2\delta(\omega + E_\lambda) + \frac{1}{Z(0)}\sum_{\lambda'}|M_{0,\lambda'}|^2\delta(\omega - E_{\lambda'}).$$

In order to evaluate this from the information which we actually obtain from an iterative diagonalization of H, we consider the impurity spectral densities corresponding to the sequence of Hamiltonians H_N, $N = 0, 1, \ldots$,

$$\rho_{d,\mu}^N(\omega, T = 0) = \frac{1}{Z_N(0)}\sum_{\lambda}|M_{\lambda,0}^N|^2\delta(\omega + E_\lambda^N) + \frac{1}{Z_N(0)}\sum_{\lambda'}|M_{0,\lambda'}^N|^2\delta(\omega - E_{\lambda'}^N).$$

From the discussion on the spectrum of H_N in the previous section, it follows that the groundstate excitations of H_N which are representative of the infinite-system H are those in the range $D_N \leq \omega \leq K(\Lambda)D_N$. Lower energy excitations and eigenstates are calculated more accurately at subsequent iterations, and higher energy excitations are not contained in H_N due to the elimination of the higher energy states at each N. Hence, for fixed N,

$$\rho^N_{d,\mu}(\omega, T=0) \approx \rho_{d,\mu}(\omega, T=0) \tag{27}$$

provided that we choose $\omega \approx \omega_N \equiv k_B T_N$. A typical choice is $\omega = 2\omega_N$ for $\Lambda = 2$. This allows $\rho_{d,\mu}(\omega, T=0)$ to be calculated at a sequence of decreasing frequencies $\omega = 2\omega_N, N = 0, 1, \ldots$ from the quantities $\rho^N_{d,\mu}$. In practice

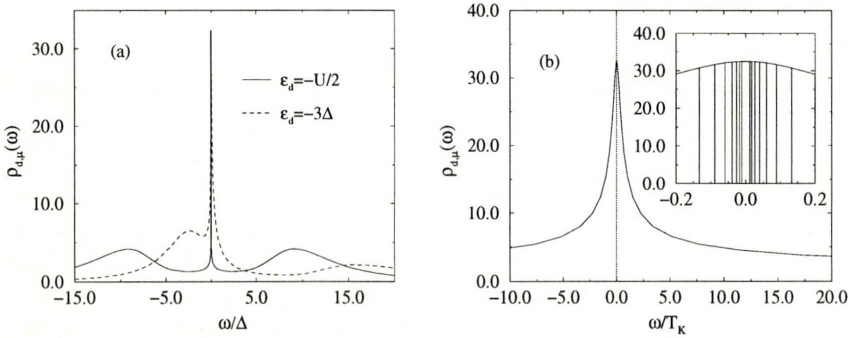

Fig. 7. (**a**) The impurity spectral density for the Anderson impurity model for $U/\pi\Delta = 6$ and different local level positions. The Kondo resonance for the case $\varepsilon_d = -U/2$ is shown in more detail in (**b**). The vertical lines in the inset show the sequence of energies $\omega = 2\omega_N$ at which the spectral density is calculated and demonstrates the ability of the method to resolve low-energy scales.

we are not interested in the discrete spectra $\rho^N_{d,\mu}(\omega) = \sum_\lambda w^N_\lambda \delta(\omega - E^N_\lambda)$ of the Hamiltonians H_N but in continuous spectra which can be compared with experiment. Smooth spectra can be obtained from the discrete spectra by replacing the delta functions $\delta(\omega - E^N_\lambda)$ by smooth distributions $P_N(\omega - E^N_\lambda)$. A natural choice for the width η_N of P_N is D_N, the characteristic scale for the energy-level structure of H_N. Two commonly used choices for P are the Gaussian and the Logarithmic Gaussian distributions [36,40]. A peak of intrinsic width Γ at frequency Ω_0 will be well resolved by the above procedure provided that $\Omega_0 \ll \Gamma$, which is the case for the Kondo resonance and other low-energy resonances. In the opposite case, the low (logarithmic) resolution at higher frequencies may be insufficient to resolve the intrinsic widths and heights of such peaks. Usually such higher frequency peaks are due to single-particle processes and can be adequately described by other methods (exceptions include interaction dominated features in the Ohmic two-state system, see below, and in strongly correlated lattice models in high dimensions [53]). In both cases, $\Omega_0 \ll \Gamma$ and $\Omega_0 \gg \Gamma$, the positions and intensities of such peaks is given correctly. An alternative procedure for obtaining smooth spectra, which in principle resolves finite-frequency peaks with the same resolution as the low-energy peaks, has been proposed in [56]. This involves a modified discretization of the conduction band with energies $\pm 1, \pm \Lambda^{-z}, \pm \Lambda^{-z-1}, \ldots$ instead of the usual discretization $\pm 1, \pm \Lambda^{-1}, \pm \Lambda^{-2}, \ldots$. By considering all z between 0 and 1 one recovers a continuous spectrum without the need to use a broadening function. The procedure requires diagonalizing H for many values of z. It has also proved useful for carrying out thermodynamic calculations at large Λ [57].

How accurate is the NRG for dynamic properties? In Fig. 7 we show results for $T = 0$ spectral densities of the Anderson impurity model [40]. A good measure of the accuracy of the procedure is given by the Friedel sum rule, a Fermi-liquid relation which states that [58]

$$\rho_{d,\mu}(0) = \frac{1}{\pi\Delta}\sin^2(\pi n_d/2), \qquad n_d = \int_{-\infty}^{0} d\omega\, \rho_{d,\mu}(\omega) \tag{28}$$

The integrated value of n_d, for the spectral density shown in Fig. 7, is 0.991. Including the renormalization in Δ due to the discretization, as discussed in [2], gives $\rho_{d,\mu}(0) = 32.779$. The value extracted directly from Fig. 7 is $\rho_{d,\mu}(0) = 32.31$ resulting in a 1.4% error, most of which is due to using the integrated value of n_d over all energy scales. Calculating n_d solely from the low energy part of the spectrum (e.g. as the limit $n_d(T \to 0)$ in a thermodynamic calculation) further reduces this error. More important, however, is that the error remains small independent of the interaction strength $0 \leq U \leq \infty$.

Two-particle Green's functions and response functions can also be calculated. Figure 8 shows the longitudinal spin-relaxation function

$$S(\omega) = -\frac{1}{\pi}\frac{\mathrm{Im}\chi_{zz}(\omega)}{\omega}, \qquad \chi_{zz} = \langle\langle S_z; S_z \rangle\rangle$$

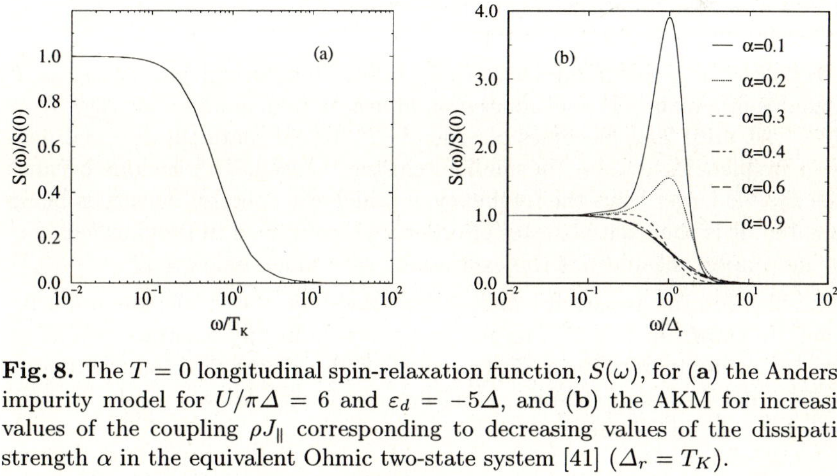

Fig. 8. The $T = 0$ longitudinal spin-relaxation function, $S(\omega)$, for (**a**) the Anderson impurity model for $U/\pi\Delta = 6$ and $\varepsilon_d = -5\Delta$, and (**b**) the AKM for increasing values of the coupling $\rho J_{\|}$ corresponding to decreasing values of the dissipation strength α in the equivalent Ohmic two-state system [41] ($\Delta_r = T_K$).

of the Anderson impurity model and of the AKM [41,42]. The former always exhibits incoherent spin dynamics. It is interesting that the latter can exhibit coherent spin dynamics for sufficiently large $\rho J_{\parallel}$. This is more easily understood in terms of the equivalent Ohmic two-state system (Sect. 1), where large $\rho J_{\parallel}$ corresponds to a sufficiently small dissipative coupling of the spin to its electronic environment leading to coherent dynamics of the spin or coherent tunneling between the two quantum-mechanical states. A subtlety here is that a quasielastic peak in $S(\omega)$ does not necessarily mean that the time-dependent dynamics of the spin is incoherent [59]. In fact the quasielastic peak in $S(\omega)$ for the AKM disappears at $\alpha = 1/3$ [41] whereas the incoherent dynamics (non-oscillatory behaviour in the *real-time* dynamics) only sets in at the Toulouse point $\alpha = 1/2$ ($\rho J_{\parallel} \approx 1$).

Transport properties

The transport properties of quantum impurity models, require knowledge of both the frequency and temperature dependence of the impurity spectral density.

The resistivity $\rho(T)$ of conduction electrons scattering from a single Anderson impurity, for example, is given by the expression

$$\rho(T)^{-1} = -e^2 \int_{-\infty}^{+\infty} \tau_{tr}(\omega, T) \frac{\partial f}{\partial \omega} \, d\omega \tag{29}$$

where the transport time $\tau_{tr}(\omega, T)$ is related to the impurity spectral density by $\tau_{tr}^{-1}(\omega, T) = \Delta \rho_{d,\mu}(\omega, T)$ and Δ is the hybridization strength. Similar expressions hold for the other transport coefficients.

The procedure for calculating finite-temperature dynamical quantities, like $\rho_{d,\mu}(\omega, T)$, required as input for calculating transport properties, is similar to that for $T = 0$ dynamics described above. The spectral density $\rho_{d,\mu}(\omega, T)$ at fixed temperature T is evaluated as above at frequencies $\omega \approx 2\omega_N, N = 0, 1, \ldots, M$ until $2\omega_M$ becomes of order $k_B T$. To calculate the spectral density at frequencies $\omega < k_B T$ a smaller "cluster" is used. This is done because when $k_B T$ is larger than the frequency at which the spectral density is being evaluated, it is the excited states of order $k_B T$ contained in previous clusters that are important and not the excitations very much below $k_B T$.

Results for the resistivity and thermopower of the Anderson impurity model are shown in Fig. 9. The method gives uniformly accurate results at high and low temperatures, as well as correctly describing the crossover region $T \approx T_K$ (detailed comparisons of the resistivity with known results at high and low temperature can be found in [40,45]).

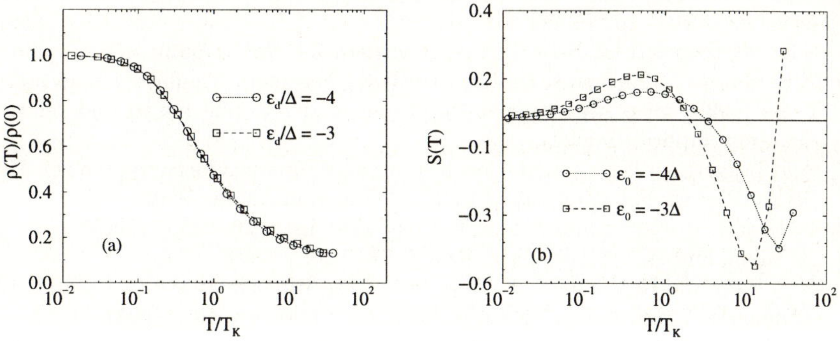

Fig. 9. (**a**) The scaled resistivity $\rho(T)/\rho(0)$ and (**b**) the thermopower $S(T)$ of the Anderson impurity model for $U/\pi\Delta = 4$ and two values of the local level position in the Kondo regime [40].

4 Conclusions

The NRG transformation for the Kondo model has developed into a powerful tool for the study of quantum impurity models. It gives information on the many-body eigenvalues and eigenstates of such models on all energy scales and thereby allows the direct calculation of their thermodynamic, dynamic and transport properties.

There is room for further improvement and extensions of the method. The use of a logarithmic discretization of the conduction band, for example, gives rise to insufficient resolution at higher energies. This limits the accuracy with which certain models can be investigated, for examples those arising in the dynamical mean field theory of correlated lattice systems [53]. New approaches [56] for overcoming these difficulties are therefore of interest.

The NRG also has potential to give information on the non-equilibrium transport through correlated impurity systems such as quantum dots [51]. However, away from equilibrium, the absence of a groundstate requires new criteria other than energy for eliminating unimportant states. Ideas based on the DMRG method described in this book are being explored and may prove useful in this context.

References

1. K.G. Wilson Rev. Mod. Phys. **47**, 773 (1975)
2. H.B. Krishnamurthy, J.W. Wilkins and K.G. Wilson, Phys. Rev. B **21**, 1044 (1980)
3. S.-T. Chui and J.W. Bray, Phys. Rev. **18**, 2426 (1978); J.W. Bray and S.-T. Chui, Phys. Rev. **19**, 4876 (1979)

4. S.R. White, Phys. Rev. Lett. **69**, 2863 (1992); Phys. Rev. B **48**, 10345 (1993); Phys. Rep. **301**, 187 (1998)
5. S.R. White and R.M. Noack, Phys. Rev. Lett. **68**, 3487 (1992)
6. A.C. Hewson, The Kondo Problem to Heavy Fermions, Cambridge University Press (1993) gives an extensive discussion of the Kondo model and related quantum impurity models.
7. A.J. Leggett, in Percolation, Localization and Superconductivity, NATO ASI Series: Physics Vol. **109**, edited by M. Goldman and S. A. Wolf
8. F. Lesage, H. Saleur and S. Skorik, Phys. Rev. Lett. **76**, 3388 (1996)
9. R. Konik and A. LeClair, Phys. Rev. B **58**, 1872 (1998)
10. K.S. Ralls and R.A. Buhrman, Phys. Rev. Lett. **60**, 998 (1988); B. Golding, N. Zimmermann and S. Coppersmith, Phys. Rev. Lett. **68**, 998 (1992)
11. A.J. Leggett, S. Chakravarty, A.T. Dorsey, M.P.A. Fisher, A. Garg and W. Zwerger, Rev. Mod. Phys. **59**, 1 (1987); **67**, 725 (1995)
12. U. Weiss, Series in Modern Condensed Matter Physics, Vol. 2, 2nd edition, World Scientific, Singapore (1999)
13. F. Guinea, V. Hakim and A. Muramatsu, Phys. Rev. B **32** 4410, (1985)
14. T.A. Costi and G. Zaránd, to appear in Phys. Rev. B **59** (1999)
15. A.M. Tsvelick and P.B. Wiegmann, Adv. Phys. **32**, 453 (1983)
16. P.W. Anderson, Phys. Rev. **124**, 41 (1961)
17. J.R. Schrieffer and P.A. Wolff, Phys. Rev. **149**, 491 (1967)
18. P.W. Anderson and G. Yuval, Phys. Rev. Lett. **23**, 89 (1969); Phys. Rev. B **1**, 4464 (1970)
19. P.W. Anderson, J. Phys. C **3**, 2439 (1970)
20. L.N. Oliveira and J.W. Wilkins, Phys. Rev. **23**, 1553 (1981)
21. B.A. Jones, in Field Theories in Condensed Matter, Ed. Z. Tesanovic, Addison-Wesley, p 87 (1990)
22. O. Sakai and Y. Shimizu, J. Phys. Soc. Japan, **61**, 2333, (1991); ibid **61**, 2348 (1991)
23. D.M. Cragg, P. Lloyd and P. Nozieres, J. Phys. C **13**, 803 (1980)
24. H.B. Pang and D.L. Cox, Phys. Rev. B **44**, 9454 (1991); H.B. Pang, Phys. Rev. Lett. **73**, 2736 (1994)
25. T.A. Costi and A.C. Hewson, Physica C. **185–189**, 2649 (1991); T.A. Costi and A.C. Hewson, J. Magn. Magn. Mat. **108**, 129 (1992)
26. I.E. Perakis, C.M. Varma and A.E Ruckenstein, Phys. Rev. Lett. **70**, 3467 (1993); I.E. Perakis and C.M. Varma, Phys. Rev. **49**, 9041 (1994)
27. K. Ingersent, B.A. Jones and J.W. Wilkins, Phys. Rev. Lett. **69**, 2594 (1992)
28. H. Kusunose, K. Miyake, Y. Shimizu and O. Sakai, Phys. Rev. Lett. **76**, 271 (1996)
29. K. Satori, H. Shiba, O. Sakai and Y. Shimizu, J. Phys. Soc. Japan, **61**, 3239 (1992); ibid, **62**, 3181 (1993)
30. K. Chen and C. Jayaprakash, J. Phys. Cond. Matt. **7**, 491 (1995)
31. K. Ingersent, Phys. Rev. B **54**, 11936 (1996); C. Gonzalez-Buxton and K. Ingersent, preprint cond-mat/9803256, (1998)
32. R. Bulla, A.C. Hewson and G.M. Zhang, Phys. Rev. B **56**, 1171 (1997)
33. R. Bulla, Th. Pruschke and A.C. Hewson, J. Phys. Cond. Matt. **9**, 10463 (1997)
34. L.N. Oliveira and J.W. Wilkins, Phys. Rev. B **24**, 4863 (1981); ibid **32**, 696 (1985); D.L. Cox, H.O. Frota, L.N. Oliveira and J.W. Wilkins, Phys. Rev. B **32**, 555 (1985)

35. H.O. Frota and L.N. Oliveira, Phys. Rev. B **43**, 7871 (1986)
36. O. Sakai, Y. Shimizu and T. Kasuya, J. Phys. Soc. Jap. **58** 3666 (1989)
37. O. Sakai, Y. Shimizu and T. Kasuya, Physica B **163**, (1990)
38. T.A. Costi and A.C. Hewson, Physica B **163**, 179-181 (1990)
39. J.J.S. Brito and H.O. Frota, Phys. Rev. B **42**, 6378 (1990)
40. T.A. Costi and A.C. Hewson, Phil. Mag. B **65**, 1165 (1992)
41. T.A. Costi and C. Kieffer, Phys. Rev. Lett. **76**, 1683 (1996)
42. T.A. Costi, Phys. Rev. Lett. **80**, 1038 (1998)
43. T.A. Costi, P. Schmitteckert, J. Kroha and P. Wölfle, Phys. Rev. Lett. **73**, 1275 (1994); T.A. Costi, P. Schmitteckert, J. Kroha and P. Wölfle, Physica C **235–240**, 2287 (1994); T.A. Costi, J. Kroha and P. Wölfle, Phys. Rev. B **53**, 1850 (1996)
44. O. Sakai, Y. Shimizu and N. Kaneko, Physica B **186-188**, 323, (1993); Y. Shimizu and O. Sakai, Physica B **186-188**, 891 (1993)
45. T.A. Costi and A.C. Hewson, J. Phys. Cond. Matt. **30**, 361 (1993)
46. T.A. Costi, A.C. Hewson and V. Zlatić J. Phys. Cond. Matt. **6**, 2251 (1994)
47. S. Suzuki, O. Sakai and Y. Shimizu, J. Phys. Soc. Japan **65**, 4034 (1996)
48. D. Goldhaber-Gordon, J. Göres, M.A. Kastner, H. Shtrikman, D. Mahalu and U. Meirav, Phys. Rev. Lett. **81**, 5225 (1998)
49. H.O. Frota and K. Flensberg, Phys. Rev. B **46**, 15207 (1992)
50. W. Izumida, O. Sakai and Y. Shimizu, J. Phys. Soc. Japan **66**, 717 (1997); ibid **67**, (1998)
51. T.A. Costi, Phys. Rev. B**55**, 3003 (1997)
52. R. Bulla, Th. Pruschke, J. Low Temp. Phys. **99**, 619 (1995)
53. W. Metzner and D. Vollhardt, Phys. Rev. Lett. **62**, 324 (1989); A. Georges, G. Kotliar, W. Krauth and M.J. Rozenberg, Rev. Mod. Phys. **68**, 13 (1996)
54. O. Sakai and Y. Shimizu, Sol. St. Commun. **89**, 307 (1994)
55. R. Bulla and Th. Pruschke and A.C. Hewson, J. Phys. Cond. Matt. **10**, 8365 (1998)
56. M. Yoshida, M. A. Whitaker and L.N. Oliveira, Phys. Rev. B **41**, 9403 (1990)
57. W.C. Oliveira and L.N. Oliveira, Phys. Rev. B **49**, 11992 (1994); S.C. Costa, C.A. Paula, V.L. Líbero and L.N. Oliveira, Phys. Rev. B **55**, 30 (1997)
58. D.C. Langreth, Phys. Rev. **150**, 516 (1966)
59. K. Völker, Phys. Rev. B **58**, 1862 (1998)

2 The Density Matrix Renormalization Group

Reinhard M. Noack[1] and Steven R. White[2]

[1] Institut de Physique Théorique, Université de Fribourg
CH-1700 Fribourg, Switzerland
[2] Department of Physics and Astronomy, University of California
Irvine, CA 92697–4575, USA

The Density Matrix Renormalization Group [1] (DMRG) is a numerical technique for finding accurate approximations to the ground state and the low-lying excited states of strongly interacting quantum lattice systems such as the Heisenberg, t-J, and Hubbard models. DMRG traces it roots to Wilson's numerical renormalization group (RG) treatment of impurity problems [2] [see also Chap. 1(I)] and is related to real space renormalization groups. DMRG is remarkable in the accuracy that can be achieved for one-dimensional systems. For example, the ground state energy of the spin-one Heisenberg chain on lattices of hundreds of sites can be calculated to an accuracy of order 10^{-10} with a modest amount of computational effort. In addition, up to a dozen or so excited states with specified conserved quantum numbers as well as virtually any equal-time observable can be calculated. The principal limitation of the method at the present time is dimensionality, or, relatedly, range of the interaction, so that the majority of systems treated up to now have been one-dimensional or quasi-one-dimensional, i.e. strips of finite width. The computational effort necessary to achieve a given accuracy grows rapidly with the width of the system.

In this chapter we will attempt to introduce DMRG in a pedagogical manner. We will first briefly discuss some simpler methods to which DMRG is closely related, including exact diagonalization methods and Wilson's original Hamiltonian-based numerical method. Recently, there have been extensions of the method to the calculation of higher energy properties such as dynamical correlation functions [3], to two-dimensional classical systems [4], to one-dimensional quantum systems at finite temperature [5–7], and to quantum chemical calculation of the states of molecules [8]. Many of these extensions are discussed at length in other chapters of this book. In this chapter, we will concentrate on the original formulation for the ground state properties of quantum lattice systems.

The outline of this chapter is as follows: we first briefly introduce exact diagonalization in Sect. 1, and then discuss Wilson's original numerical renormalization group procedure as applied to quantum lattice systems in Sect. 2. We illustrate the reason for the failure of the Wilson RG for most quantum lattice models using the noninteracting tight-binding particle in Sect. 3. Two algorithms that overcome these shortcomings for the noninteracting system are also introduced in this section. In Sect. 4, we discuss how to generalize one of these, the superblock procedure, to interacting systems via a projection using the reduced density matrix. The two classes of DMRG algorithms, the infinite system algorithm and the finite system algorithm, are discussed in Sect. 5, along with additional details of the algorithms for interacting systems.

This chapter contains two appendices. Appendix A describes the form of the density matrix for single particle systems, such as the tight-binding chain. Appendix B describes the DMRG algorithm for the tight-binding chain and gives a C++ program to carry out the calculation. We hope that using this program to carry out a simple calculation will provide an excellent introduction to the ideas in DMRG.

1 Exact Diagonalization

Let us assume that we want to find the low-energy properties of strongly interacting quantum lattice models such as the Heisenberg model,

$$H = J \sum_{\langle ij \rangle} \boldsymbol{S}_i \cdot \boldsymbol{S}_j \ , \tag{1}$$

or the Hubbard model,

$$H = -t \sum_{\langle ij \rangle, \sigma} \left(c^\dagger_{i\sigma} c_{j\sigma} + c^\dagger_{j\sigma} c_{i\sigma} \right) + U \sum_i n_{i\uparrow} n_{i\downarrow} \ , \tag{2}$$

where $c^\dagger_{i\sigma}$ creates an electron of spin $\sigma = \uparrow$ or $\downarrow$ on site i, $n_{i\sigma} = c^\dagger_{i\sigma} c_{i\sigma}$, t is the hopping matrix element between neighboring sites, $\langle ij \rangle$ denotes a sum over pairs of nearest neighbor sites, and U is the energy cost due to the Coulomb repulsion of two electrons on the same site. The Heisenberg model can be obtained from the strong coupling (large U) limit of the half-filled Hubbard model with the quantum mechanical spin operator $\boldsymbol{S}_i = c^\dagger_{i\sigma} \boldsymbol{\tau}_{\sigma\sigma'} c_{i\sigma'}$, where $\boldsymbol{\tau}_{\sigma\sigma'}$ is a vector of Pauli matrices, and $J = 4t^2/U$. Here we will consider primarily the one-dimensional version of these models in order to illustrate the DMRG. The sum over nearest neighbors $\langle ij \rangle$ then reduces to a sum over lattice sites i with $j = i + 1$. Although these models can be written down in compact form, they involve many degrees of freedom. The spin-1/2 Heisenberg model on an L-site lattice has 2^L degrees of freedom, and the Hubbard model 4^L.

Unlike classical models whose ground states are usually trivial, but whose thermodynamics are interesting (e.g. the Ising model), the properties of the ground states of these quantum lattice models are difficult to calculate, and are, in general, poorly understood. For example, the question of whether or not the ground state of the doped two-dimensional Hubbard model on a square lattice is superconducting with $d_{x^2-y^2}$ pairing symmetry and therefore relevant to the high-T_c superconductors has not yet been definitively answered.

Both the Hubbard and the Heisenberg models do have exact solutions in one dimension via the Bethe Ansatz. However, the solution is somewhat unwieldy and only a limited number of properties of the system can be easily extracted. Comparison with the exact solution will nevertheless provide an opportunity to test the performance of the DMRG. In more than one dimension, there are no well-controlled analytical methods to treat these models. Therefore, numerical techniques have become an important means for studying them. The most important numerical methods are quantum Monte Carlo, exact diagonalization, and the numerical renormalization group.

Perhaps the simplest approach to numerically treating a quantum lattice system is to diagonalize the Hamiltonian for a finite-size lattice. A convenient basis for the Hilbert space for such a system has states of the form

$$|1\rangle \otimes |2\rangle \otimes \ldots \otimes |n\rangle \; , \tag{3}$$

where $|i\rangle$ labels a state of a single site of the system, and $\otimes$ denotes a direct product. For example, for a spin-1/2 system, $|i\rangle = \uparrow$ or $\downarrow$. The size of the Hilbert space grows exponentially in the number of sites n. If A is a single-site operator belonging to site i, its matrix elements have the simple form

$$\langle n| \ldots \langle 1|A|1'\rangle \ldots |n'\rangle = \delta_{11'} \ldots \delta_{i-1,i-1'} \delta_{i+1,i+1'} \ldots \delta_{n,n'} A_{i,i'} \; , \tag{4}$$

where $A_{i,i'}$ is the matrix elements of the operator for a single site. Similar expressions hold for products of single-site operators.

The δ-functions in these expressions make the matrix for the Hamiltonian operator in the many-site Hilbert space extremely sparse. Consequently, the usual dense matrix diagonalization methods, such as the Jacobi method or Householder transformation with diagonalization of the resulting tridiagonal matrix, are not the most efficient algorithms for these systems. We are primarily interested in the ground state and possibly a few low-lying excited states. Two widely used and very efficient methods for finding a few eigenstates of a large, sparse matrix are the *Lanczos method* and the *Davidson method*. These methods build up a small set of basis vectors, and minimize the energy within this basis. The reduced set of basis vectors is systematically expanded until convergence is reached. Typically, the original sparse Hamiltonian H only comes into play in the multiplication of a vector $\boldsymbol{v}$ by H, in which case only the nonzero elements of H are relevant. The Davidson method [9], also utilizes the diagonal elements of H in an attempt to generate an improved reduced basis. Consequently, if the Hamiltonian is at all dominated by its diagonal elements, the Davidson method will probably converge more quickly than the Lanczos method.

In efficient exact diagonalization programs, the Hamiltonian matrix is usually not explicitly generated. Instead, a procedure to multiply a vector by H is used which generates the Hamiltonian matrix elements as they are needed.

Although these diagonalization algorithms work remarkably well for finding the lowest eigenvalues and eigenvectors of the large, sparse matrices found in quantum lattice problems, the maximum system size that can be treated is still severely limited by the exponential growth of the Hilbert space. The principal limitations are then that the computer memory required to store a Hilbert space vector becomes too large to handle, and operations on these vectors become too expensive. In practice, the spin-1/2 Heisenberg model can be diagonalized on up to about 36 sites and the Hubbard model on up to about 20 sites. It would therefore be advantageous to formulate a variational diagonalization scheme that also truncates the Hilbert space used to represent H in a controlled way. This can be done using the *numerical renormalization group*.

2 Wilson's Numerical Renormalization Group

In this section, we will sketch out the basic ideas of Wilson's numerical renormalization group [2], using a notation that will be useful later for discussions of the DMRG. See Chap. 1(I) for a more complete discussion of Wilson's numerical RG as applied to impurity problems. We will postpone discussion of the mechanics of keeping and transforming the operators to a later section, since they are the same as for the DMRG.

The basic idea of the renormalization group is to integrate out unimportant degrees of freedom progressively using a succession of renormalization group transformations. In order to investigate the behavior of the single-impurity Kondo problem, Wilson implemented the RG in a purely *numerical* way [2]. The procedure starts with a numerical representation of the Hamiltonian in a particular basis, then adds degrees of freedom by carrying out a renormalization group transformation, typically by increasing the size of the finite system, and finally numerically transforms the representation of the Hamiltonian to a reduced basis.

To be more concrete, let us consider a one-dimensional quantum lattice model such as the Heisenberg or Hubbard model. In fact, Wilson was able to map the spherically symmetric Kondo problem onto such a one-dimensional quantum lattice model. In the Kondo case, the impurity is represented by the first site and the spherically symmetric momentum shells (logarithmically discretized) of the conduction electrons are represented by a semi-infinite lattice with only near-neighbor interactions.

Wilson's numerical RG procedure then proceeds as follows:

1. Isolate a portion of the system containing L sites. Here L is chosen to be small enough so that the Hamiltonian H_L can be diagonalized exactly.
2. Diagonalize H_L numerically, obtaining the m lowest eigenvalues and eigenvectors.

3. Transform H_L and other operators in the "block" of length L to a new basis consisting of the m lowest eigenvectors of H_L, i.e. form $\bar{H}_L = O_L^\dagger H_L O_L$, $\bar{A}_L = O_L^\dagger A_L O_L$, etc., where the columns of O_L contain the m lowest eigenvectors of H_L, and A_L is an arbitrary operator in the block. Note that $\bar{H}_L$ is a diagonal matrix with m elements.
4. Add a site to $\bar{H}_L$ to form H_{L+1}. In order to do this, the interaction between the block of length L and the additional site added must be reconstructed. We will discuss in more detail how this is done in Sect. 5.
5. Repeat starting with step 2, substituting H_{L+1} for H_L.

This scheme is depicted pictorially in Fig. 1. For a Hamiltonian represented in a real-space basis, the RG step is a real-space blocking scheme. Typically, the number of states m kept at each step is held constant, so the time and memory required for each diagonalization stays the same, and the computer time needed is linear in L.

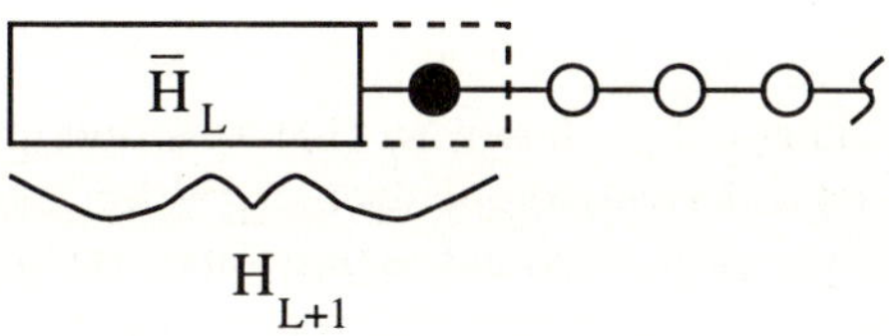

Fig. 1. A pictorial depiction of the Wilson numerical RG procedure.

The basic idea of this scheme is that only the low-energy eigenstates obtained for a system of size L will be important in making up the low-energy states of a system of size $L + 1$. Note that in isolating the block of length L, one has to decide how to treat the boundaries of the block. The simplest thing to do is to neglect connections to surrounding sites, which corresponds to applying open boundary conditions to the system being diagonalized.

This procedure worked well in Wilson's original work on the single impurity Kondo problem and, with minor variations, is still used today for a variety of single and two Kondo and Anderson impurity problems [see Chap. 1(I)]. In his original paper [2], Wilson very carefully justified the truncation by perturbatively calculating the error, and compared the numerical results with analytical analysis of the behavior near the fixed points. Due to the transformation into an energy basis and the logarithmic discretization of the conduction electron "sites", each site corresponds to a successively lower energy scale, and the coupling between successive sites decreases exponentially.

However, when this procedure is applied to other systems for which the lattice model does not include an intrinsic separation of energy scales, such as the one-dimensional Heisenberg or Hubbard models, the accuracy becomes quite poor after just a few iterations [10].

3 Numerical RG for the Particle on a Chain

In order to understand why the Wilson numerical renormalization group procedure breaks down for interacting quantum lattice systems, it is useful to consider first its application to a simple noninteracting quantum lattice problem, a single particle on a tight-binding chain. The Hamiltonian we will consider is

$$H = -\sum_{i=1}^{L-1} (\, |i\rangle\langle i+1| + |i+1\rangle\langle i| \,) + 2\sum_{i=1}^{L} |i\rangle\langle i| \, , \tag{5}$$

where the state $|i\rangle$ corresponds to a localized tight-binding orbital on site i. The matrix elements $H_{ij} = \langle i|H|j\rangle$ are then

$$H_{ij} = \begin{pmatrix} 2 & -1 & 0 & 0 \dots \\ -1 & 2 & -1 & 0 \dots \\ 0 & -1 & 2 & -1 \dots \\ \vdots & \vdots & \vdots & \vdots \end{pmatrix} . \tag{6}$$

The value 2 on the diagonal is chosen so that this operator is just the discretization of the second derivative operator, $-\partial^2/\partial x^2$.

The Wilson procedure described in the previous section can be carried out on this system with just a few minor differences. First, since this is a one-particle problem, the dimension of the Hilbert space for a lattice of length L is L, rather an exponential of L as in an interacting system. Since the Hilbert space grows less rapidly with system size for the noninteracting systems, we will add two equal-sized blocks in the real-space blocking step, rather than adding a site at a time. Secondly, the mechanics of putting the blocks together is a little simpler.

In step 1 of the Wilson procedure, we isolate a block of length L. We can understand how this is done for the noninteracting system by considering a semi-infinite system broken up into blocks of length L. The Hamiltonian can then be written

$$H = \begin{pmatrix} H_L & T_L & 0 & 0 & \dots \\ T_L^\dagger & H_L & T_L & 0 & \dots \\ 0 & T_L^\dagger & H_L & T_L & \dots \\ \vdots & \vdots & \vdots & \vdots \end{pmatrix} . \tag{7}$$

For $L = 1$, the H_1 is a 1×1 matrix with value 2 and T_1 is a 1×1 matrix with value -1. For larger L, H_L has the form of (6), and T_L connects only sites on the block boundaries, i.e. has all zero elements except for a -1 in the lower left corner. Isolating a block then consists of neglecting the T_L, and therefore applies fixed boundary conditions to H_L. Here we use the term "fixed" boundary conditions rather than the equivalent "open" boundary conditions to emphasize that the single-particle wavefunction vanishes at the boundary. (On the lattice, the wavefunction is actually zero only at the sites *beyond* the boundary, i.e. site 0 and $L+1$, see Fig. 2.)

In step 2, we diagonalize H_L and form

$$O_L = \begin{pmatrix} | & & | \\ \boldsymbol{v}_1 & \cdots & \boldsymbol{v}_m \\ | & & | \end{pmatrix} , \tag{8}$$

where $\boldsymbol{v}_1 \ldots \boldsymbol{v}_m$ are the eigenvectors corresponding to the $m \leq L$ lowest eigenvalues of H_L. In step 3, we form the diagonal $m \times m$ matrix $\bar{H}_L = O_L^\dagger H_L O_L$ and transform the connection between H_L and the rest of the system, T_L to the the new basis by forming $\bar{T}_L = O_L^\dagger T_L O_L$.

We now increase the size of the system, step 4, by putting two blocks of size L together to form a system of size $2L$ with

$$H_{2L} = \begin{pmatrix} \bar{H}_L & \bar{T}_L \\ \bar{T}_L^\dagger & \bar{H}_L \end{pmatrix} \tag{9}$$

and

$$T_{2L} = \begin{pmatrix} 0 & 0 \\ \bar{T}_L & 0 \end{pmatrix} . \tag{10}$$

The procedure can then be repeated starting with step 2 by substituting H_{2L} and T_{2L} for H_L and T_L. The size of the system therefore doubles at each step, but the size of the matrix to be diagonalized is at most $2m \times 2m$. Notice that since we transform to a truncated the basis at each step, the matrix elements of $\bar{H}_L$ and $\bar{T}_L$ can no longer be easily related to the original real-space basis. However, if m were equal to L at each step, the procedure would be exact; it would just be a complicated reshuffling of the original Hamiltonian.

As illustrated in the column labeled "Wilson" in Table 1, this procedure performs quite badly as soon as $m < L$. There are large errors in the energies of the lowest few states after only the first few truncating steps. This failure was pointed out by Wilson at an informal seminar at Cornell University in 1986 as an example of a numerical RG procedure which does not work. He also pointed out that in this simple system it is easy to understand why the procedure fails.

Table 1. Lowest energies after 10 blocking transformations for the noninteracting single particle on a 1-D chain with fixed boundary conditions.

	Exact	Wilson	Fixed-Free
E_0	2.3508×10^{-6}	1.9207×10^{-2}	2.3508×10^{-6}
E_1	9.4032×10^{-6}	1.9209×10^{-2}	9.4032×10^{-6}
E_2	2.1157×10^{-5}	1.9214×10^{-2}	2.1157×10^{-5}
E_3	3.7613×10^{-5}	1.9217×10^{-2}	3.7613×10^{-5}

In the continuum limit, the tight-binding model with fixed boundary conditions describes a particle in a box of length L with an infinitely high potential at the walls. The eigenfunctions are therefore particle-in-a-box eigenfunctions, $\psi_n(x) \sim \sin n\pi x/L$ with n a positive integer, and vanish at the boundaries of the box. In the RG procedure, the lowest few eigenstates of a system of length L are combined to form the low-lying eigenstates of a system of length $2L$. The lowest eigenstates of a system of length L and of length $2L$ are plotted in Fig. 2. Clearly, a combination of the ground states of two systems of length L is a bad approximation to the ground state for a system of length $2L$. Since the wave vectors of the discrete system take on small but finite values on the first and last sites of the system, the "kink" at the boundary between the blocks can be removed, but only by using almost all of the eigenstates of the smaller block.

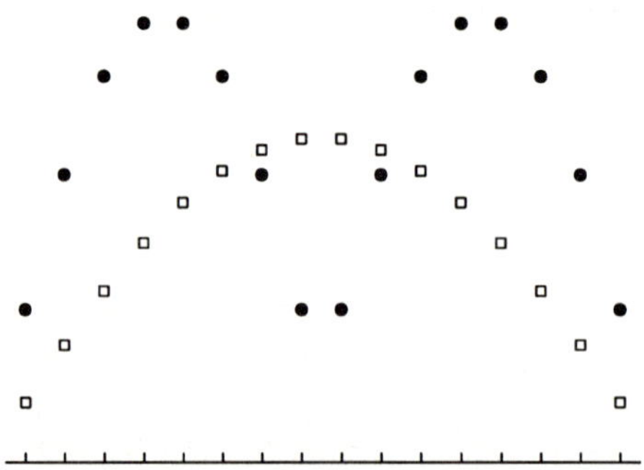

Fig. 2. The lowest eigenstates of two 8-site blocks (solid circles) and a 16-site block (open squares) for the one-dimensional tight-binding model with fixed boundary conditions.

The lesson that is learned is that the treatment of the boundaries of the blocks is crucial in formulating an accurate RG procedure. One could try to fix this problem by applying some other boundary conditions such as periodic boundary conditions to the block rather than fixed boundary conditions. While it is possible to formulate such an RG procedure, one finds that this procedure is also quite inaccurate after the first few truncations. The reason for this is that a periodic wave vector must have the same value at both boundaries of a block, so that the wave vector of the larger block is broken up into "steps", leading to an inaccurate representation of the low-lying states.

White and Noack [11] formulated two types of RG procedures which solve these problems and work quite well for the single-particle problem. Both are based on choosing a new basis for $\bar{H}_L$ which is *not* the eigenbasis of H_L. In the first procedure, called the *combination of boundary conditions* (CBC) method, the new basis is formed from the low-lying eigenstates of several different block Hamiltonians. The Hamiltonians are formed by applying a number of *different* boundary conditions to the edge of the block. For example, fixed and free (for free boundary conditions the derivative of the wave function vanishes at the boundary) boundary conditions can be applied. One can form $H_L^{bb'}$ with b, b' = fixed or free. For example,

$$H_{L=2}^{\text{free,fixed}} = \begin{pmatrix} 1 & -1 \\ -1 & 2 \end{pmatrix} . \tag{11}$$

We diagonalize $H_L^{bb'}$ for all 4 combinations of boundary conditions, and then form O_L from the $m/4$ eigenvectors associated with the lowest eigenvalues from each combination of boundary conditions. Since the columns of O_L are not orthogonal, we must explicitly orthogonalize them using the Gram-Schmidt procedure. We then form $\bar{H}_L^{bb'} = O_L^\dagger H_L^{bb'} O_L$ and $\bar{T}_L = O_L^\dagger H_L^{bb'} O_L$. Note that $\bar{H}_L^{bb'}$ is not diagonal. The Hamiltonian of the system of size $2L$ is then

$$H_{2L}^{bb'} = \begin{pmatrix} \bar{H}_L^{b,\text{fixed}} & \bar{T}_L \\ \bar{T}_L^\dagger & \bar{H}_L^{\text{fixed,b}'} \end{pmatrix} . \tag{12}$$

That fixed boundary conditions must be used where the blocks are joined together can be understood by considering a procedure in which all the states are kept at each step and requiring it to be exact. The matrix T_{2L} is formed as in (10).

The fixed-free CBC procedure works amazingly well. As can be seen by comparing the "Exact" and "Fixed-Free" columns of Table 1, the ground-state energy is obtained almost exactly (the error is in the tenth digit) even after 10 iterations ($L = 2048$) when $m = 8$ states are kept. The CBC procedure can also be formulated with other combinations of boundary conditions such as periodic and antiperiodic. While the periodic-antiperiodic CBC procedure is not as accurate as the fixed-free procedure, it performs much better than the Wilson procedure with periodic or antiperiodic boundary conditions.

The second type of procedure developed in Ref. [11], called the *superblock* method, chooses a new basis for $\bar{H}_{2L}$ and $\bar{T}_{2L}$ based on the idea that they will eventually be used to make up part of a larger system. In order to do this, a "superblock" (with periodic boundary conditions) made up of $p > 2$ blocks is formed and diagonalized. For example,

$$H_{2L}^{p=4} = \begin{pmatrix} \bar{H}_L & \bar{T}_L & 0 & \bar{T}_L^\dagger \\ \bar{T}_L^\dagger & \bar{H}_L & \bar{T}_L & 0 \\ 0 & \bar{T}_L^\dagger & \bar{H}_L & \bar{T}_L \\ \bar{T}_L & 0 & \bar{T}_L^\dagger & \bar{H}_L \end{pmatrix} . \tag{13}$$

The transformation O_{2L} is then made up by projecting the m lowest-lying eigenstates of H_{2L}^p onto the coordinates of the first two blocks, and then orthonormaling its columns. In other words, if u_j^α (with $j = 1, \ldots, 4m$) is an eigenvector of H_{2L}^4, then a nonorthonormalized column vector of O_{2L} is composed of the first $2m$ elements, $j = 1, \ldots, 2m$, of u_j^α, assuming $\bar{H}_L$ is an

$m \times m$ matrix. This new basis is used to transform $\bar{H}_{2L} = O_{2L}^{\dagger} H_{2L} O_{2L}$ and $\bar{T}_{2L} = O_{2L}^{\dagger} T_{2L} O_{2L}$, as defined in (9) and (10).

The idea is that the fluctuations in the additional blocks surrounding the portion of the system to be transformed effectively apply general boundary conditions, or equivalently, provide the conditions at the boundaries that the transformed blocks would see as part of a larger system. As p becomes large, this procedure becomes exact because it reduces to an exact diagonalization of the complete final system. Another interesting feature of this procedure is that the diagonalization step is decoupled from the real-space blocking step: a different size system is diagonalized than is blocked together. This procedure yields accurate results for the tight-binding particle eigenstates, although not quite as accurate as the fixed-free CBC procedure.

Of course, we are interested in developing an RG procedure for interacting quantum lattice systems, not the single-particle problem, so the crucial question is whether these procedures can be generalized to work on interacting systems. The CBC method is difficult to generalize because it is difficult to find a general enough set of boundary conditions for interacting systems. To see this, consider the many-body wavefunction for a system of noninteracting fermions, i.e. for the Hubbard model at $U = 0$. An arbitrary many-body state is composed of the Slater determinant of the single-particle wavefunctions of the individual particles, some of which may have nodes and some of which may have antinodes at the boundaries of a block. It is easy to find boundary conditions for which every particle on the block has a node or every particle has an antinode at the boundaries, but is is difficult to find boundary conditions which produce different behavior for different particles. However, a general representation for a block that is part of a larger system must provide a complete range of boundary behavior for each particle individually.

The superblock method seems more promising for application to interacting systems, since the general behavior at the boundaries is provided automatically by embedding the block of interest in a larger superblock. However, the projection of the wavefunction of the superblock onto the system block, which is a simple, single-valued coordinate projection in the noninteracting system, becomes multivalued for the interacting system: one state of the superblock can, in general, project onto many states of the system block. The Density Matrix Renormalization Group is based on choosing an optimal way to do this projection and on combining it with a version of the superblock procedure.

4 The Density Matrix Projection

In this section, we will discuss how to generalize the projection of the superblock described in the previous section for the noninteracting system to interacting systems. The procedure involves forming the reduced density matrix for the system block as part of the superblock. We will show that the basis obtained using this density matrix projection is the *optimal* basis in a particular sense.

First, let us briefly review the properties of density matrices. An excellent treatment is given in Feynman's book on statistical mechanics [12]. The term "density matrix" is used to refer to a number of different, but related mathematical objects, both in quantum mechanics and quantum statistical mechanics. Here we consider a quantum mechanical system in a definite pure state, and consider the properties of a part of that system. Since we will later use this procedure as part of a superblock algorithm, we will label the entire system the *superblock*, the part that we are interested in constructing a basis for the *system block*, and the remainder of the system the *environment block*, as depicted in Fig. 3. Let $|i\rangle$ label the states of the system block, and $|j\rangle$ label the states of the environment block, i.e. the rest of the superblock. If ψ is a state of the superblock,

$$|\psi\rangle = \sum_{ij} \psi_{ij} |i\rangle |j\rangle \ . \tag{14}$$

The reduced density matrix for the system block is defined as

$$\rho_{ii'} = \sum_{j} \psi^*_{ij} \psi_{i'j} \ . \tag{15}$$

By normalization, $\mathrm{Tr}\rho = 1$. The density matrix contains all the information needed from the wavefunction ψ to calculate any property restricted to the system block. If operator A acts only on the system block, then

$$\langle A \rangle = \sum_{ii'} A_{ii'} \rho_{i'i} = \mathrm{Tr}\ \rho A \ . \tag{16}$$

Now let us diagonalize the density matrix. Let ρ have eigenstates $|u^\alpha\rangle$ and eigenvalues $w_\alpha \geq 0$. Since $\mathrm{Tr}\rho = 1$, $\sum_\alpha w_\alpha = 1$. Then for any system block operator A,

$$\langle A \rangle = \sum_{\alpha} w_\alpha \langle u^\alpha | A | u^\alpha \rangle \ . \tag{17}$$

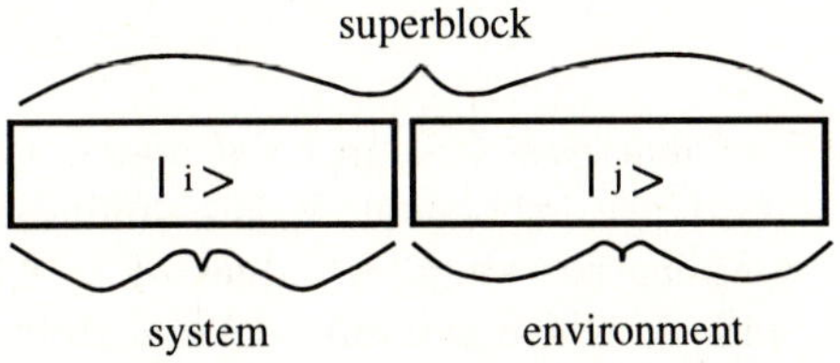

Fig. 3. A superblock divided into a system block and an environment block.

Equation (17) will apply immediately to our numerical renormalization group procedure. Suppose we wish to throw away some states from the system block. If for a particular α, $w_\alpha \approx 0$, we make no error in $\langle A\rangle$, for any A, if we discard $|u^\alpha\rangle$. We have found a way to find which states to keep (those with significant w_α) and which to discard.

This argument can be made much more precise. In particular, we can show that keeping the most probable eigenstates of the density matrix gives the most accurate representation of the state of the superblock, i.e., the system block plus the environment block. Let us assume we have diagonalized the superblock and obtained one particular state $|\psi\rangle$, typically the ground state. We wish to define a procedure for producing a set of states of the system block $|u^\alpha\rangle$, $\alpha = 1, \ldots, m$, with $|u^\alpha\rangle = \sum_i u_i^\alpha |i\rangle$, which are optimal for representing ψ in some sense. Because we allow only m states, we cannot represent $|\psi\rangle$ exactly if $\ell > m$, where ℓ is the number of system block states $|i\rangle$. We wish to construct an accurate expansion for $|\psi\rangle$ of the form

$$|\psi\rangle \approx |\bar{\psi}\rangle = \sum_{\alpha,j} a_{\alpha,j} |u^\alpha\rangle |j\rangle \ . \tag{18}$$

In other words, we wish to minimize

$$\mathcal{S} = \big| |\psi\rangle - |\bar{\psi}\rangle \big|^2 \tag{19}$$

by varying over all $a_{\alpha,j}$ and u^α, subject to $\langle u^\alpha | u^{\alpha'}\rangle = \delta_{\alpha\alpha'}$. Without loss of generality, we can write

$$|\bar{\psi}\rangle = \sum_\alpha a_\alpha |u^\alpha\rangle |v^\alpha\rangle \tag{20}$$

where $v_j^\alpha = \langle j | v^\alpha \rangle = N_\alpha a_{\alpha,j}$, with N_α chosen to set $\sum_j |v_j^\alpha|^2 = 1$. Switching to matrix notation, we have

$$\mathcal{S} = \sum_{ij} (\psi_{ij} - \sum_{\alpha=1}^{m} a_\alpha u_i^\alpha v_j^\alpha)^2 \ , \tag{21}$$

and we minimize $\mathcal{S}$ over all u^α, v^α, and a_α, given the specified value of m. The solution to this minimization problem is known from linear algebra. We now think of ψ_{ij} as a rectangular matrix. The solution is produced by the singular value decomposition [13] of ψ,

$$\psi = U D V^T \ , \tag{22}$$

where U and D are $\ell \times \ell$ matrices, V is an $\ell \times J$ matrix (where $j = 1, \ldots, J$, and we assume $J \geq \ell$), U is orthogonal, V is column-orthogonal, and the diagonal matrix D contains the singular values of ψ. Linear algebra tells us that the u^α, v^α, and a_α which minimize $\mathcal{S}$ are given as follows: the m largest-magnitude diagonal elements of D are the a_α and the corresponding columns of U and V are the u^α and v^α. (We emphasize that the singular value decomposition is not being used here as a numerical method, only as a convenient factorization which allows us to use a theoretical result from linear algebra.)

These optimal states u^α are also eigenvectors of the reduced density matrix of the block as part of the system. This reduced density matrix for the block depends on the state of the system, which in this case is a pure state $|\psi\rangle$. (The system could also be in a mixed state [see below] or at finite temperature.) The density matrix for the block in this case, where ψ_{ij} is assumed real, is given by

$$\rho_{ii'} = \sum_j \psi_{ij}\psi_{i'j} . \qquad (23)$$

We see that

$$\rho = UD^2U^T , \qquad (24)$$

i.e. U diagonalizes ρ. The eigenvalues of ρ are $w_\alpha = a_\alpha^2$ and the optimal states u^α are the eigenstates of ρ with the largest eigenvalues. Each w_α represents the probability of the block being in the state u^α, with $\sum_\alpha w_\alpha = 1$. The deviation of $P_m \equiv \sum_{\alpha=1}^m w_\alpha$ from unity, i.e. the "discarded weight" of the density matrix eigenvalues, measures the accuracy of the truncation to m states.

To summarize, in the previous two paragraphs we have shown that when the superblock is assumed to be in a pure state, the optimal states to keep are the m most significant eigenstates of the reduced density matrix of the system block, obtained from the wavefunction of the superblock via (23).

We can also consider the superblock to be in a mixed state. This is the natural assumption for a system at finite temperature, and it is also useful to assume a mixed state when one wishes to obtain several of the lowest lying states: if we put the superblock with equal probability into each of several states, then the system block states obtained from the density matrix will equally well represent each of these superblock states. We represent the mixed case by saying that the superblock has probability W_k to be in state $|\psi^k\rangle$. If the superblock is at a finite temperature, then the W_k are normalized Boltzmann weights. In this case the appropriate definition for the error in the representation is

$$S = \sum_k W_k \sum_{ij} (\psi_{ij}^k - \sum_{\alpha=1}^m a_\alpha^k u_i^\alpha v_j^{k,\alpha})^2 . \qquad (25)$$

Note that we are interested in determining a single set of optimal u^α, whereas we allow the rest of the system additional freedom to choose a different v^α for each state k. Minimizing over the u^α, $v^{k,\alpha}$, and a_α^k, we find

$$\rho u^\alpha = w_\alpha u^\alpha \qquad (26)$$

with

$$\rho_{ii'} = \sum_k W_k \sum_j \psi^k_{ij} \psi^k_{i'j} \tag{27}$$

and

$$w_\alpha = \sum_k W_k (a^k_\alpha)^2 \,. \tag{28}$$

This equation for ρ is the definition of the reduced density matrix when the superblock is in a mixed state, and the u^α are the eigenstates of ρ.

Thus the conclusion when the superblock is in a mixed state is identical to the result for a pure state: the optimal states to keep are the eigenvectors of the reduced density matrix with the largest eigenvalues.

The effectiveness of the truncation of the Hilbert space of the system block via the density matrix depends crucially on the distribution of the density matrix eigenvalues w_α. For certain exactly solvable systems, it is possible to determine the density matrix eigenvalues exactly. For these integrable systems, the distribution of the density matrix eigenvalues can be shown to decay exponentially, see Chap. 3.1(II).

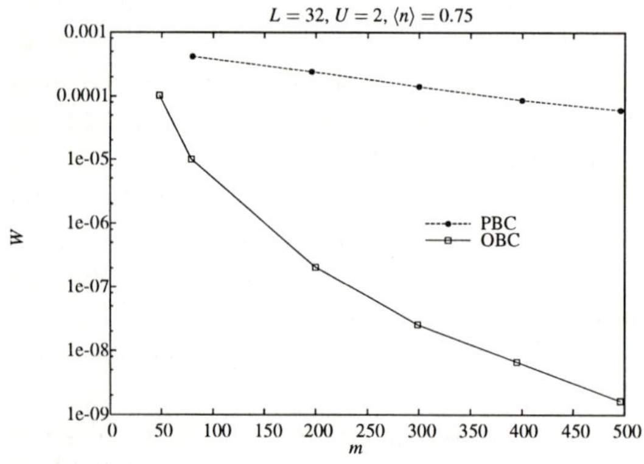

Fig. 4. Sum of the discarded weight of the density matrix eigenvalues for the finite system DMRG algorithm, as a function of the number of states kept m.

As a realistic example of the distribution of the density matrix eigenvalues, we give numerical results for the one-dimensional Hubbard model obtained from the finite size DMRG algorithm, which will be discussed in the next section. In Fig. 4, we plot the sum of the discarded density matrix eigenvalues, $\sum_{\alpha=m+1}^{m_{\max}} w_\alpha$, where $m_{\max}$ is the size of the density matrix. This discarded weight is plotted in Fig. 4 as a function of m on a logarithmic scale for the 32-site Hubbard chain with open and periodic boundary conditions at $n = 0.75$. As can be seen, the discarded weight falls off rapidly with the number of states kept. For open boundary conditions, the decay is clearly slower than exponential for small m, in contrast to the original results found for the spin-one Heisenberg chain [1]. This could be because the system is in the Luttinger liquid regime and therefore has no gap in either spin or charge excitations. It is possible that the decay becomes exponential at large enough m, when the error in the energy becomes less than the finite energy level spacing always found on a finite system. This discarded weight is strongly correlated with the error in the ground state energy and is often used as a measure of the error. It has been used to to make an $m \to \infty$ extrapolation of the energy [14]. Also depicted in Fig. 4 is the discarded weight for the same system with periodic boundary conditions. The convergence of the discarded weight (and thus the ground state energy) with m is much slower than for open boundary conditions. It is generally found that the accuracy of the energy for a given m is many orders of magnitude worse for periodic than for open boundary conditions. It is therefore usually better to treat systems with open boundary conditions on much larger lattices rather than systems with periodic boundary conditions.

5 DMRG Algorithms

In this section, we will describe how to combine the superblock procedure with the density matrix projection in order to define efficient DMRG algorithms. There are three main ingredients needed to form a DMRG algorithm: first, we have to decide how to add degrees of freedom to the system, i.e. how to build up the system block; second, we have to determine the configuration of the superblock; and finally, we must choose which superblock eigenstate or eigenstates to use to construct the density matrix.

For interacting systems, it is clear that one wants to add the minimum number of degrees of freedom at once to the system block in order keep as large a fraction of the system block states as possible, and to keep the size of the Hilbert space of the superblock as small as possible. Therefore, one usually wants to build up the system block one site at a time in a procedure similar to that described for the Wilson numerical RG in Sect. 2.

The algorithms then fall into two classes, depending on how the environment block is chosen to form the superblock: the infinite system algorithm and the finite system algorithm. We will discuss these algorithms in detail below.

We will call the superblock state or states used to form the reduced density matrix for the system block *target states.* If only ground state properties are desired, it is most accurate to target just the ground state of the superblock. (The Hamiltonian is usually block diagonal in particular quantum numbers such as S_z; by ground state we will mean ground state for a particular quantum number.) If excited states or matrix elements between different states are required, more than one target state can be used. However, for fixed number of states kept m, the accuracy with which the properties of each individual state can be determined goes down as more states are targeted. For simplicity, we will assume that only the ground state is targeted in the following.

The infinite system algorithm

The infinite system algorithm is the most straightforward extension of the Wilson procedure described in Sect. 2 that incorporates the superblock concept. We build up the system block one site at a time, just as in the Wilson procedure, but must choose some sort of environment block. The simplest way of forming the environment block is to use a reflection of the system block. The superblock configuration is shown in Fig. 5. Here $\bar{H}_\ell$ is the Hamiltonian for the system block in the reduced basis, as before, and the solid dots represent single sites. The right block, $\bar{H}_\ell^R$, is formed by relabeling the sites in the system block so that they are reflected onto the right part of the lattice.

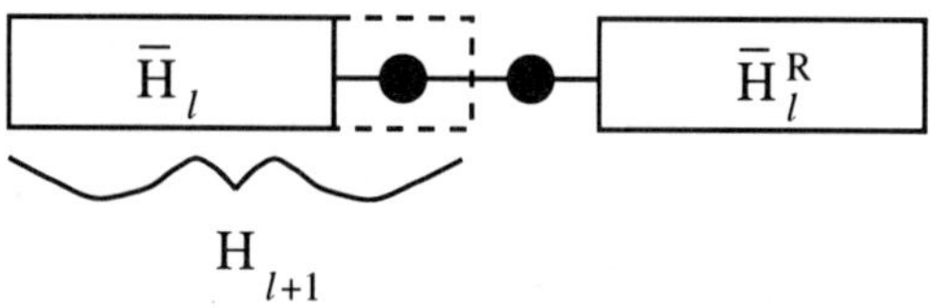

Fig. 5. The superblock configuration for the infinite-system algorithm.

The infinite system algorithm then proceeds as follows:

1. Form a superblock containing L sites which is small enough to be exactly diagonalized.
2. Diagonalize the superblock Hamiltonian H_L^{super} numerically, obtaining *only* the ground state eigenvalue and eigenvector $|\psi\rangle$ using the Lanczos or Davidson algorithm.
3. Form the reduced density matrix $\rho_{ii'}$ for the new system block from $|\psi\rangle$ using (15). Note that $\ell' = \ell = L/2 - 1$.
4. Diagonalize $\rho_{ii'}$ with a dense matrix diagonalization routine to obtain the m eigenvectors with the largest eigenvalues.
5. Construct $H_{\ell+1}$ and other operators in the new system block and transform them to the reduced density matrix eigenbasis using $\bar{H}_{\ell+1} = O_L^\dagger H_{\ell+1} O_L$, $\bar{A}_{\ell+1} = O_L^\dagger A_{\ell+1} O_L$, etc., where the columns of O_L contain the m highest eigenvectors of $\rho_{ii'}$, and $A_{\ell+1}$ is an operator in the system block.
6. Form a superblock of size $L+2$ using $\bar{H}_{\ell+1}$, two single sites and $\bar{H}_{\ell+1}^R$.
7. Repeat starting with step 2, substituting H_{L+2}^{super} for H_L^{super}.

It is clear that this algorithm is very much in the spirit of the original Wilson procedure in that the system being diagonalized grows at each step. There are, however, a few important differences. First, as in the noninteracting superblock procedure, the diagonalization step and the real-space blocking step take place on different size systems. Therefore, the energy and various expectation values are calculated during the superblock diagonalization, while the density matrix diagonalization rather than a Hamiltonian diagonalization is used to determine the new basis for the system block. Second, the size of the superblock grows by two sites rather than one site at every step. Third, we have assumed that the system is reflection symmetric. It is possible to formulate algorithms that do not assume reflection symmetry, but this is done most easily in the context of the finite system algorithm described below.

The finite system algorithm

In the finite size algorithm, the environment block is chosen in a different way: it is chosen so that the size of the superblock is kept fixed at each step. Suppose that we have run the infinite system algorithm until the superblock reaches size L, but have stored all the $\bar{H}^R_{\ell'}$ for $\ell = 1, \ldots, L/2-2$ as well as the all the additional operators needed to connect the blocks at each step. We can then continue to build up the system block, but keep $L = \ell + \ell' + 2$ fixed by using the appropriate previously stored $\bar{H}^R_{\ell'}$. The finite size algorithm then proceeds as follows:

0. Carry out the infinite system algorithm until the superblock reaches size L, storing $\bar{H}_\ell$ and the operators needed to connect the blocks at each step.
1. Carry out steps 3-5 of the infinite system algorithm to obtain $\bar{H}_{\ell+1}$. Store it. (Now $\ell \neq \ell'$.)
2. Form a superblock of size L using $\bar{H}_{\ell+1}$, two single sites and $\bar{H}^R_{\ell'-1}$. The superblock configuration is given by Fig. 6, where $\ell' = L - \ell - 2$.
3. Repeat steps 1-2 until $\ell = L - 3$ (i.e. $\ell' = 1$). This is the *left to right* phase of the algorithm.
4. Carry out steps 3-5 of the infinite system algorithm, reversing the roles of $\bar{H}_\ell$ and $\bar{H}^R_{\ell'}$, i.e. switch directions to build up the right block and obtain $\bar{H}^R_{\ell'+1}$. Store it.
5. Form a superblock of size L using $\bar{H}_{\ell-1}$, two single sites and $\bar{H}^R_{\ell'+1}$.
6. Repeat steps 4-5 until $\ell = 1$. This is the *right to left* phase of the algorithm.
7. Repeat starting with step 1.

A useful analogy is to think of this procedure as being like running a zipper repeatedly from left to right and then right to left through a superblock that is always the same size. Each time the zipper changes direction, a new set of stored blocks is used as the environment block. In this way, the representations of the stored blocks are iteratively improved and the zipping can be repeated until convergence is reached.

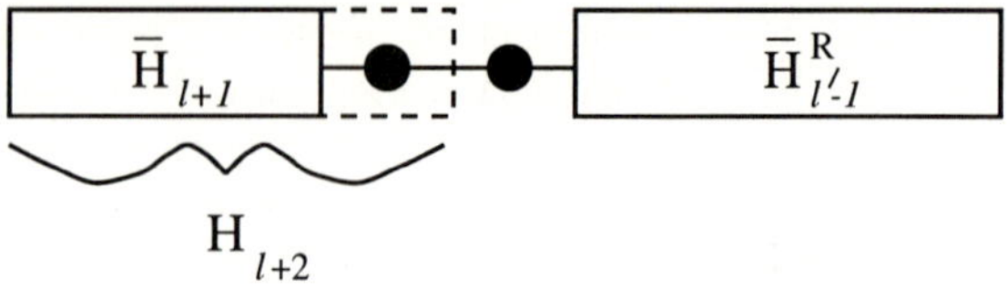

Fig. 6. A typical superblock configuration for the left-to-right phase of the finite-system algorithm.

Note that since the left block and the right blocks are stored independently, we do not have to assume that the lattice is reflection symmetric (at least after step 0). Since the same size superblock is always diagonalized, the algorithm is less dependent than the infinite system algorithm on translational invariance, i.e. on the optimum representation of different size superblocks being similar.

If reflection symmetry is present, it can be used at the point at which $\ell = \ell'$ to shorten the length of the zips. One way of formulating the algorithm in this case is to build up the left blocks from $\ell = 1$ to $\ell = L/2 - 1$, and build up the right blocks from $\ell' = L/2$ to $\ell' = L - 3$, i.e. to only zip from the left side of the superblock to the middle and then back to the left side. The fact that we have used reflection symmetry in the infinite system phase, step 0, is usually not important. However, it is also possible to formulate infinite system algorithms that do not use reflection symmetry. This issue will be discussed in more detail below in the context of algorithms for two-dimensional and fermion systems.

For a given system size L, the finite system algorithm almost always gives substantially more accurate results than the infinite system algorithm, and is therefore usually preferred unless there is a specific reason to go to the thermodynamic limit.

It is instructive to examine how the infinite and finite system algorithms work in detail on a particular system. Here we will utilize the one-dimensional Hubbard model, (2), as an example. This model is fairly complicated, containing both spin and fermionic degrees of freedom, but has a Bethe Ansatz exact solution [15], which has been extended to the case of open boundary conditions [16]. Here we will investigate how the energy obtained from the finite-system DMRG algorithm converges to the exact Bethe Ansatz energy on a particular system. This is done for a fractionally filled 32-site system with open boundary conditions in Fig. 7. The ordinate "Iteration" refers to the iteration of reflection symmetric finite system algorithm, for which each iteration consists of a right-to-left "zip" from center (the reflection symmetric configuration) to the left edge, followed by a left-to-right "zip" back to the center.

As can be seen from the curves which were made for different runs with different numbers of states m kept in the system block, there is virtually no

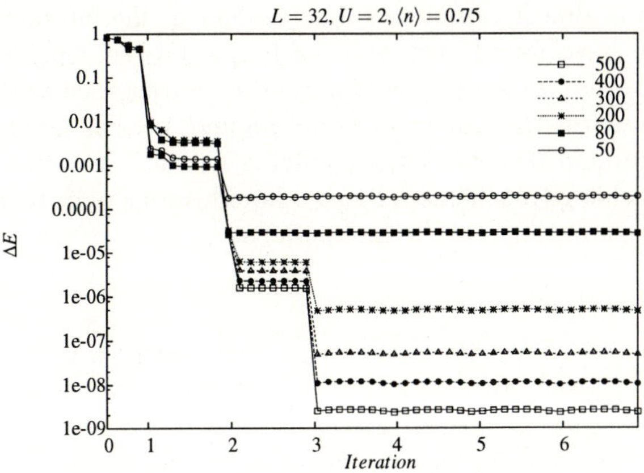

Fig. 7. The difference between the ground-state energy obtained using the finite system DMRG and the exact Bethe Ansatz energy for a 32-site system at $U = 2$ and filling $\langle n \rangle = 0.75$, plotted on a logarithmic scale as a function of iteration of the finite system algorithm. Each point represents a Davidson diagonalization of the superblock (every fourth diagonalization is plotted), and the different symbols represent different runs made with a fixed number $m = 50, \ldots, 500$ of states kept.

dependence of the accuracy of the energy on m in the zeroth iteration, the infinite system phase. This behavior is typical for fractionally filled fermion systems and for finite-width systems. It is due to the poor job the infinite system algorithm does at representing systems with an intrinsic length, such as the wavelength of local density oscillations or the width of a two-dimensional system. After two to three iterations, the error in the energy saturates for a given m, with the saturation occurring after a larger number of iterations for larger m. Therefore, it is only helpful to increase m once sufficient convergence in the number of iterations has been achieved. The optimum strategy is then to increase m fairly rapidly at each iteration. If this is done, the majority of time is spent doing the last iteration, and the finite system algorithm only takes a factor of two to three longer than the infinite system algorithm for the same number of states, but can be orders of magnitude more accurate. Note that the rate of convergence with the number of iterations can vary strongly from model to model, so the optimum strategy will also vary.

Once the optimum convergence is achieved, the crucial question is how the accuracy of the energy depends on the maximum number of states kept in a run, i.e. on how the height of the plateaus in Fig. 7 depends on m. This dependence is illustrated in Fig. 8 for different U values. The form of the decay in the error in the ground state energy with m is very similar to the form of decay of the discarded weight of the density matrix eigenvalues, Fig. 4. The error in the energy is usually proportional to the discarded weight, once the algorithm has converged sufficiently. Another interesting point is that the rate of convergence is almost completely independent of the interaction strength U. In fact, we have found that even adding a nearest-neighbor Coulomb interaction does not substantially change the convergence with m [17], as long as the system is the Luttinger liquid regime, i.e. has no spin or charge gap. The nature and density of the low-lying excitations in the superblock, which does not change much in a Luttinger liquid, seems to be more important than the nature and range of the interactions.

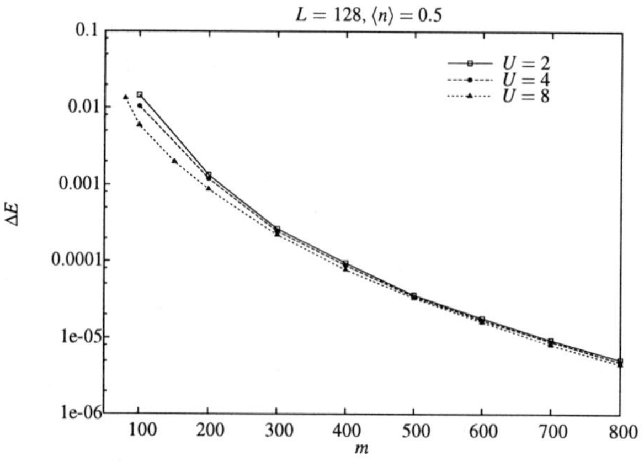

Fig. 8. Convergence of the finite system algorithm as a function of the number of states kept m for different U.

Details for interacting systems

Up to now, we have not considered in detail how to store and transform the operators necessary to carry out the renormalization group transformation for an interacting system. In this section, we will discuss how to do this efficiently.

In order to construct the Hamiltonian of the system, a block must have various operators stored as matrices connecting these states. For example, for the Heisenberg model with exchange terms

$$\boldsymbol{S}_i \cdot \boldsymbol{S}_{i+1} = S_i^z S_{i+1}^z + \frac{1}{2}(S_i^+ S_{i+1}^- + S_i^- S_{i+1}^+) \tag{29}$$

one needs to store $m \times m$ matrix representations of S_i^z, S_i^+, and S_i^- for i equal to the left or right end sites of the block. (In practice, one need not store S_i^-, since it can be obtained by taking the Hermitian conjugate of S_i^+). For a Hubbard model one would have to store matrices for $c_{i\sigma}$, with $\sigma = \uparrow$ and $\downarrow$, in order to reconstruct the hopping term $\sum_\sigma (c_{i+1\sigma}^\dagger c_{i\sigma} + c_{i\sigma}^\dagger c_{i+1\sigma})$.

Consider joining two blocks B_1 and B_2 together in a Heisenberg system. In the Wilson procedure, B_2 will typically consist of a single site, and for the DMRG algorithms, it is sufficient to consider how to compose two blocks. If B_1 has m_1 states, and B_2 has m_2 states, the combined block B_1B_2 has m_1m_2 states. We label the combined states by two indices, ij. The matrix representing the Hamiltonian of B_1B_2 is then given by

$$[H_{B_1B_2}]_{ij;i'j'} = [H_{B_1}]_{ii'}\,\delta_{jj'} + [H_{B_2}]_{jj'}\,\delta_{ii'} + [S^z_\ell]_{ii'}\,[S^z_{\ell+1}]_{jj'} \\ +\tfrac{1}{2}\,[S^+_\ell]_{ii'}\,[S^-_{\ell+1}]_{jj'} + \tfrac{1}{2}\,[S^-_\ell]_{ii'}\,[S^+_{\ell+1}]_{jj'} \tag{30}$$

where H_{B_1} is the Hamiltonian matrix of block B_1, and ℓ is its rightmost site. In order for the connections in the Hamiltonian to be restored when the two blocks are combined, each block must contain each of the matrices appearing in (30).

The most time-consuming part of a DMRG calculation is the diagonalization of the system Hamiltonian, which occurs once for every step. Only the ground state or a few low-lying states are needed, and so the Lanczos or Davidson iterative algorithms should be used. These algorithms both require repeated multiplications of superblock vectors ψ by the superblock Hamiltonian H^{super}. However, the actual Hamiltonian matrix should not be constructed and stored. Instead, the following procedure to directly multiply $H^{\text{super}}\psi$ uses much less memory and is also much faster. Consider again, for simplicity, a system formed from two blocks. The superblock Hamiltonian can be written in the general form

$$[H^{\text{super}}]_{ij;i'j'} = \sum_\alpha A^\alpha_{ii'} B^\alpha_{jj'} \ . \tag{31}$$

Then the product $H^{\text{super}}\psi$ can be written as

$$\sum_{i'j'} [H^{\text{super}}]_{ij;i'j'}\,\psi^\alpha_{i'j'} = \sum_\alpha \sum_{i'} A^\alpha_{ii'} \sum_{j'} B^\alpha_{jj'} \psi_{i'j'} \ . \tag{32}$$

For each α, the last sum is performed first, as a matrix-matrix multiplication of B^α and ψ^T, to form a temporary matrix $C^\alpha_{ji'}$. Then a matrix-matrix multiplication of A^α and $[C^\alpha]^T$ forms a partial result, which is added into the result vector, giving a sum on α.

Whenever a site is added onto a block, or more generally two blocks are added, the operator matrices must be updated. The eigenstates of the density matrix can be written in the form u^α_{ij}, which we write as $O_{ij;\alpha}$, $\alpha = 1, \ldots, m$. Here i and j represent state indices of the two blocks that are being added together. Then for each operator A that is needed, $A_{ij;i'j'}$ is replaced by $A_{\alpha\alpha'}$, where

$$A_{\alpha\alpha'} = \sum_{iji'j'} O_{ij;\alpha} A_{ij;i'j'} O_{i'j';\alpha'} \ . \tag{33}$$

The terms appearing in (30) show the various ways operators $A_{ij;i'j'}$ can be formed from single-block operators $A_{ii'}$.

Any efficient DMRG program must make use of quantum numbers to speed up the calculation and reduce storage. For example, in order to construct the system Hamiltonian it may be necessary to store for a block the matrix form of the operator $\hat{S}^+_\ell$, where ℓ is the right-most site of the block. If there are m states in the block, this is an $m \times m$ matrix. However, if states are labeled and grouped by block quantum number S^z, then this matrix is mostly zeroes, with the nonzero parts in rectangular blocks. These blocks connect states with specific quantum numbers, e.g. the states corresponding to the left index of the matrix may have $S_z = 0$, and for the right index $S_z = -1$. It is essential to store only the nonzero elements of this matrix. Although this can done using sparse matrices, the best way to do it is as a set of dense matrices, one for each nonzero rectangular block. The multiplication of $H^{\text{super}}\psi$ described above takes place as described in the previous paragraph, except that now there is an additional sum or loop over quantum numbers, and the dense matrices which are multiplied are much smaller. Keeping track of all the matrices, each of a different size, can be very well organized in C++ by defining classes to represent operator matrices, submatrices, etc. The fact that these matrices are all different sizes and have different dimensions at each DMRG step makes it somewhat more difficult to use storage efficiently in Fortran 77, which does not have dynamic allocation of memory.

It is useful at this point to mention typical maximum numbers of states kept, m, for various systems on current computers. For the one-dimensional Hubbard model, $m = 800$ for a system of up to a few hundred sites can be treated on typical workstation, using a few hundred megabytes of main memory [17,18]. For the Heisenberg model, up to $m = 1100$ states have been kept for the spin-one chain with an impurity [19] and $m = 1700$ for the spin-two chain [20]. These Heisenberg-chain calculations have been carried out using a highly optimized Fortran code.

Measurements

Measurements are made using the superblock wavefunction $|\psi\rangle$ to evaluate expectation values of the form $\langle\psi|A|\psi\rangle$. Rather complicated operators can be evaluated fairly easily, but dynamical information is more difficult to obtain. In order to measure A, one must have kept operator matrices for the components of A. For example, to measure the on-site spin-density S^z_j for all sites ℓ, one must keep track of matrices $[S^z_\ell]_{ii'}$, for all each site ℓ in each of the blocks. These operators must be updated using (33) at every step of each iteration. We will once again divide superblock into two parts with states labeled by $|i\rangle$ and $|j\rangle$. One then obtains the expectation value using

$$\langle\psi|S^z_\ell|\psi\rangle = \sum_{i,i',j} \psi^*_{ij}\, [S^z_\ell]_{ii'}\, \psi_{i'j} \; , \tag{34}$$

etc. This procedure gives *exact* evaluations of $\langle\psi|A|\psi\rangle$ for the *approximate* eigenstate $|\psi\rangle$.

For a correlation function such as $\langle\psi|S^z_\ell S^z_m|\psi\rangle$, the evaluation depends on whether ℓ and m are on the same block or not. If they are on different blocks, then one need only have kept track of $[S^z_\ell]_{ii'}$ and $[S^z_m]_{jj'}$, and one has

$$\langle\psi|S^z_\ell S^z_m|\psi\rangle = \sum_{i,i',j,j'} \psi^*_{ij}\,[S^z_\ell]_{ii'}[S^z_m]_{jj'}\,\psi_{i'j'}\ . \tag{35}$$

If ℓ and m are on the same block, one *should not* use

$$\langle\psi|S^z_\ell S^z_m|\psi\rangle \approx \sum_{i,i',i'',j} \psi^*_{ij}\,[S^z_\ell]_{ii'}[S^z_m]_{i'i''}\,\psi_{i''j}\ . \tag{36}$$

This expression does not evaluate the correlation function exactly within the approximate state $|\psi\rangle$. The sum over i' should run over a complete set of states, but does not, whereas the sums over the other variables need run only over those states needed to represent $|\psi\rangle$, since they appear as a subscript in either the $|\psi\rangle$ on the left or on the right.

To evaluate this type of correlation function, one needs to have kept track of $[S^z_j S^z_k]_{ii'}$ throughout the calculation. One then evaluates

$$\langle\psi|S^z_\ell S^z_m|\psi\rangle = \sum_{i,i',j} \psi^*_{ij}\,[S^z_\ell S^z_m]_{ii'}\,\psi_{i'j} \tag{37}$$

to obtain the correlation function.

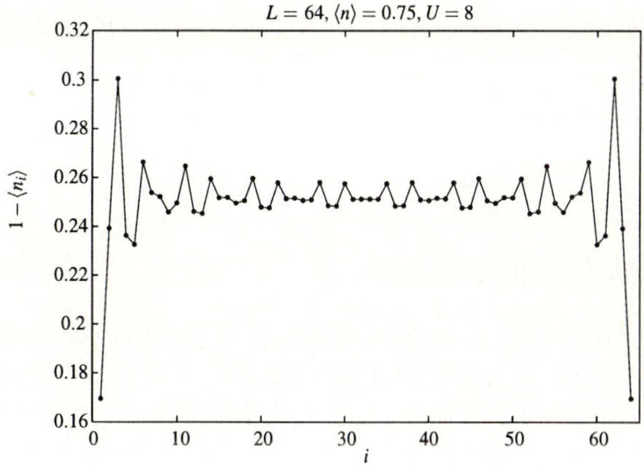

Fig. 9. Local hole density $1 - \langle n_i\rangle$ plotted as a function of i.

As an example of measurements of the first type, (34), we display the local hole density $1 - \langle n_i\rangle$, where $n_i \equiv n_{i\uparrow} + n_{i\downarrow}$ for the one-dimensional

Hubbard model in Fig. 9. The oscillations are Friedel oscillations due to the open boundary conditions.

In Fig. 10, we show the nearest-neighbor spin-spin correlations for the half-filled system as an example of a correlation function between different sites, calculated using (37). This correlation function also has oscillations due to the open boundaries. Since the correlation functions cannot be calculated directly from the Bethe Ansatz, is is not possible to directly calculate the error, as for the ground state energy. The issue of how to form a good approximation to correlation functions in the thermodynamic limit using these finite-lattice correlation functions, as well as a comparison with the asymptotic results obtain from a combination of bosonization and the Bethe Ansatz [21] is discussed in more detail in Chap. 1.1(II).

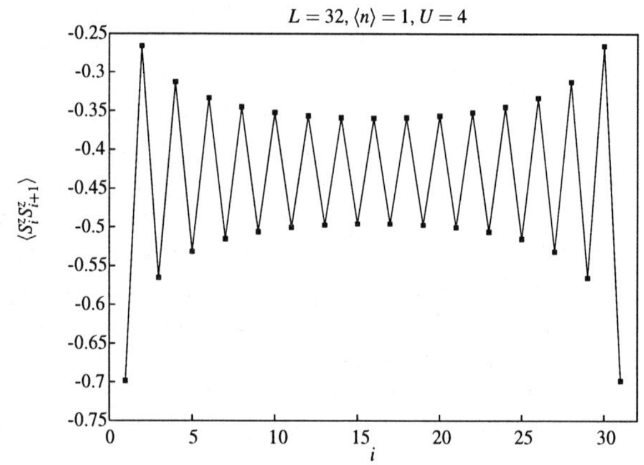

Fig. 10. Nearest neighbor spin-spin correlation function $\langle S_i^z S_{i+1}^z \rangle$ plotted as a function of i for the half-filled system.

Wave function transformations

An important improvement in the efficiency of DMRG comes from keeping track of the wavefunction from step to step [22]. The step referred to here is the process of adding a site to a block and requires the diagonalization of a system configuration of two blocks and two sites [1]. In each DMRG step, an iterative sparse matrix algorithm, such as the Davidson method, is used to find the ground state of the system. Thus far, we have not specified a starting point for the Davidson procedure. To ensure that the DMRG procedure is always stable and convergent, the system ground state usually has to be determined to rather high accuracy. (One diagonalization which converges to a low-lying eigenstate other than the ground state can ruin the accuracy of the entire DMRG sweep.) Consequently, a substantial number of Davidson steps are necessary to converge to sufficient accuracy, typically 40-100. The total calculation time is proportional to the average number of Davidson steps.

If a very good initial guess is available for the Davidson procedure, the number of Davidson steps can be reduced substantially. An ideal initial guess, for the case of the finite system DMRG algorithm, is the final wavefunction from the previous DMRG step. This wavefunction, however, is in a different basis, corresponding to a different system configuration, but it can be transformed into the basis corresponding to the current configuration, as described below. Use of this transformation to obtain the initial state in a Davidson diagonalization can reduce the number of Davidson steps by one half, typically, assuming that one iterates Davidson until it converges to high accuracy. Use of this initial guess has an even more important advantage: it is not necessary to converge to high accuracy, since there is no danger of converging to an incorrect low-lying eigenstate. The initial guess not only has low energy, it approximately describes the correct eigenstate, as obtained in the previous step. Hence a lower accuracy convergence is possible without damaging the stability of DMRG. In fact, the algorithm can be made completely stable even if the number of Davidson steps is restricted to two or three! Thus one saves a factor of 20-50 in the time required by the Davidson procedure. The overall speedup is somewhat reduced from this factor because the calculation time to perform other parts of the DMRG procedure, such as diagonalizing the density matrix, becomes significant.

A DMRG step adds a site onto a block, constructing an appropriate basis for the new block. The new basis is defined by the eigenvectors of the density matrix, which is determined from the current system wavefunction. The eigenvectors defining the new basis (which we rewrite as matrices L and R below) can then be used to transform the current system wavefunction into the appropriate basis for the *next* step. Here we consider a finite-system DMRG step, in which we move from left to right, adding a site onto the left block, in the standard configuration with two sites between the left and right blocks. Let $|\alpha_\ell\rangle$ be the states of left block ℓ, where ℓ is the rightmost site of the block. The two sites in the middle are $\ell+1$ and $\ell+2$. Let $|s_\ell\rangle$ be the states of site ℓ, $|s_{\ell+1}\rangle$ for site $\ell+1$, etc. Then the basis states for the new left block are given by

$$|\alpha_{\ell+1}\rangle = \sum_{s_{\ell+1},\alpha_\ell} L^{\ell+1}[s_{\ell+1}]_{\alpha_{\ell+1},\alpha_\ell} |\alpha_\ell\rangle \otimes |s_{\ell+1}\rangle \, . \tag{38}$$

This notation is similar to that of Östlund and Rommer [23]. For more details, see Chap. 3(I). The transformation matrix $L^{\ell+1}[s_{\ell+1}]_{\alpha_{\ell+1},\alpha_\ell}$ is a slightly rewritten form of the truncated matrix of density matrix eigenvectors $u^{\alpha_{\ell+1}}$: specifically, $L^{\ell+1}[s_{\ell+1}]_{\alpha_{\ell+1},\alpha_\ell} = u^{\alpha_{\ell+1}}_{s_{\ell+1}\alpha_\ell}$. L includes only the eigenvectors which are retained, i.e. whose corresponding eigenvalues are above a cutoff. Let the states of the right block be $|\beta_{\ell+3}\rangle$, where we note that $\ell+3$ is the leftmost site of the block. These states were formed at an earlier right-to-left DMRG step, using a different set of density matrix eigenvectors, which we write in terms of a transformation matrix $R^{\ell+3}$:

$$|\beta_{\ell+3}\rangle = \sum_{s_{\ell+3},\beta_{\ell+4}} R^{\ell+3}[s_{\ell+3}]_{\beta_{\ell+3},\beta_{\ell+4}} |s_{\ell+3}\rangle \otimes |\beta_{\ell+4}\rangle \ . \tag{39}$$

Note that reflection symmetry is not assumed here: the L and R matrices are independent.

The wavefunction is written in a basis for the two block plus two site superblock. This superblock basis has basis states of the form

$$|\alpha_\ell s_{\ell+1} s_{\ell+2} \beta_{\ell+3}\rangle = |\alpha_\ell\rangle \otimes |s_{\ell+1}\rangle \otimes |s_{\ell+2}\rangle \otimes |\beta_{\ell+3}\rangle \ . \tag{40}$$

A system wavefunction $|\psi\rangle$ is written in this basis as

$$|\psi\rangle = \sum_{\alpha_\ell s_{\ell+1} s_{\ell+2} \beta_{\ell+3}} \psi(\alpha_\ell s_{\ell+1} s_{\ell+2} \beta_{\ell+3}) |\alpha_\ell s_{\ell+1} s_{\ell+2} \beta_{\ell+3}\rangle \ . \tag{41}$$

One needs to transform this wavefunction into the basis appropriate for the next DMRG step, in which the superblock is shifted by one site, with basis states of the form $|\alpha_{\ell+1} s_{\ell+2} s_{\ell+3} \beta_{\ell+4}\rangle$. However, the transformation between the two bases cannot be exact, since there is a truncation in going from $|\alpha_\ell s_{\ell+1}\rangle$ to $|\alpha_{\ell+1}\rangle$. However, the states $|\alpha_{\ell+1}\rangle$ are formed using the density matrix to be ideally adapted for representing $|\psi\rangle$. This means that the wavefunction can be transformed in an approximate but controlled fashion, with the error in the transformation depending on the truncation error in the DMRG step. Since the error in the density matrix is given by the truncation error, and since the density matrix is, roughly speaking, the square of the wavefunction, the error in the wavefunction transformation should be roughly the square root of the truncation error.

The simplest way to derive the transformation is to assume, based on the above argument, that for the transformation of the wavefunction only, one can approximate

$$\sum_{\alpha_{\ell+1}} |\alpha_{\ell+1}\rangle\langle\alpha_{\ell+1}| \approx 1 \ . \tag{42}$$

With this approximation one readily obtains

$$\begin{aligned} &\psi(\alpha_{\ell+1} s_{\ell+2} s_{\ell+3} \beta_{\ell+4}) \approx \\ &\quad \sum_{\alpha_\ell s_{\ell+1} \beta_{\ell+3}} L^{\ell+1}[s_{\ell+1}]_{\alpha_{\ell+1},\alpha_\ell} \psi(\alpha_\ell s_{\ell+1} s_{\ell+2} \beta_{\ell+3}) R^{\ell+3}[s_{\ell+3}]_{\beta_{\ell+3},\beta_{\ell+4}} \ . \end{aligned} \tag{43}$$

This is the desired transformation.

The most efficient way to implement this transformation numerically is to first form the intermediate wavefunction

$$\psi(\alpha_{\ell+1} s_{\ell+2} \beta_{\ell+3}) = \sum_{\alpha_\ell s_{\ell+1}} L^{\ell+1}[s_{\ell+1}]_{\alpha_{\ell+1},\alpha_\ell} \psi(\alpha_\ell s_{\ell+1} s_{\ell+2} \beta_{\ell+3}) \ , \tag{44}$$

and then form the final result

$$\psi(\alpha_{\ell+1} s_{\ell+2} s_{\ell+3} \beta_{\ell+4}) = \sum_{\beta_{\ell+3}} \psi(\alpha_{\ell+1} s_{\ell+2} \beta_{\ell+3}) R^{\ell+3}[s_{\ell+3}]_{\beta_{\ell+3},\beta_{\ell+4}} \, . \qquad (45)$$

In this form, the transformation requires very little computer time compared to other parts of the calculation.

This transformation is used for one half of the DMRG steps, when a site is being added to the left block. An analogous transformation is used for adding a site to the right block.

Implementing this transformation requires saving all the matrices L and R, which was not necessary in the original formulation of DMRG. The storage for these matrices is typically 20-30% of the storage required for the blocks themselves, so the extra storage is not a major concern. In an efficient DMRG implementation for a typical single processor workstation, both the blocks and the transformation matrices should be stored on disk.

Extension to two dimensions

The issue of how to optimally extend the DMRG to two or more dimensional quantum systems is difficult and is still a subject of active development. It is instructive to consider first how to extend the one-dimensional algorithm to quasi-one-dimensional systems with a finite width, e.g. to the Heisenberg or Hubbard model on a "ladder". One straightforward way to do this would be to replace the single sites added between the blocks with a row of sites. However, for wide systems the extra degrees of freedom in the two center "sites" would make the size of the system's Hilbert space prohibitively large. It is usually better to add single sites by mapping the 2D system onto a 1D system, simply by tracing a path through the lattice. A typical superblock configuration for a ladder system is shown in Fig. 11. The site added to the system block is enclosed by a dashed line and the dotted line shows the order in which sites are added in a sweep. An up-down-up-down path is shown; an up-up-up path would be just as efficient. One can see that it is not possible to reflect the left block onto a right block of the proper geometry at every step, so the finite system algorithm must used. A similar difficulty appears in fermion systems at fractional filling because it is not possible to set the number of particles so that the average density stays constant at all system sizes. The denominator of the fraction of the filling is analogous to the row size for two-dimensional lattices. With the mapping onto a 1D system, the two-dimensional procedure differs from the one-dimensional finite system procedure only in that there are additional connections between the system and environment blocks along the boundary.

Several infinite system approaches can be used in the warmup sweep [24]. The simplest is just to use several sites for the environment block, without truncation. Usually the accuracy of the warmup sweep is not critical; a few sweeps of the finite-system procedure can make up for a poor quality warmup sweep.

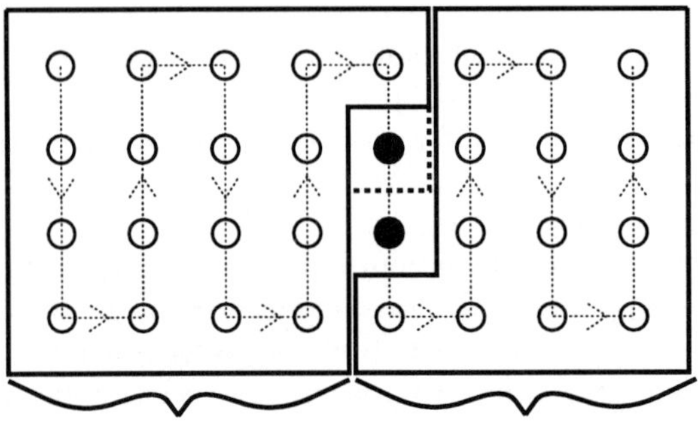

Fig. 11. The superblock configuration for the two-dimensional algorithm. The order in which sites are added to the system block on a series of iterations is given by the dotted line, and the site added to the approximate system block Hamiltonian is outlined by the dashed line.

The number of states needed to maintain a certain truncation error in the density matrix projection procedure depends strongly on the number of operators connecting the two parts of the system. Best accuracy is obtained when the number of connections between the system and environment blocks is minimized. Therefore, we usually study systems with open rather than periodic or antiperiodic boundary conditions. Also, we find that the number of states m needed to maintain a given accuracy depends strongly on the width and weakly on the length of the system.

Just how rapidly the truncation error increases with the width of the system is not clear in general. Liang and Pang [25] studied the error in the energy as a function of width for a gas of noninteracting spinless fermions and found that the number of states needed to maintain a given accuracy grew exponentially with the width of the system. In an interacting system such as the Hubbard model, the detailed structure of the energy spectrum seems to be important. For example, in the two chain Hubbard model at half-filling, where there is a spin and pairing gap, the truncation error for a given m is much smaller than away from half-filling, where the spin gap is reduced and the gap to pairing excitations is no longer present. For Heisenberg ladders, the presence or absence of a gap in the spin spectrum depends on whether the number of chains is even or odd, so the truncation error for a given m depends on the number of chains in a complicated way. In any case, for systems of more than one dimension, it is important to be able to keep as many states m per block as possible. Efficient algorithms are therefore crucial in this case.

6 Remarks

In this chapter, we have described the Density Matrix Renormalization Group as originally developed to determine the properties of low-lying states of interacting quantum systems. We have followed the historical development of the method in order to motivate the ingredients that go into the algorithms. We have also tried to strike a balance between a pedagogical explanation of the basic ideas and the inclusion of sufficient detail to allow the reader to produce efficient implementations. In order to provide a good starting point for understanding the basic ideas and limitations of various numerical RG schemes, the simplest problem, the tight-binding chain, was presented in considerable detail. As the best way to learn the technique is to implement a scheme on the computer and to play with it, we have included, in Appendix B, a program which treats this problem as an example.

The DMRG is still under quite active development. Since the basic ideas seem to be quite general and the algorithms well-suited for experimenting, there is hope that significant improvements will be made to both performance and applicability in the future. As we have described, the most serious limitation of the original ground-state algorithm is the strong dependence of its convergence on the degree of two-dimensionality and on the complexity and range of the interaction. Up to now, progress has been made through improved efficiency of the implementations and better performance of the computers. A variation of the algorithm that scales more favorably for higher dimensional systems would, however, be quite useful. The momentum-space formulation discussed in Chap. 6(I) could be a step in this direction.

There has also been much recent activity on extending the DMRG and in applying it to new problems. In particular, there have been interesting developments in extending the DMRG to calculate dynamic and finite-temperature properties of interacting quantum systems, in applying it to two-dimensional classical systems, and in studying its behavior analytically by examining its behavior near the infinite-system fixed point and by making contact with variational states. These developments are the subject of the remainder of Part I of this book, and a wide variety of applications are discussed in Part II.

A Density Matrix for Single Particle Systems

Here we show, in the case of a single particle, that the density matrix is equivalent to a simple projection of the wavefunction, used in the superblock method for the tight-binding chain described in Sect. 4.

Consider a wavefunction $\psi(k)$, where k runs over the sites of the system, $k = 1, \ldots, L$. We will call sites $1, \ldots, \ell$ the left block, labeled by i, and sites

$\ell+1,\ldots,L$ the right block, labeled by j. In order to write a single-particle wavefunction in a product form

$$|\psi\rangle = \sum_{ij} \psi_{ij} |i\rangle |j\rangle \,, \tag{A.1}$$

it is necessary to construct an enlarged basis which includes zero and two-particle states. Specifically, we use the basis

$$\begin{aligned} &|0\rangle_{\mathrm{L}} \\ |1\rangle_{\mathrm{L}} &= c_1^\dagger |0\rangle_{\mathrm{L}} \\ &\vdots \\ |\ell\rangle_{\mathrm{L}} &= c_\ell^\dagger |\ell\rangle_{\mathrm{L}} \end{aligned} \tag{A.2}$$

for the left block, and similarly for the right block. Here $c_i^\dagger$ creates a particle at site i. Then the wavefunction $\psi(k)$ can be written in matrix form as

$$\psi = \begin{pmatrix} 0 & \psi(\ell+1) \ldots \psi(L) \\ \psi(1) & \\ \vdots & 0 \\ \psi(\ell) & \end{pmatrix} . \tag{A.3}$$

In this matrix, the upper left zero represents the amplitude in the state $|0\rangle_{\mathrm{L}} \otimes |0\rangle_{\mathrm{R}}$. This state is included in the basis but since there is one particle, its coefficient is always zero. The rest of the first column represents the states $|i\rangle_{\mathrm{L}} \otimes |0\rangle_{\mathrm{R}}$, and similarly the rest of the first row represents $|0\rangle_{\mathrm{L}} \otimes |j\rangle_{\mathrm{R}}$. The large lower right block of zeros represents two particle states.

Then the density matrix $\rho_{ii'}$ in matrix form is

$$\rho = \psi\psi^\dagger = \begin{pmatrix} w_{\mathrm{R}} & 0 & 0 & \ldots & 0 \\ 0 & \psi_1\psi_1 & \psi_1\psi_2 & \ldots & \psi_1\psi_\ell \\ 0 & \psi_2\psi_1 & \psi_2\psi_2 & \ldots & \psi_1\psi_\ell \\ \vdots & \vdots & \vdots & \ddots & \vdots \\ 0 & \psi_\ell\psi_1 & \psi_\ell\psi_2 & \ldots & \psi_\ell\psi_\ell \end{pmatrix} , \tag{A.4}$$

where $w_{\mathrm{R}} = 1 - w_{\mathrm{L}} = 1 - \sum_{i=1}^{\ell} |\psi_i|^2$ is the probability that the particle is in the right block. This density matrix has two nonzero eigenvalues, with corresponding eigenvectors $(1\ 0\ \ldots\ 0)^T$ and $w_{\mathrm{L}}^{-1/2}(\psi_1 \ldots \psi_\ell)^T$. The first eigenvector does not need to be explicitly treated; the second is equivalent to the projection of the wavefunction considered in Sect. 4.

B A Program in C++ for the Particle on a Chain

General

In order to illustrate the DMRG method more explicitly, we describe here the DMRG algorithm and a complete DMRG C++ program to obtain the ground state energy and wave function for the particle on-a-chain problem numerically. We hope that after writing or playing with a program based on this appendix, the main part of this chapter should be much easier to understand. The algorithm which we will describe here is a simplified version of the finite system algorithm described in Sect. 5.

Our goal is to solve the particle on a chain problem with Hamiltonian (5) numerically. To make the program more like an efficient DMRG program for an interacting system, in which the dimensions of the matrices are much larger, we assume that our computer has only enough memory to diagonalize a 4×4 matrix. We will assume that it has enough "disk" storage to hold about two vectors of length L. By this we mean that the innermost loops involve manipulations of only a small amount of data, and access to the main data of size $\sim L$ occurs in outer loops, a little at a time. We will assume we do not know the analytic solution to the problem, and the algorithm we present will work equally well if the nonzero elements of H are perturbed slightly, either randomly or not. The solution we present is the DMRG method, in simplified form only because it is applied to a single particle problem.

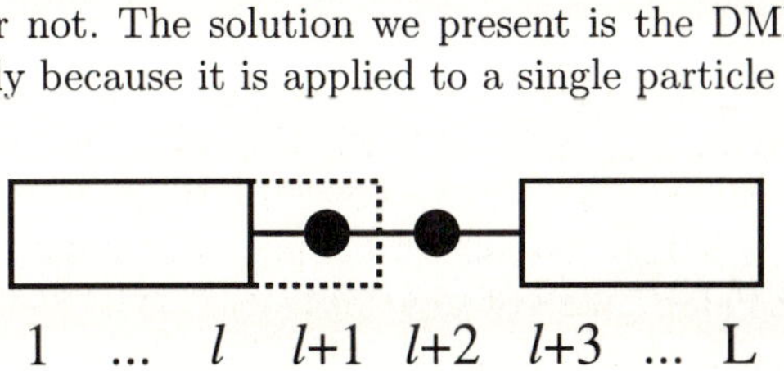

Fig. B.1. Breakup of a system, of length L, into two blocks and two sites.

Our procedure will divide the system up into four pieces, called blocks, as shown in Fig. B.1. (This is the "superblock" of Sect. 5.) Block 1 (the "left block") will consist of sites $1 \ldots \ell$; block 2 will consist of a single site, $\ell + 1$; block 3 will also consist of a single site, $\ell + 2$; and block 4 (the "right block") will consist of sites $\ell+3 \ldots L$. During the course of the algorithm, the dividing point ℓ will be moved back and forth through the system, so that every site except the first and last is represented by one of the two middle sites during a "sweep" through the system.

Each block will be represented by a single basis state. For the middle blocks, this is not a limitation; for the left and right blocks, it is a severe one. A wavefunction ψ_j, $j = 1, \ldots L$, is written as

$$\psi_j = \begin{cases} a_1 L_j & j \le \ell \\ a_2 & j = \ell + 1 \\ a_3 & j = \ell + 2 \\ a_4 R_j & j \ge \ell + 3 \end{cases} . \tag{B.1}$$

The basis states L and R are defined only within the corresponding blocks, and they are normalized, $\langle L|L\rangle = \langle R|R\rangle = 1$. Note that any state can be represented in this way if L and R are chosen appropriately for that state. Using (6) and (B.1), we find that the Hamiltonian matrix element between two states ψ and ψ' can be written as

$$\langle \psi|H|\psi'\rangle = \begin{pmatrix} a_1 \\ a_2 \\ a_3 \\ a_4 \end{pmatrix}^T \begin{pmatrix} H_{11} & T_{12} & 0 & 0 \\ T_{12} & 2 & -1 & 0 \\ 0 & -1 & 2 & T_{34} \\ 0 & 0 & T_{34} & H_{44} \end{pmatrix} \begin{pmatrix} a_1' \\ a_2' \\ a_3' \\ a_4' \end{pmatrix} \tag{B.2}$$

where

$$H_{11} = \langle L|H|L\rangle \, , \tag{B.3}$$

$$H_{44} = \langle R|H|R\rangle \, , \tag{B.4}$$

$$T_{12} = \langle L|H|l+1\rangle = -L_\ell \, , \tag{B.5}$$

and

$$T_{34} = \langle l+2|H|R\rangle = -R_{\ell+3} \, . \tag{B.6}$$

The 4×4 matrix H in (B.2) represents the Hamiltonian in our restricted basis. Its minimum eigenvalue is an upper bound for the exact ground state energy. When L and R represent the left and right parts of the exact ground state, then the minimum eigenvalue is the exact ground state energy.

In order to turn this into an algorithm for finding the ground state of the system, we will specify an iterative procedure for improving L and R. Let $L(\ell)$ and $R(\ell+3)$ represent the bases for step ℓ. Then we want to improve the set of all basis states, $\{L(\ell)\}$ and $\{R(\ell)\}$. Our basic step in this procedure will be to get an improved basis state $L(\ell+1)$, given $L(\ell)$ and $R(\ell+3)$. Given this improved basis state, we proceed through the lattice from left the right, improving the set $\{L(\ell)\}$. When we reach the right end, we will reverse the procedure, generating an improved $R(\ell+2)$, given $L(\ell)$ and $R(\ell+3)$, as we proceed through the lattice in the reverse direction, eventually improving all elements of the set $\{R(\ell)\}$.

The basic left-to-right step is this: First, we diagonalize our 4×4 Hamiltonian matrix, (B.2), obtaining a ground state wavefunction in this basis, a vector of length 4, which we write as (a_1, a_2, a_3, a_4). We normalize a_1 and a_2

as $a'_1 = a_1/N$ and $a'_2 = a_2/N$, where $N = (a_1^2 + a_2^2)^{1/2}$. Then the new basis state, properly normalized, is

$$L(\ell+1)' = \begin{pmatrix} a'_1 L(\ell)_1 \\ \vdots \\ a'_1 L(\ell)_\ell \\ a'_2 \end{pmatrix} . \tag{B.7}$$

The new Hamiltonian matrix element, needed to construct H for the next step, is

$$\langle L(\ell+1)'|H|L(\ell+1)'\rangle = {a'_1}^2 \langle L(\ell)|H|L(\ell)\rangle + 2{a'_2}^2 - 2a'_1 a'_2 L(\ell)_\ell \, . \tag{B.8}$$

If $L(\ell)$ and $R(\ell+3)$ are the exact left and right pieces of the true ground state of the system, then $L(\ell+1)'$ obtained in this way is the exact left piece of the true ground state, when the system is divided at $\ell+1$ rather than ℓ. Thus if our procedure ever finds the correct ground state at some step ℓ, then after one additional sweep through the lattice, all the $\{L(\ell)\}$ and $\{R(\ell)\}$ will represent exact parts of the true ground state and the procedure will have converged. Furthermore, assume $|\psi(\ell)\rangle$ is the state at step ℓ. Then, with $L(\ell+1)'$ as specified in (B.7), in the new basis at step $\ell+1$, $|\psi(\ell)\rangle$ can be represented exactly. Therefore the energy at step $\ell+1$ cannot be higher than at step ℓ. However, the basis in step $\ell+1$ generally has additional degrees of freedom not in the basis at step ℓ, and consequently, the energy at step $\ell+1$, obtained from diagonalizing the 4×4 Hamiltonian matrix, will generally be lower than at step ℓ. Consequently, we expect a monotonic convergence to the true ground state with this procedure.

In order to initialize this iterative procedure, we need to generate a set of approximate initial blocks $\{L(\ell)\}$ and $\{R(\ell)\}$. A good initial approximation comes from applying our procedure to a set of systems of sizes 4, 6, ..., L. Starting with the system of size 4, the left and right blocks are just single sites, with trivial L and R. We use this system to get an approximate $L(2)$. Proceeding to the system of size 6, this $L(2)$ can be used, and for the block R we use the reflection of $L(2)$, switching the left and right sides of the basis state. At each additional step, a reflection of the left block is used for the right block. In this way, we gradually grow the system up to the full size L, generating an initial set of $\{L(\ell)\}$ and $\{R(\ell)\}$.

Program

We now present a complete C++ program to solve the particle on a chain problem using the steps outlined above. The object-oriented nature of C++ means that the main routine, for a particle on a chain, can be reused with rather little change for a non-trivial interacting problem. The program utilizes the "`MatrixRef`" matrix library, written by the authors. Both the MatrixRef matrix library and a complete version of the program described here are currently available on the world-wide web at http://hedrock.ps.uci.edu. In what follows, we will list and discuss the important parts of the program.

Blocks: We first define a class for a Block.

```
class Block
    {
public:
    Real H11, L_inner;
    Block()         // Default: construct a one-site block
        : H11(2.0), L_inner(1.0) { }
    Block Reflect() const { return *this; }
    };
```

Here type (Real is defined in the matrix library as a double, equivalent to a Fortran real*8). The two variables H11 and L_inner store the values of either $\langle L|H|L\rangle$ or $\langle R|H|R\rangle$, and either L_ℓ or $R_{\ell+3}$, depending on whether the block is a left or right block. The constructor function Block() initializes the block in the way appropriate for a single site block, with $H_{11} = 2$ and $L_\ell = 1$. The Reflect() function, which reflects a block, does nothing in this simple case because of the dual usage of L_inner, but is included to illustrate the more general case.

System: Next we define a class System, which contains the four Blocks making up the lattice, for a particular step ℓ.

```
class WaveFunction;

class System
    {
public:
    const Block &b1, &b2, &b3, &b4;
    System(const Block& bb1,const Block&bb2,
                const Block&bb3, const Block&bb4)
        : b1(bb1), b2(bb2), b3(bb3), b4(bb4) { }

    Real GetGroundState(WaveFunction& p);
    };
```

A System has four data members, references to four blocks, b1, b2, b3, and b4. The function GetGroundState, returns a Real number, the ground state energy, and finds a ground state wavefunction. The definition of WaveFunction is

```
class WaveFunction
    {
public:
```

```
Vector v;
WaveFunction() : v(4) {}
};
```

Wave function: A `WaveFunction` here is simply a `Vector` of length 4. Vectors are defined in the matrix library. The function `GetGroundState` is

```
Real System::GetGroundState(WaveFunction& p)
    {
    Matrix H(4,4), evecs(4,4);
    Vector evals(4);
    H = 0.0;
    H(1,1) = b1.H11;    H(2,2) = b2.H11;
    H(3,3) = b3.H11;    H(4,4) = b4.H11;
    H(1,2) = H(2,1) = -b1.L_inner;
    H(2,3) = H(3,2) = -1.0;
    H(3,4) = H(4,3) = -b4.L_inner;
    EigenValues(H,evals,evecs);
    p.v = evecs.Column(1);
    if(p.v.sumels() < 0.0) p.v *= -1.0;
    Real energy = evals(1);
    return energy;
    }
```

This function defines a 4×4 matrix H, a matrix of eigenvectors evecs, and a vector of eigenvalues evals. H is first initialized to contain all 0's, and then the non-zero elements are defined according to (B.2), (B.5), and (B.6). Then, the Matrix library routine `EigenValues` is called, which diagonalizes H. The `WaveFunction` p is set to the ground state vector, and the ground state energy is returned.

Density matrix: The last class we will define is a `DensityMatrix`. Thus far, in our description of the single particle DMRG algorithm, density matrices have not appeared. However, as shown in Appendix A, the elements a_1 and a_2 of the wavefunction ψ are equivalent to the reduced density matrix for the first two blocks as part of the system as a whole.

```
enum LR {Left, Right};

class DensityMatrix
   {
public:
   Real a,b;
   DensityMatrix(const WaveFunction& psi,LR lr)
      {
      if(lr == Left)
          { a = psi.v(1); b = psi.v(2); }
```

```
    else
        { a = psi.v(4); b = psi.v(3); }
    }
  Vector NewBasis()
    {
    Vector res(2);
    Real norm = sqrt(a*a+b*b);
    res(1) = a / norm;
    res(2) = b / norm;
    return res;
    }
  };
```

A `DensityMatrix` has two data members, `a` and `b`, representing a_1 and a_2 or a_4 and a_3, depending on the whether one is forming a density matrix for the left or right half of the system. The first line defines a new type, `LR`, which takes on only two symbolic values, `Left` and `Right`, indicating which half of the system we want. One makes a `DensityMatrix` out of a `WaveFunction` `psi` simply by assigning the appropriate values to `a` and `b`. The function `NewBasis` forms a new basis for $L(\ell+1)$, which simply involves normalizing the two-element vector (a,b).

In order to utilize the new basis to form a new block, there are functions `NewLeft` and `NewRight`, which combine two blocks (either the left two blocks [`NewLeft`], or the right two blocks [`NewRight`]) using a new basis, and using (B.8). Here we show `NewLeft`; `NewRight` is quite similar.

```
Block NewLeft(const Block& b1, const Block& b2, const Vector& bas)
  {
  Block res;
  res.H11 = bas(1) * bas(1) * b1.H11 + 2 * bas(2) * bas(2)
       - 2 * bas(1) * bas(2) * b1.L_inner;
  res.L_inner = bas(2);
  return res;
  }
```

Main program: The main program is relatively straightforward, using these classes and routines. First, we read in the length of the system and the number of sweeps to be performed, and initialize an array of blocks which will hold both $\{L(\ell)\}$ and $\{R(\ell)\}$.

```
int main()
  {
  Block siteblock;
  cerr << "Input length, number of iterations: ";
  int i, length, nsweeps;
  cin >> length >> nsweeps;
  cout << "length, nsweeps = " << length SP nsweeps << endl;
```

```
    cout << "Exact energy = " << exacten(length) << endl;
    exlen = length;
    Array1<Block> allblocks(length);
    WaveFunction psi;
    Real energy;
```

Next, we perform the warmup sweep, which uses the infinite-size algorithm to build up the system from four sites initially to the full system size.

```
// Warmup sweep, using the infinite-size algorithm
    allblocks[1] = siteblock;
    for(i = 1; i < length/2; i++)
       {
       Block rightblock = allblocks(i).Reflect();
       System S(allblocks(i),siteblock,siteblock,rightblock);
       energy = S.GetGroundState(psi);
       cout << i+1 SP psi.v(2) SP energy SP 0 << endl;
       DensityMatrix rho(psi,Left);
       Vector basis = rho.NewBasis();
       allblocks[i+1] = NewLeft(allblocks(i),siteblock,basis);
       }
```

The macro SP in the cout statement makes a space, and is defined in the matrix library. The final 0 to be printed out on each line indicates this is the warmup sweep. Once the system has reached its full size, nsweeps sweeps are performed to converge to the ground state. Each sweep consists of a right-to-left part, and a left-to-right part. In this program, because the system is symmetric under reflections, the sweeps start or end at the middle, symmetric configuration, rather than the far right site. The right-hand blocks in the middle configuration are obtained as reflections of the left hand blocks.

```
// Finite System sweeps
    for(int swp = 1; swp <= nsweeps; swp++)
       {
// We assume reflection symmetry:
       allblocks[length/2 + 2] = allblocks(length/2 - 1).Reflect();
       cout << endl;

// Right to left
       for(i = length/2+2; i > 3; i--)
          {
          System S(allblocks(i-3),siteblock,siteblock,allblocks(i));
          energy = S.GetGroundState(psi);
          cout << i-1 SP psi.v(3) SP energy SP swp - 0.5 << endl;
          DensityMatrix rho(psi,Right);
          Vector basis = rho.NewBasis();
          allblocks[i-1] = NewRight(siteblock,allblocks(i),basis);
          }
```

```
// Left to right
       cout << endl << 1 SP psi.v(1) SP energy SP swp << endl;
       for(i = 1; i < length/2-1; i++)
          {
          System S(allblocks(i),siteblock,siteblock,allblocks(i+3));
          energy = S.GetGroundState(psi);
          cout << i+1 SP psi.v(2) SP energy SP swp << endl;
          DensityMatrix rho(psi,Left);
          Vector basis = rho.NewBasis();
          allblocks[i+1] = NewLeft(allblocks(i),siteblock,basis);
          }
       }
    return 0;
    }
```

At each step i, the energy and value of the wavefunction at site $i+1$ is printed out. (We use i, rather than l, as the step variable because ℓ looks too much like 1 in program listings.)

Results

Results from this program are shown in Fig. B.2. The ground state energy is plotted as a function of the index of the site ℓ just added, for the warmup and first three sweeps. Note that during the warmup, the system length is

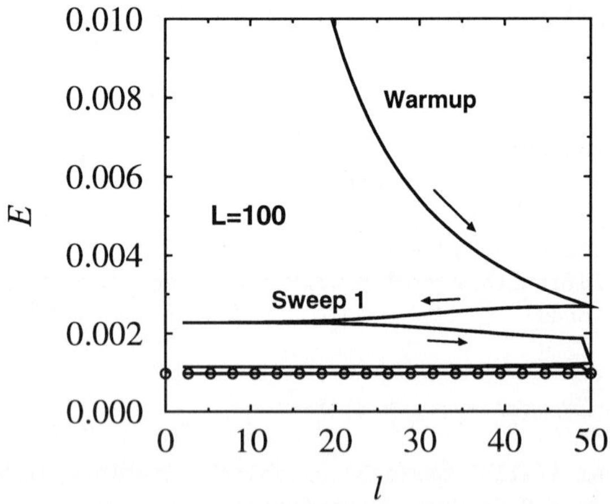

Fig. B.2. Results for the ground state energy for a particle on a chain of length $L = 100$.

less than the eventual system size, which accounts for most of difference in energy from the exact result. (Here $\ell = i + 1$ for left to right sweeps and $\ell = i - 1$ for right to left sweeps.) During the warmup sweep, the system size is still growing, and at each step the length is 2ℓ. Starting with the beginning of the first sweep, the length is fixed at L. Each sweep starts in the middle, progresses to the left side, and returns. There is a discontinuous jump in the energy between sweeps because at those points a reflection of the left block is used to replace the first right block of the new sweep. During the warmup, this is done every step, but it occurs only once per sweep during the other sweeps. The open circles represent the exact energy, $E = 0.000967435$. (The general formula is $E = 2[1 - \cos(\frac{\pi}{L+1})]$.) Good convergence is obtained after three sweeps.

Results for the wavefunction are shown in Fig. B.3. The wavefunction at the site just added is plotted versus the site index for the first three sweeps. Each half-sweep is labeled by an integer, starting at 1. The open circles represent the exact wavefunction, which is given (in unnormalized form) by

$$\psi_j = \sin(qj) \tag{B.9}$$

with $q = \pi/(L+1)$, and $j = 1, \ldots L$. Convergence of the wavefunction is reasonable after three sweeps; another sweep or two would be needed for precise agreement with the exact results.

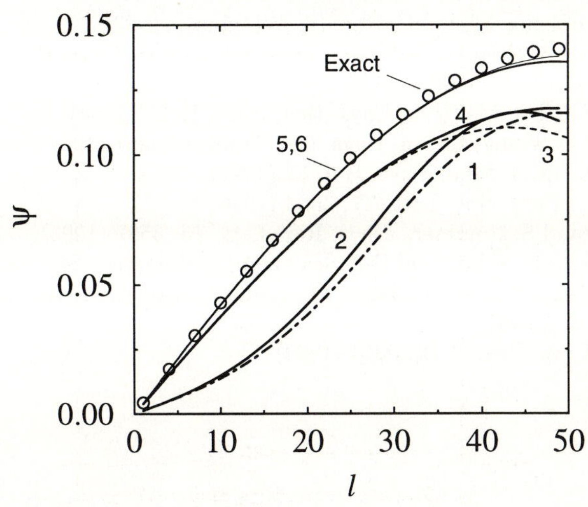

Fig. B.3. Results for the wavefunction for a particle on a chain of length $L = 100$.

The above program can be easily modified to find more than just one state, to deal with modifications to the Hamiltonian, including randomness, etc. A useful exercise would be to modify this program to include randomness or another local potential, to include longer range hopping, etc. Another useful exercise would be to translate this program into another programming language.

References

1. S.R. White, Phys. Rev. Lett. **69**, 2863 (1992), Phys. Rev. B **48**, 10345 (1993)
2. K.G. Wilson, Rev. Mod. Phys. **47**, 773 (1975)
3. K. Hallberg, Phys. Rev. B **52**, 9827 (1995); H. Pang, H. Akhlaghpour and M. Jarrell, Phys. Rev. B **53**, 5086 (1996)
4. T. Nishino, J. Phys. Soc. Jpn. **64**, 3598 (1995)
5. S. Moukouri and L.G. Caron, Phys. Rev. Lett. **77**, 4640 (1996)
6. R.J. Bursill, T. Xiang and G.A. Gehring, J. Phys. Cond. Matt. **8**, 583 (1996)
7. X.Q. Wang and T. Xiang, Phys. Rev. B **56**, 5061 (1997)
8. S.R. White and R.L. Martin, unpublished, cond-mat/9808118
9. E.R. Davidson, J. Comp. Phys. **17**, 87 (1975); E.R. Davidson, Computers in Physics **7**, No. 5, 519 (1993)
10. J.W. Bray and S.T. Chui, Phys. Rev. B **19**, 4876 (1979); T. Xiang and G.A. Gehring, Phys. Rev. B **48**, 303 (1993)
11. S.R. White and R.M. Noack, Phys. Rev. Lett. **68** 3487, (1992)
12. R.P. Feynman, Statistical Mechanics: A Set of Lectures, Benjamin (1972)
13. W.H. Press, B.P. Flannery, S.A. Teukolsky, and W.T Vetterling, Numerical Recipes: The Art of Scientific Computing, Cambridge (1986)
14. E. Jeckelmann and S.R. White, Phys. Rev. B **57**, 6376 (1998)
15. E.H. Lieb and F.Y. Wu, Phys. Rev. Lett. **20**, 1445 (1968)
16. H. Schulz, J. Phys. C: Solid State Phys. **18**, 581 (1985)
17. S. Kneer, Diploma Thesis, Universität Würzburg (1997)
18. G. Bedürftig, B. Brendel, H. Frahm and R.M. Noack, Phys. Rev. B **58**, 10225 (1998)
19. X.Q. Wang and S. Mallwitz, Phys. Rev. B **48**, R492 (1996)
20. S.J. Qin, X.Q. Wang, Lu Yu, Phys. Rev. B **56**, R14251 (1997)
21. H.J. Schulz, Int. J. Mod. Phys. B **5**, 57 (1991)
22. S.R. White, Phys. Rev. Lett. **77**, 3633 (1996)
23. S. Östlund and S. Rommer, Phys. Rev. Lett. **75**, 3537 (1995)
24. R.M. Noack, S.R. White and D.J. Scalapino, in Computer Simulations in Condensed Matter Physics VII, Eds. D.P. Landau, K.K. Mon and H.B. Schüttler Spinger, p 85 (1994)
25. S. Liang, Phys. Rev. B **49**, 9214 (1994)

3 Thermodynamic Limit and Matrix-Product States

Stefan Rommer** and Stellan Östlund

Institute for Theoretical Physics, Chalmers University of Technology,
SE - 412 96 Göteborg, Sweden

In this lecture we study the nature and underlying principles of the density matrix renormalization group (DMRG) method. Our primary goal is not to reach best possible accuracy by doing full scale DMRG calculations. Instead we will keep a very modest number of states in the DMRG truncation and analytically analyze the renormalization process. Many of the results that will be reviewed in this lecture were originally presented in [1,2]. For background information on the DMRG method see Chap. 2 (I) and the original articles by White [3,4].

In Sect. 1 of this lecture we show that if the DMRG algorithm converges to a fixed point in the thermodynamic limit, the DMRG ground state leads to a special form for the wave function. Guided by this form we define a set of ansatz wave functions in matrix product (MP) form. To make things more concrete we apply our ideas to the antiferromagnetic Heisenberg spin-1 chain with bilinear and biquadratic interactions, defined by

$$H = \sum_{i=1}^{n} \mathbf{S}_i \cdot \mathbf{S}_{i+1} - \beta(\mathbf{S}_i \cdot \mathbf{S}_{i+1})^2 \, . \tag{1}$$

Section 2 contains a brief discussion of some aspects of this model that are of relevance for later sections.

In Sects. 3 and 4 we analyze the matrix product ansatz states and find that the DMRG ground state can actually be rederived by doing a variational calculation in the matrix product ansatz states. This demonstrates a relation of the DMRG to a variational calculation. In Sect. 4 we calculate several properties of the matrix product states. In particular we determine correlation functions in the MP states and find that the states are intrinsically either long range ordered or finitely correlated.

In Sect. 5 we extend the ansatz to include a set of Bloch states that describe elementary excitations in both finite and infinite systems. We show numerical results for the low-lying excitation spectrum of the spin-1 chain. The last section, Sect. 6, contains a summary. This lecture also contains two

** Present address: Department of Physics and Astronomy, University of California, Irvine, CA 92697-4575

short appendices. Appendix A deals with parity of the MP wavefunction and App. B contains details on how to calculate an orthonormal basis in the MP states. These appendices give details on how calculations in the MP states can be done.

matrix-product states have been extensively used in both classical and quantum mechanics and it is out of the scope of this lecture to give a detailed account on this issue. Before we discuss their connection to the DMRG we will however briefly discuss some other uses of MP states for quantum systems that are closely related to this lecture.

When it comes to increasing numerical accuracy, the variational calculations that will be discussed in this lecture are probably not a serious alternative to big DMRG calculations. However, the matrix product states give us some insight into how the DMRG works. We can also better understand the limitations of the DMRG method. In this lecture we will see an example of this when we look at correlation functions in the MP states.

The matrix-product states are also interesting in their own right. For example they can be used as a convenient representation of valence bond ground states for spin-1 chains [5,6], like the ground state of the Affleck-Kennedy-Lieb-Tasaki (AKLT) model. The AKLT model, i.e. the Hamiltonian in (1) with $\beta = -1/3$, is one simple, but non-trivial, example of a model with MP ground state. This ground state will be introduced in Sect. 2.

MP states have also been extensively used to construct new classes of spin Hamiltonians with exact MP ground states [6–9]. The task is here in a sense opposite to the usual when solving a Hamiltonian; construct MP states and then try to find the Hamiltonians for which these states are the exact ground states. MP states have also been used as approximate trial states for spin chains [10,11] and spin ladder systems [12,13]. Even though the numerical results of, for example, the ground-state energy of the MP ansatz for the pure Heisenberg chain cannot compete in accuracy with a full scale DMRG calculation, the MP ansätze give us a picture of how the ground states of non-solvable spin chains might look like. As an example, the AKLT state is believed to be a good caricature of the true ground state of the ordinary spin-1 Heisenberg model ((1) with $\beta = 0$).

Klümper *et al.* [7] as well as Kolezhuk *et al.* [8,11] use a somewhat different MP formalism from the one that will be presented in this lecture. It is however not difficult to see that their MP states can be written in the form that we will use.

1 Thermodynamic Limit and the Matrix-Product State

In a renormalization scheme like the DMRG one typically starts with a very short 1D chain whose properties can be calculated exactly. The scheme then proceeds by iteratively adding a single site. When the chain gets longer we do not use the full set of basis states for describing the system. Instead a reduced number of basis states is used to represent a system in each renormalization step. In this way we keep the Hilbert space at a manageable size as the system size grows.

In the literature, the DMRG algorithm is usually described by how the Hamiltonian and other operators of a system are constructed as the chain grows by one site and how these operators are calculated as the basis truncation is done. The basis vectors themselves, expressed in the original complete basis, are however not explicitly constructed. This is also not necessary and would not be numerically feasible. After all, it is the large dimension of the full Hilbert space we want to avoid. In this section we will however analyze a renormalization step and look at how the basis states change.

Assume we have a block that represents a chain with $n-1$ sites. Let m_s be the number of possible states of a single lattice site. If we would treat this system exactly there would be m_s^{n-1} states in the Hilbert space basis for this system. In the case of a spin-1 chain, we could label the site with the z-component of the single spin-1, so that $m_s = 3$. The number of states in this complete basis rapidly becomes too large to handle when n is increased. Assume therefore that an approximation is made and our chain is represented by a smaller set of states labeled by $\{|\beta\rangle_{n-1}\}$. This set of states has been chosen by the previous iterations of the renormalization with the aim to describe the low energy physics. Assume there are m states in this basis, where $m \leq m_s^{n-1}$. If this is the first iteration, $\{|\beta\rangle_{n-1}\}$ is the complete basis.

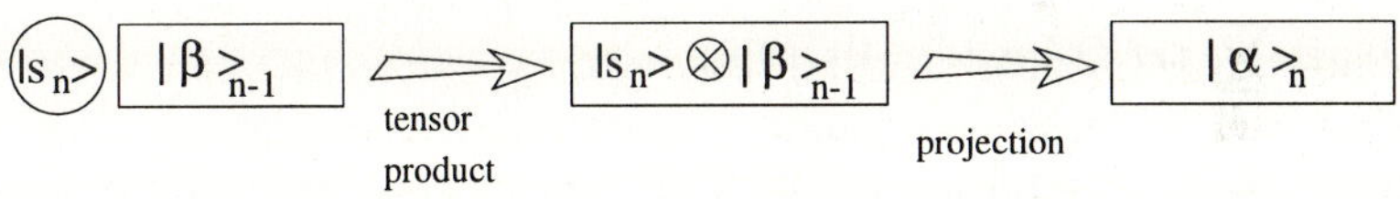

Fig. 1. A renormalization step. First the old block representing $n-1$ sites is joined to a single site labeled by s_n. The tensor product then "glues" the different parts together. Finally there is a projection to a new block representing n sites.

We now add a single site, labeled by s_n, the z-component of spin, to the left hand side of our block resulting in a new block with n sites and $m_s * m$ states in its basis. The basis states are now generated by the product representation $\{|s_n\rangle \otimes |\beta\rangle_{n-1}\}$. See Fig. 1 for an illustration of this step. We now use a projection operator A_n to generate a new truncated basis with typically m states that represent the "important" states of the longer block. This whole process is written

$$|\alpha\rangle_n = \sum_{\beta, s_n} A_n^{\alpha,(\beta,s_n)} |s_n\rangle \otimes |\beta\rangle_{n-1} \, , \tag{2}$$

where we have indexed A by the chain length n and its matrix indices α and (β, s_n). Note that (β, s_n) is thought of as a single index labeling a tensor product of the states $|s_n\rangle$ and $|\beta\rangle_{n-1}$. In the DMRG, a specific algorithm is used to calculate A, but this is not important in the present discussion.

We now make two crucial observations.

1. First we perform a simple change in notation: $A_n^{\alpha,\beta}[s_n] \equiv A_n^{\alpha,(\beta,s_n)}$, thus writing the $m \times (m_s * m)$ matrix as a set of m_s matrices with dimension $m \times m$.
2. Second, we assume that the recursion leads to a fixed point for the projection operator so that we can write $A_n[s] \to A[s]$, as $n \to \infty$.

By repeatedly applying the renormalization step in (2) we now find that

$$|\alpha\rangle_n = \sum_{s_n \dots s_1, \beta} (A[s_n]A[s_{n-1}] \dots A[s_1])^{\alpha,\beta} |s_n s_{n-1} \dots s_1\rangle \otimes |\beta\rangle_0 \,, \tag{3}$$

where $|\beta\rangle_0$ represents some initial state. We thus see that the renormalization procedure results in a wave function that can be written in a matrix product form.

Equation (3) now suggests a natural form for trial wavefunctions. We can in a compact way write linear combinations of the m^2 states

$$|\alpha, \beta\rangle_n = \sum_{s_n \dots s_1} (A[s_n]A[s_{n-1}] \dots A[s_1])^{\alpha,\beta} |s_n s_{n-1} \dots s_1\rangle$$

by introducing a matrix Q. For every $m \times m$ matrix Q we define the (unnormalized) state $|Q\rangle_n$

$$|Q\rangle_n \equiv \sum_{\{s\}} \mathrm{tr}\,(\, QA[s_n] \dots A[s_1]\,) |s_n \dots s_1\rangle \,. \tag{4}$$

The state $|Q\rangle_n$ can be viewed as a state that is uniform in the bulk, but with a linear combination of boundary conditions defined by $|\alpha\rangle_n$ on the left and $|\beta\rangle_0$ on the right. The special case of $Q = \mathbb{1}$, the identity matrix, leads to a translationally invariant state with periodic boundary conditions. This $Q = \mathbb{1}$ state we will later on use as our trial ground state.

If we now demand that the projection of (2) preserves orthonormal bases, $\langle\alpha|\alpha'\rangle = \delta_{\alpha,\alpha'}$, we can use the recursion formula (2) and the orthogonality of the local spin states and previous block states to find

$$\begin{aligned} \delta_{\alpha,\alpha'} &= \sum_{\beta,\beta',s,s'} (A^{\alpha',\beta'}[s'])^* A^{\alpha,\beta}[s] \, \langle s'|s\rangle \langle \beta'|\beta\rangle \\ &= \sum_s (A[s]A^\dagger[s])^{\alpha,\alpha'} \,. \end{aligned} \tag{5}$$

Hence in matrix form we have $\sum_s A[s]A^\dagger[s] = \mathbb{1}$. This constraint will be used later to reduce the number of free parameters in A.

2 The Model

We will now take a closer look at the spin model defined in (1) and review some of its properties that will be of interest in later sections.

The behavior of the model in (1) as a function of β has been extensively studied [14–22]. Depending on the ratio between the bilinear and the biquadratic terms in the Hamiltonian, this model is believed to exhibit several different phases. See e.g. [20] for a review of the phase diagram.

In this lecture we apply our methods to (1) with $\beta = -1/3$ and $\beta = 0$. It is believed that the model exhibits a single phase in the range $-1 < \beta < 1$, usually called the Haldane phase. At $\beta = 0$ we find the ordinary Heisenberg model, which Haldane conjectured [23] to have exponential decaying spin-spin correlations and a finite energy gap to the first excited state. This conjecture has been very well supported by numerical calculations [14,24,25]. At the point $\beta = 1$ the model is exactly solvable using the Bethe ansatz. The spectrum was calculated by Takhtajan [26] and Babujian [27] and is gapless with power-law decay of correlations. From field theory arguments [28] it is suggested that the gap opens up on both sides of the point $\beta = 1$. The numerical evidence for the opening of the gap is, however, inconclusive [19,20]. At $\beta = -1$ we find the Lai-Sutherland model which is again solvable by Bethe ansatz techniques [29–31]. The model is gapless with algebraically decaying correlations.

The model with $\beta = -1/3$ is the AKLT Hamiltonian with a unique ground state that can be constructed as an exact nearest-neighbor valence bond state (VBS), as was shown by Affleck *et al.* [32]. The AKLT Hamiltonian has exponential decay of correlations and a finite energy gap.

The VBS state can easily be constructed in a pictorial way [32]. Imagine that instead of a spin-1 object on each site of a chain there are two spin-1/2 objects. These two spin-1/2 can be represented in the s_z-basis by the z-component of each of the spins. The two spins can also be represented in the total spin basis by their total spin and the z-component of total spin. The transformation between these two bases is given by the Clebsch-Gordan coefficients [33]. In the total spin basis the two spin-1/2 form a singlet (spin-0) or a triplet (spin-1). Now let the two spin-1/2 be combined into a triplet and let this triplet represent a single spin-1 site in a spin-1 chain. To construct the VBS state, all we now have to do is to combine neighboring spin-1/2 on *different* sites into singlets (see Fig. 2). As we will see later, this state can

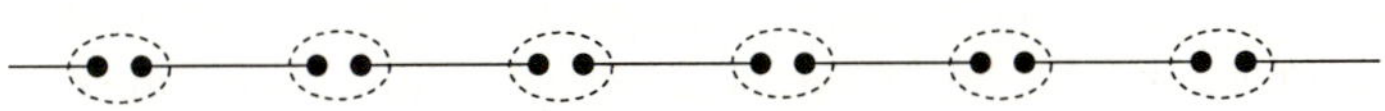

Fig. 2. The exact ground state of the AKLT Hamiltonian. The dots represent spin-1/2 variables. The dashed circles represent triplet bonds between the two spins on the same site and will thus be spin-1 objects. The solid lines represent singlet bonds between spin-1/2 objects on different sites.

also simply be written as a matrix-product state.

The valence-bond state is the exact ground state only for $\beta = -1/3$ but it is considered to give a reasonable picture of the ground state also for other values of β in this phase. For $-1 < \beta < 1$ our model is thus believed to have a gap, a unique ground state and exponentially decaying correlations.

From the valence-bond caricature of the ground state in the Haldane phase, there should be spin-1/2 degrees of freedom at each end of an open spin-1 chain. That these spin-1/2 edge states exist in the $\beta = -1/3$ model is clear from the construction of the exact ground state (see Fig. 2). For the ordinary Heisenberg model with $\beta = 0$, numerical calculations support the spin-1/2 edge states and thus the valence-bond picture [34].

Let us now look at the spin-1 chain from a total spin basis point of view. We will, in a DMRG fashion, start with a single site and then add a single spin-1 site at a time to the chain. Motivated by the spin-1/2 edge excitations of the open spin-1 chain, we will take a single spin-1/2 site as our first site. After adding the first spin-1 to our system, it thus consists of one spin-1/2 and one spin-1 site. In the total spin basis, the basis states of this system is a sum of two irreducible representations of total spin, one spin-1/2 and one spin-3/2,

$$\frac{1}{2} \otimes 1 = \frac{1}{2} \oplus \frac{3}{2} \, . \tag{6}$$

Note that since we have a spin-1/2 on the edge, the irreducible representations of total spin of our chain will be half-odd integers.

After adding another spin-1 site to the chain we find that the irreducible representations are

$$\left(\frac{1}{2} \oplus \frac{3}{2}\right) \otimes 1 = \frac{1}{2} \oplus \frac{3}{2} \oplus \frac{1}{2} \oplus \frac{3}{2} \oplus \frac{5}{2} \, . \tag{7}$$

In this way we can continue to add sites to our chain. We note that, as we add more spins to the chain, we do not only get higher values of total spin among our basis states, we also get multiple copies of irreducible representations of the same total spin. See Fig. 3 for an illustration of this process.

If we now, in a DMRG fashion, want to represent a finite chain by a reduced number of basis states, which basis states should we choose? The Hamiltonian of (1) is spin rotationally invariant since it commutes with all three components of the total spin $\mathbf{S}_{tot} = \sum_i \mathbf{S}_i$. In order to preserve this symmetry, we should pick complete representations of total spin to build up our basis. By this we mean that we want the basis to consist of complete multiplets of total spin, i.e. if one of the basis states for example has total spin $j = 3/2$ and z-component of total spin $m_z = 1/2$, then we also must have the states with $m_z = -3/2$, $m_z = -1/2$ and $m_z = 3/2$ of the same multiplet as basis states. Since the ground state of the spin-1 chain (without the spin-1/2 edge spin) is a singlet of total spin [35], we could guess that to describe this state the best representations to keep in each step would

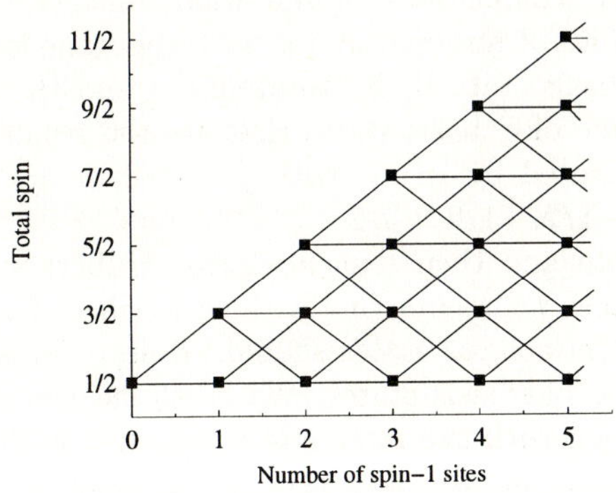

Fig. 3. Illustrates how the total spin representations of the spin-1 chain with a spin-1/2 edge spin evolve as we add sites to the system. Each square denotes one or more copies of an irreducible representation of total spin.

be those with small values of total spin. In the next section we will see an example of this.

3 Construction of the Matrices *A*

We now analyze the projection matrix A. We want our effective Hamiltonian, after the truncation of the basis, to have as many symmetries preserved as possible from the original Hamiltonian. As in the previous section, in order that the projection in each step preserves the spin rotational symmetry, our basis states of a block must form a representation of total spin.

Since we keep basis states with many different values of total spin as well as many states with the same total spin in each iteration, all the basis states together must form a sum of irreducible representations of total spin. Adding a spin one does not mix even or half-odd spin representations, thus the basis states must form a sum of either all half-odd or all integer spin representations. Most naturally for the spin-1 chain one would work with integer spin representations, but as we saw in the previous section, by placing a single spin-1/2 on the side of the entire chain one could use half integer spin representations to represent the blocks instead. This is consistent with the existence of a spin-1/2 edge state [34,32] and we have found that working with half-odd integer representations give better numerical results.

We now discuss how the projection operator A can be constructed for the case were we have kept 6 basis states in each iteration. A more complicated example with 12 basis states can be found in [2]. We have used the half-odd spin representations.

By doing a DMRG calculation on the spin-1 chain we have found that when approximately 6 states are kept the blocks are represented by one spin-

1/2 and one spin-3/2 irreducible representation. Since we in this case do not have multiple copies of representations with the same total spin, we can uniquely label each basis state by its total spin j and the z-component of total spin, m. The six "old" basis states that are the result of the previous iteration are thus denoted by $|j,m\rangle$, with $j = 1/2$, $m = -1/2, +1/2$ and $j = 3/2$, $m = -3/2, -1/2, +1/2, +3/2$

After adding a site and then truncating the Hilbert space we get the "new" basis states. Since we assume that we have reached the thermodynamic limit fixed point, the new basis states should fall into the same irreducible representations as the old basis states. We label the six basis states that represent the new block with one more site by $|j',m'\rangle$, with $j' = 1/2$, $m' = -1/2, +1/2$ and $j' = 3/2$, $m' = -3/2, -1/2, +1/2, +3/2$.

Let us now examine what happens in our example when going from the old $|j,m\rangle$ to the new $|j',m'\rangle$. When adding a single spin-1 to the old block of six states we get $6 \times 3 = 18$ "intermediate" states in the product representation of the old block states with a spin 1. This procedure is the same as is described by (7) and these 18 states thus fall into 5 irreducible representations. Since we in the intermediate states have more than one representation with a given total spin we cannot just label them with the total spin. If we however use the "old" total spin j as an additional label we can uniquely label the intermediate states by $|j,j'',m''\rangle$, where j'' is the total spin of the intermediate state and m'' is the corresponding z-component. These five intermediate reps are shown in the j''-column in Fig. 4. We then project from these 5 reps back down to the two reps that we have chosen to keep. This projection must preserve the total spin symmetry, i.e. it cannot mix different j'' and it cannot depend on total m''. We thus get only a few nonzero projection terms. Since there is exactly one intermediate spin-1/2 and one intermediate spin-3/2 for each of the two "old" representations j there is one projection term from each of the "old" j to each of the "new" j'. There are thus four nonzero projection terms which we denote by $P^{j',j}$. The $P^{j',j}$ are indicated by lines in Fig. 4. These projection terms are in fact not independent, but are related by the requirement that the new states are normalized.

Expressing all this mathematically, we let, as above, j uniquely label a rep of total spin of the old block. Each state is thus labeled by $|j,m\rangle$ where m is the z-component of total spin. The single spin to be added is labeled by $|s\rangle$, where s is the z-component of the spin-1.

The 18 intermediate states can be explicitly constructed in the total spin basis using the Clebsch-Gordan coefficients of the form $\langle (j_1,m_1)(j_2,m_2)|j,m\rangle$ as

$$|j,j'',m''\rangle = \sum_{m,s} \langle (j,m)(1,s)|j'',m''\rangle \ (|s\rangle \otimes |j,m\rangle)\,. \tag{8}$$

From these 18 states $|j,j'',m''\rangle$ we now do a projection to 6 "new" basis states $|j',m'\rangle$, i.e. we define 6 new states by taking linear combinations of the 18 intermediate states. As above we demand that the projection preserves total

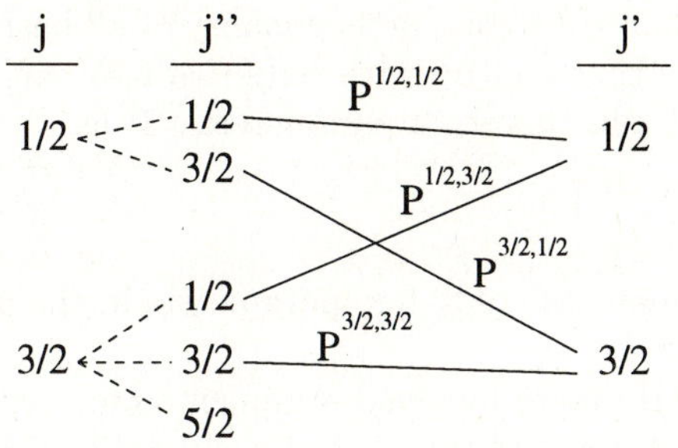

Fig. 4. Construction of the block states is shown when six states are kept in the basis. Old representations are on the left and new representations on the right. Intermediate representations are denoted by j''. The z-component of total spin is not explicitly written out. Each solid line represents a nonzero projection $P^{j',j}$ of basis representations.

spin and the z-component of total spin. The projection (the linear combinations) that define our new states will thus be very restrictive and we will have $j'' = j'$ and $m'' = m'$ in the projection. The projection is thus determined by a few projection parameters (coefficients) $P^{j',j}$ and is written

$$|j', m'\rangle = \sum_j P^{j',j} |j, j', m'\rangle \,. \tag{9}$$

Note that $j'' = j'$ and $m'' = m'$ so that $|j, j', m'\rangle$ in (9) denotes the 18 intermediate states. Inserting (8) into (9) we find that (C.f. (2))

$$|j', m'\rangle = \sum_{s,(j,m)} A^{(j',m'),(j,m)}[s](|s\rangle \otimes |j, m\rangle) \,,$$

where

$$A^{(j',m'),(j,m)}[s] = P^{j',j} \langle (j,m)\ (1,s) | j', m' \rangle \,. \tag{10}$$

Thus, although the projection matrices A contain a total of $3 \times 6 \times 6 = 108$ numbers, they are in fact generated by the relatively few degrees of freedom available in $P^{j',j}$.

For this case with six basis states there are naively 4 parameters $P^{j',j}$ and we could write them in a 2×2 matrix $\mathbf{P}$

$$\mathbf{P} \equiv \begin{pmatrix} P^{\frac{1}{2},\frac{1}{2}} & P^{\frac{1}{2},\frac{3}{2}} \\ P^{\frac{3}{2},\frac{1}{2}} & P^{\frac{3}{2},\frac{3}{2}} \end{pmatrix} . \tag{11}$$

If we demand normalization of all basis states, $\langle j_1, m_1 | j_2, m_2 \rangle = \delta_{j_1,j_2} \delta_{m_1,m_2}$, a simple calculation using (5) and (10) yields the condition that the diagonal elements of $\mathbf{P}^T\mathbf{P}$ are all 1, where the superscript T denotes transpose. This gives two constraints. Since the spin-1/2 basis states are automatically orthogonal to the spin-3/2 states, orthogonality of all basis states gives us no new constraints. We thus end up with only two free parameters. Under the given constraints we can choose to parameterize $\mathbf{P}$ as

$$\mathbf{P} = \begin{pmatrix} \cos(\phi) \cos(\theta) \\ \sin(\phi) \sin(\theta) \end{pmatrix} , \tag{12}$$

where ϕ and θ represent the only free parameters in the projection matrices $A[s]$ in the case $m = 6$.

Similar but slightly more involved arguments are used in [2] for 12 basis states. In that case it is found that out of the $3 \times 12 \times 12 = 432$ numbers in A, there are only eight free parameters.

Recently, Dukelsky *et al.* have done a related study but with a slightly different choice of independent parameters [36]. They considered matrix-product states with up to six irreducible representations in the truncated basis.

With only a few free parameters we can use a variational principle for the energy to determine these. At this point it is clear that the DMRG plays no essential role in the construction aside from providing a guide to which representations to keep. Even this choice could be done variationally. In the next section we will make an ansatz for a ground state wavefunction and perform the variational calculation to find the ground-state energy and the projection matrices $A[s]$. For $m = 6$ this amounts to finding the parameters ϕ and θ of (12) that minimize the energy of our ground-state ansatz.

The Hamiltonian in (1) with $\beta = -1/3$ is a very special point in this context. As can be seen from the construction of the ground state of the AKLT-model (Fig. 2) the total spin of the ground state does not increase as we add sites to the system. After we have added a site, only the spin-1/2 representation on the right hand side of (6) will contribute to the ground state. Thus only a single spin-1/2 representation at each step is needed to represent the ground state. From the arguments of this section we now see that if there is only a single irreducible representation in our truncated basis we have no freedom at all in the projection matrix $P^{j',j}$. Without having to do a variational calculation, the matrix product state with a single spin-1/2 irreducible representation gives us the exact ground state of the AKLT-model.

The fact that the DMRG in the thermodynamic limit produces a ground state in matrix product form also suggests an explanation of the fact that the DMRG is exact for models that have exact matrix product ground states. This is also studied in Chap. 3.1 (II).

4 Expectation values

To do the variational calculation that was discussed in Sect. 3 we take as an ansatz for the ground-state wave function the translationally invariant state $Q = \mathbb{1}$ from (4) which we denote by $|1\rangle$. Thus

$$|1\rangle \equiv \sum_{\{s_j\}} \mathrm{tr}\,(\, A[s_n] \ldots A[s_1]\,) |s_n \ldots s_1\rangle \,. \tag{13}$$

Note that although it is not explicitly written out, $|1\rangle$ has a definite number of lattice sites n. For the AKLT model [32] ($\beta = -1/3$) our ground-state ansatz is exact as are the matrix-product states of [5–7].

We will now show how expectation values in the matrix-product states can be calculated. We will do all calculations for the ground-state ansatz $|1\rangle$ but generalization to the general state $|Q)$ in (4) is straightforward. We will calculate several properties of the states. In particular we will determine the form of correlation functions in the matrix-product states. The calculations done in this section will also be of help in the next section were we will look at excited states.

The expectation value of an operator $\mathcal{O}$, e.g. energy or correlation function, in this state is given by

$$\langle 1|\mathcal{O}|1\rangle = \sum_{\{s_j\},\{s'_j\}} \mathrm{tr}\big(A^*[s'_n]\dots A^*[s'_1]\big)\,\mathrm{tr}\big(A[s_n]\dots A[s_1]\big) \times\langle s'_n\dots s'_1|\mathcal{O}|s_n\dots s_1\rangle\,. \tag{14}$$

To write this expression in a simpler form we define the tensor product matrix $(B\otimes C)$ by $(B\otimes C)^{(\alpha,\beta),(\tau,\nu)} = B^{\alpha,\tau}C^{\beta,\nu}$. We will in the rest of the lecture interchangeably use ordinary matrix indices α,β and composite indices (α,β), where composite indices are written with a parenthesis around them. This means that we can write a $m\times m$ "matrix" A as either a matrix $A^{\alpha,\beta}$ or as a m^2 vector $A^{(\alpha,\beta)}$. When the indices are not explicitly written out, the matrix or vector character of the symbol is assumed to be clear from the context. We now use the trace and matrix product identities $\mathrm{tr}\,(\,B\,)\,\mathrm{tr}\,(\,C\,) = \mathrm{tr}\,(\,B\otimes C\,)$ and $(BCD)\otimes(EFG) = (B\otimes E)(C\otimes F)(D\otimes G)$ to find

$$\langle 1|\mathcal{O}|1\rangle = \sum_{\{s_j\},\{s'_j\}} \mathrm{tr}\Big((A^*[s'_n]\otimes A[s_n])\dots(A^*[s'_1]\otimes A[s_1])\Big) \times\langle s'_n\dots s'_1|\mathcal{O}|s_n\dots s_1\rangle\,. \tag{15}$$

To write this in a more compact form we define a mapping $\widehat{M}$ from 3×3 spin matrices M to $m^2\times m^2$ matrices $\widehat{M}$ by

$$\widehat{M} \equiv \sum_{s',s} M_{s',s}\,(A^*[s']\otimes A[s])\,. \tag{16}$$

We denote by $S \equiv (S^x, S^y, S^z)$ the spin-1 representation of total spin and thus by $\hat{S} \equiv (\hat{S}^x, \hat{S}^y, \hat{S}^z)$ the "hat" mapping of the 3×3 spin matrices S. By $\widehat{1}$ we denote the "hat" mapping of the 3×3 identity matrix. We now see from (15) that the norm and the expectation value of the spin at the site j is given by

$$\begin{aligned}\langle 1|1\rangle &= \mathrm{tr}\,(\,\hat{1}^n\,),\\ \langle 1|S_j|1\rangle &= \mathrm{tr}\left(\,\hat{1}^{n-1}\hat{S}\,\right),\end{aligned} \tag{17}$$

where we in the last equation have used the cyclicity of the trace. Other expectation values are also easily obtained. Since we can factorize matrix elements like

$$\begin{aligned}\langle s_j', s_i' | \mathbf{S}_i \cdot \mathbf{S}_j | s_j, s_i \rangle &\equiv (\mathbf{S}_i \cdot \mathbf{S}_j)_{s_j', s_i', s_j, s_i} \\ &= (\mathbf{S})_{s_i', s_i} \cdot (\mathbf{S})_{s_j', s_j} ,\end{aligned}$$

we find that expectation values of energy and spin-spin correlation function are given by

$$\begin{aligned}\langle 1 | \mathbf{S}_j \cdot \mathbf{S}_{j+1} | 1 \rangle &= \mathrm{tr}\left(\hat{1}^{n-2} \hat{S} \hat{S} \right), \\ \langle 1 | \mathbf{S}_j \cdot \mathbf{S}_{j+l} | 1 \rangle &= \mathrm{tr}\left(\hat{1}^{n-l-1} \hat{S} \hat{1}^{l-1} \hat{S} \right) .\end{aligned} \tag{18}$$

A similar but a more complicated formula can be derived also for the operator $(\mathbf{S}_j \cdot \mathbf{S}_{j+1})^2$.

Formulas similar to (17) and (18) for expectation values in MP states have also been derived by Fannes *et al.* [6].

A possible problem with the construction of the projection operator is that parity is not built into the construction of the ground state since the projectors operate from the left to the right. There is therefore the possibility that parity is violated in the ground state $|1\rangle$. In App. A it is shown that the state $|1\rangle$ has parity $(-1)^n$, where n is the number of sites. This appendix also serves as another example for how calculations in the matrix product states can be done.

The ground-state energy

For $m = 6$ the freedom in the projection operators $A[s]$, and thus also in the ground-state ansatz $|1\rangle$, can be parameterized by the parameters ϕ and θ in (12). The energy of the state $|1\rangle$ is then a function of these two parameters,

$$E_{\mathrm{MP}}(\phi, \theta) = \langle 1 | H | 1 \rangle . \tag{19}$$

Note that $\langle 1 | 1 \rangle = 1$ due to (5). Note also that although it is not explicitly written out, E_{MP} depends on the number of lattice sites n. The convergence of E_{MP} to the limit $n \to \infty$ is however rapid and the value in the thermodynamic limit is easily calculated.

We can now use the variational principle for the energy and determine the parameters ϕ and θ, and thus also the projection matrices $A[s]$ by minimizing E_{MP} over all values of ϕ and θ. For $m = 12$ a similar minimization can be done but with eight parameters instead of two.

For both cases the computational effort is modest and we implemented the minimization procedure in Mathematica on an ordinary desktop workstation. We used a conjugate gradient method [37] to do the minimization in the two- and eight-dimensional spaces.

The projection matrices $A[s]$ obtained by this variational technique were found to agree up to high numerical accuracy with the projection operator obtained from similar DMRG calculations. Note that to compare the projection matrices from the variational calculation with the projection matrices from the DMRG calculation, the DMRG has to be done so as to preserve the total spin symmetry. We used the infinite system DMRG algorithm with a spin-1/2 edge spin. The ground-state energy was estimated from the difference in total energy of two systems from consecutive iteration steps.

Table 1. ground-state energy per site

β	E_{MP}	E_{DMRG}	E_{MP}	E_{DMRG}	exact	best numerical
	$(m = 6)$		$(m = 12)$			
$-1/3$	-0.666667	-0.666667	-0.666667	-0.666667	$-2/3$	
0	-1.39966	-1.39962	-1.40138	-1.40138	-	$-1.401484038971(4)$

The result for the lowest energy state for $\beta = -1/3$ and $\beta = 0$ is found in Table 1. The best result known to us for $\beta = 0$ comes from DMRG calculations in [14]. The exact result at the AKLT point $\beta = -1/3$ can be found in [32].

The small difference for the ground-state energy in Table 1 between the variational calculation and the DMRG calculation for $\beta = 0$ can be explained by the fact that the MP ground state ansatz is by construction uniform while the DMRG method will not produce a uniform ground state, since the projection operators A_n of (2) will depend on the site index n.

Correlation functions

The matrix $\hat{1}$ is central in this formulation of matrix product states. In (18) it appears as a kind of transfer matrix for the spin-1 chain. We will now look at the ground-state correlation functions in more detail. It turns out that the matrix product states are either long range ordered or have exponentially decaying correlations and that the correlation lengths for any operator is given by the eigenvalues of $\hat{1}$. Generally the correlation function of two operators is given by (C.f. (18))

$$\begin{aligned} \langle 1|\mathcal{O}_{i+l}\mathcal{O}_i|1\rangle &= \mathrm{tr}\left(\hat{1}^{n-l-i}\hat{\mathcal{O}}_{i+l}\hat{1}^{l-1}\hat{\mathcal{O}}_i\hat{1}^{i-1}\right) \\ &= \mathrm{tr}\left(\hat{1}^{n-l-1}\hat{\mathcal{O}}_{i+l}\hat{1}^{l-1}\hat{\mathcal{O}}_i\right) . \end{aligned} \tag{20}$$

Now diagonalize $\hat{1}$,

$$\hat{1} = UXU^{-1} , \tag{21}$$

where U is the matrix containing the eigenvectors of $\hat{1}$ and X is the diagonal matrix with the eigenvalues on the diagonal. It is easy to show that

$\hat{1}$ must have an eigenvalue equal to one [1,2,36]. To do this, we take the m^2-dimensional vector $v^{(\beta',\beta)} \equiv \delta_{\beta,\beta'}$. By using (5), we now find

$$\begin{aligned}
\sum_{\beta',\beta} (\hat{1})^{(\alpha',\alpha),(\beta',\beta)} v^{(\beta',\beta)} &= \sum_{\beta',\beta} \sum_{s} (A^*[s] \otimes A[s])^{(\alpha',\alpha),(\beta',\beta)} v^{(\beta',\beta)} \\
&= \sum_{\beta',\beta} \sum_{s} (A^*[s])^{\alpha',\beta'} (A[s])^{\alpha,\beta} \delta_{\beta,\beta'} \\
&= \sum_{\beta,s} (A^*[s])^{\alpha',\beta} (A[s])^{\alpha,\beta} \\
&= \delta_{\alpha,\alpha'} \\
&\equiv v^{(\alpha',\alpha)} . \qquad (22)
\end{aligned}$$

The vector v is thus an eigenvector to $\hat{1}$ with eigenvalue 1. Further, it is found numerically that all other eigenvalues have absolute values less than one. Now insert (21) into (20) and let us assume that n is large. We then see that only the eigenvalue one will survive in the term $\hat{1}^{n-l-1}$ and we get

$$\langle 1|\mathcal{O}_{i+l}\mathcal{O}_i|1\rangle = \langle \lambda_1|\hat{\mathcal{O}}_{i+l} U X^{l-1} U^{-1} \hat{\mathcal{O}}_i|\rho_1\rangle , \qquad (23)$$

where $\langle \lambda_1|$ and $|\rho_1\rangle$ are the left and right eigenvectors of $\hat{1}$ corresponding to the eigenvalue one. The correlation function will thus be

$$\langle 1|\mathcal{O}_{i+l}\mathcal{O}_i|1\rangle = \sum_{k=1}^{m^2} \langle \lambda_1|\hat{\mathcal{O}}_{i+l}|\rho_k\rangle \, \langle \lambda_k|\hat{\mathcal{O}}_i|\rho_1\rangle \, x_k^{l-1} , \qquad (24)$$

where x_k is the kth eigenvalue of $\hat{1}$, with $x_1 = 1$ and $|x_k| < 1$ for $k = 2 \ldots m^2$. If $l-1$ is large, we can extract the leading behavior of (24). However, it turns out that for some operators, the expectation value between $\langle \lambda_1|$ and $|\rho_1\rangle$ is zero. The leading behavior will thus be governed by the largest eigenvalue for which the expectation values on the right hand side of (24) does not vanish. We thus find that, if both $\langle \lambda_1|\hat{\mathcal{O}}_{i+l}|\rho_1\rangle$ and $\langle \lambda_1|\hat{\mathcal{O}}_i|\rho_1\rangle$ are nonzero, the state has long range order since the eigenvalue $x_1 = 1$ will prevent decay of the correlations. If however the rows of $\hat{\mathcal{O}}_{i+l}$ or the columns of $\hat{\mathcal{O}}_i$ are orthogonal to $\langle \lambda_1|$ and $|\rho_1\rangle$ respectively, we will have exponential decay of correlations. We get for the correlation length

$$\xi = -\frac{1}{\ln |x_k|} , \qquad (25)$$

where x_k is the largest eigenvalue of $\hat{1}$ for which the expectation values in (24) does not vanish. It turns out that the spin-spin correlation function, $\langle S^z_{i+l} S^z_i\rangle$ is governed by the second largest eigenvalue and thus decays exponentially. The so called string correlation function,

$$g(l) = \langle S^z_{i+l} \left(\prod_{j=i+1}^{i+l-1} e^{i\pi S^z_j} \right) S^z_i \rangle , \qquad (26)$$

however turns out to be long ranged.

Numerically we find for $m = 12$ the next leading eigenvalue of $\hat{1}$ to be -0.777, giving an asymptotic spin-spin correlation length from (25) of $\xi = 3.963$, compared to best estimates [14] of $\xi = 6.03(1)$. For $m = 6$ we find the next leading eigenvalue to be -0.681 giving $\xi = 2.60$. We see that the severe truncation of our basis to only six or twelve states has resulted in the asymptotic correlations being quite poor, although we have verified that intermediate length spin-spin correlations are consistent with more precise calculations [38]. For the string correlation function in (26) we find with $m = 12$ that $g(\infty) = -0.3759$, whereas best estimates are [14] $g(\infty) = -0.374325096(2)$.

It is well established that the DMRG works best for gapful systems with short ranged correlations and that the convergence of calculated eigenvalues or correlation functions with m is slower when the gap is small or the correlation length is large. Legeza and Fáth report a decrease in accuracy by several orders of magnitude at a critical point as compared to off-critical points for the Ising model in a transverse field [39]. The number of states m that has to be kept for a certain accuracy thus grows very large as a critical point is approached [40]. Efforts to study phase transitions by investigating the divergence of the correlation function or the vanishing of the gap have not been successful because the difficulties to accurately determine these properties close to a critical point [22,20,41].

How do systems with algebraically decaying correlations (infinite correlation length) fit into the matrix product picture presented in this lecture? Does the DMRG converge to a fixed point in the thermodynamic limit also for these systems? In a recent study by Andersson *et al.* [42] it has been found that for a system of spinless fermions on a lattice the DMRG converges to a fixed point in the thermodynamic limit. This system is gapless and has algebraicly decaying correlations. At intermediate range, where the correlations are given by the sum of exponentials in (24), it is found that this sum mimics the behavior of a powerlaw function. The asymptotic correlations on the other hand always turn out to be exponential. It is however found that the asymptotic correlation length defined by (25) increases with m as $m^{1.3}$ which indicates an infinite correlation length for the exact system [42].

5 Bloch States

In this section we use the states $|Q)_n$ as defined in (4), where Q is a general $m \times m$ matrix, to make an ansatz for the low-lying excited states. For a translationally invariant system we can define our states to be Bloch states. A reasonable ansatz for a Bloch state $|Q,k)_n$ defined by a matrix Q and a momentum k is given by

$$|Q,k)_n \equiv \sum_{s_n \dots s_1} \sum_{j=1}^{n} e^{ijk} \mathrm{tr}\Big(A[s_n] \dots A[s_{j+1}] Q A[s_j] \dots A[s_1]\Big) \times |s_n \dots s_1\rangle .$$

This wavefunction can be viewed as the ground state $|1\rangle$ with a disturbance Q introduced at some site, and then letting the disturbance run over all sites to form a state with a definite momentum. In this way we get a single "particle" excitation.

We will now calculate properties of these states. We will determine the Hamiltonian and normalization matrices

$$(Q', k|H|Q, k)_n \tag{27}$$

$$(Q', k|Q, k)_n \tag{28}$$

and find orthonormal eigenstates $|Q_\gamma, k\rangle$ of the matrix (27),

$$\langle Q_{\gamma'}, k|H|Q_\gamma, k\rangle = E_\gamma \delta_{\gamma',\gamma} \tag{29}$$

$$\langle Q_{\gamma'}, k|Q_\gamma, k\rangle = \delta_{\gamma',\gamma} \ , \tag{30}$$

where γ labels the different eigenstates. Note that states with different momentum k are automatically orthogonal. It should also be emphasized that there will not be any minimization done here. The projection matrices $A[s]$ are the same as for the ground state. The only freedom we have in the states $|Q, k)_n$ reside in the $m \times m$ matrix Q and we will determine the set of eigenmatrices Q_γ from an eigenvalue equation.

We will now derive expressions for matrix elements (27) and (28). The method to calculate them are the same as was presented for the state $|1\rangle$ in the previous section. The Bloch states are however more complicated and the calculations are a bit tedious. We will therefore only summarize them here. The details can be found in [2]. See also App. B for some details regarding the state $|Q)_n$ of (4).

One finds that the general structure of all matrix elements between the Bloch states is that they consist of traces with a convolution sum over matrix products inside each trace. For the overlap we find

$$(Q', k|Q, k)_n = n \ \mathrm{tr} \left((Q' \otimes \mathbb{1}) \sum_{j=0}^{n-1} e^{ijk} \hat{1}^{n-j} (\mathbb{1} \otimes Q) \hat{1}^j \right) , \tag{31}$$

with similar, but more complicated, expressions for the Hamiltonian and for the z-component of total spin. For finite-length chains, the sums in expressions of this type can be expediently calculated by a recursive scheme for the case when n is a power of two [2].

By doing a transformation which is described in some detail in App. B, we can calculate a norm matrix $G(k, n)$ defined by

$$(Q', k|Q, k)_n \equiv n \ Q' G(k, n) Q \ , \tag{32}$$

where the right hand side of (32) is an ordinary vector-matrix-vector product and we regard Q' and Q as m^2-dimensional vectors and where $G(k, n)$ is an $m^2 \times m^2$ matrix. The nice thing about (32) is that we have effectively turned the computation of the trace in (31) for all Q and Q' into a matrix inner product between Q, Q' and a single $m^2 \times m^2$ matrix $G(k,n)$, independent of Q and Q'.

Similarly we can define a $m^2 \times m^2$ Hamiltonian matrix by

$$(Q', k|H|Q, k)_n \equiv n\, Q'H(k,n)Q \,. \tag{33}$$

A matrix $S_T^z(k,n)$ representing the z-component of total spin can be defined analogously.

Rewriting the traces over matrix products as ordinary vector-matrix-vector products is very useful since now we can diagonalize matrices like $H(k,n)$ in (33) and thereby find the eigenstates in the subspace spanned by all ansatz vectors $|Q,k)_n$. The principle for calculating these matrices is described in App. B. Due to the number of terms in the expressions for the expectation values as in (31) it is however numerically cumbersome for finite length chains.

There is however an elegant way to extract the leading behavior of $H(k,n)$ and $G(k,n)$ as $n \to \infty$ [2], which we will briefly describe here.

Let us first define the z-transform (sometimes called a discrete Laplace transform) of a series $\{a_n\}_{n=0}^{\infty}$ by $F(\lambda) = \sum_{n=0}^{\infty} a_n e^{-n\lambda}$. Let us now denote the sum inside the trace in (31) by S_n, so that

$$S_n = \sum_{j=0}^{n-1} e^{ijk} \hat{1}^{n-j} (\mathbb{1} \otimes Q) \hat{1}^j \,.$$

We now define a series $\{S_n\}_{n=0}^{\infty}$, and take the z-transform of this sequence. By examining the analytical structure of the transformed series it is possible to extract the leading behavior of the sum S_n, as $n \to \infty$. In this way we can get the asymptotic form of $G(k,n)$ in the limit of large n. This procedure is then also applied to all sums in the expressions for the Hamiltonian matrix elements. This whole procedure finally results in the asymptotic forms

$$H(k,n) = n^2 H_2(k) + nH_1(k) + H_0(k) + \mathcal{O}(z)^n \tag{34}$$

$$G(k,n) = nG_1(k) + G_0(k) + \mathcal{O}(z)^n \,, \tag{35}$$

with $H(k,n)$ and $G(k,n)$ as defined in (33) and (32) and were $H_i(k)$, $i = 0,1,2$ and $G_i(k)$, $i = 0,1$ are $m^2 \times m^2$ matrices. Here z represents the next leading eigenvalue of $\hat{1}$ which numerically is found to be $|z| \approx 0.8$ for $m = 12$. There are thus very small corrections to the asymptotic form. We also find that H_2 and G_1 are non-vanishing only when the momentum k is zero. The energy of a state $|Q,k)_n$ is

$$E_Q(k,n) = \frac{QH(k,n)Q}{QG(k,n)Q} \,. \tag{36}$$

Since the normalization matrix $G(k,n)$ is not the identity matrix, the naive basis states $(Q_{i,j})^{\alpha,\beta} = \delta_{i,\alpha}\delta_{j,\beta}$, with $i = 1,\ldots,m$ and $j = 1,\ldots,m$, are not orthonormal. We thus have to consider also the matrix $G(k,n)$ in the diagonalization. The eigenvalue equation which must be solved is thus

$$H(k,n)Q(k,n) = E_Q(k,n)G(k,n)Q(k,n) \;, \tag{37}$$

where $Q(k,n)$ is an m^2-dimensional vector. Combining (34), (35) and (37) for $k \neq 0$ we have

$$\big(nH_1(k) + H_0(k)\big)Q(k,n) = \big(nE_0 + \Delta_k(n)\big)G_0(k)Q(k,n) \;, \tag{38}$$

where E_0 is the ground-state energy per site and $\Delta_k(n)$ is the excitation energy. E_0 denotes the ground-state energy per site in the limit $n \to \infty$, and is therefore independent of n. Since we are interested in the solutions to (38) when $n \to \infty$ we assume also Q and Δ_k to be independent of n. From the solution of (38) we get the ground-state energy, which is the same energy as we found from the minimization in Sect. 4. We also get the excitation spectrum Δ_k and the corresponding orthonormal eigenstates $|Q_\gamma, k\rangle$, were γ labels the different eigenstates. These states represent our "single magnon" states and will have the properties of (29) and (30). An asymptotic form for the z-component of total spin, $S^z_T(k,n)$, similar to the form for $H(k,n)$, containing terms up to order n^2 can be determined in a similar way.

We have computed the asymptotic forms when $n \to \infty$ for the Hamiltonian and norm matrices defined in (33) and (32) as well as for the total spin matrix for $\beta = 0$ and for different momenta k. We simultaneously diagonalize $H(k,n)$ and $S^z_T(k,n)$ which means that the eigenstates will be states with a definite z-component of total spin. We have numerically determined these states along with their energy and spin expectation values. For $m = 12$ we find the single-particle spectrum shown in Fig. 5.

The low-lying triplet branch defines the gap $\Delta_\pi = 0.4094$, which is very good compared to the most accurately known result [14,24,25] of 0.410502(1). Furthermore, we compute the spin wave velocity $v = 2.452$ to be compared to the calculations in [24], where $v = 2.49(1)$ was obtained. Clearly we reproduce the single-particle triplet excitations with high accuracy considering the few number of states in our basis. Our calculation also yields a detailed spectrum of lowest lying "single-magnon" excitations shown by dotted lines in Fig. 5. Our second lowest energy excitation at $k = \pi$ is a singlet shown by a dotted line in the same figure with $\Delta_\pi(singlet) = 2.348$.

As a function of k, the second lowest single-particle excitation is either a singlet or a spin-2 object, as has also been observed in exact finite size calculations [19]. Parity of each of the elementary excitations is verified by checking the relation (39) with Q as well as with the matrices A. The boundary to two particle excitations at a given value of k is computed explicitly by minimizing the sum of energies of excitations whose pseudo-momentum sums to k, and similarly for the three-particle excitations. These results are shown by the light and dark shaded regions in Fig. 5. The picture fits well with previously obtained results.

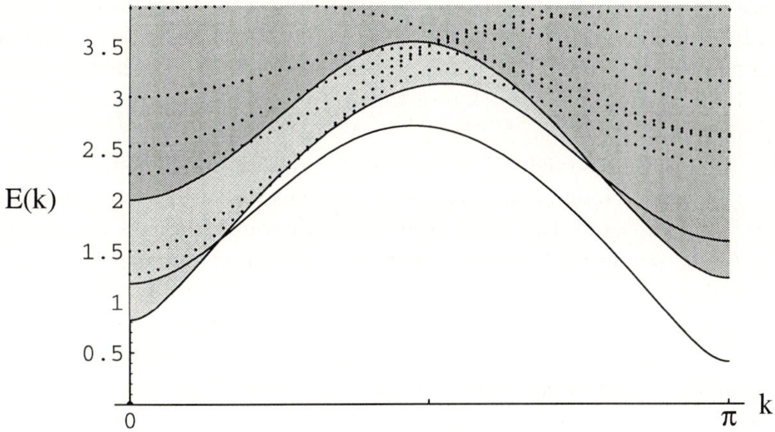

Fig. 5. The spectrum for $\beta = 0$ and $m = 12$ is shown. The lowest single particle triplet is shown as a solid line, with the lightly shaded region representing two-particle excitations and the dark region three-particle excitations. Solid lines define the boundaries to the two- and three-particle continuum. Dotted lines indicate the spectrum of higher energy single-magnon excitations. The spin of these dotted excitations are, in order of increasing energy at $k = \pi$: 0,1,2,2,3,1,1,0.

6 Summary and Conclusions

In this lecture we have studied the DMRG method as it reaches the thermodynamic limit. It was found that if the renormalization converges to a fixed point, the DMRG ground state can be written in a matrix product form. This form was used to define a set of matrix product ansatz wave functions that can be investigated on their own, with no direct connection to the DMRG. The numerical calculations were tested on the spin-1 chain. It was found that by using symmetry considerations, the number of free parameters in these ansatz wave functions can be greatly reduced. In particular, the spin rotational symmetry of the spin-1 chain was explored and the basis states in the truncated basis were chosen to be complete representations of total spin of the chain. With just a few free parameters, a variational principle for the energy was now used to find the lowest energy state in the few-parameter family of ansatz states. By performing the variational calculation it was found that the variational matrix product ground state agrees to a very high accuracy with the ground state found from similar DMRG calculations.

This suggests that the rapid convergence of the DMRG is explained by the fact that the states selected are optimally chosen eigenstates of total block spin. Properly chosen, these states are highly efficient for building wave functions with a small basis that have low total spin for all subblocks.

By further investigating the matrix-product states it was found that the asymptotic correlations in the states are either long ranged or decay exponentially. Intermediate range correlations were found to be either long ranged or to be a sum of exponentials. This suggests an explanation of the fact that the DMRG has been found to work best for gapful system with exponentially decaying correlations.

In this lecture we also defined a set of states describing elementary excitations in the spin-1 chain. These states could be explored in the thermodynamic limit and the excitation spectrum of the Heisenberg spin-1 chain was numerically estimated and found to agree well with previous results.

A Parity

In this appendix we will show in detail that the state $|1\rangle$ has a definite parity. Let $\mathcal{P}$ be the parity operator. We thus have

$$\begin{aligned}\mathcal{P}|1\rangle &= \sum_{\{s_j\}} \mathrm{tr}\,(\,A[s_n]\dots A[s_1]\,)\mathcal{P}|s_n\dots s_1\rangle \\ &= \sum_{\{s_j\}} \mathrm{tr}\,(\,A[s_n]\dots A[s_1]\,)|s_1\dots s_n\rangle\;.\end{aligned}$$

Suppose now that there exists an invertible $m \times m$ matrix $Q_\mathcal{P}$ such that

$$Q_\mathcal{P}A[s] = sign[\mathcal{P}](A[s])^T Q_\mathcal{P}\;, \tag{39}$$

where A^T denotes transpose and $sign[\mathcal{P}]$ is a proportionality constant that will be seen to be the eigenvalue of the parity operator. Then it follows that

$$\begin{aligned}\mathcal{P}|1\rangle &= \sum_{\{s_j\}} \mathrm{tr}\,(\,Q_\mathcal{P}^{-1}Q_\mathcal{P}A[s_n]\dots A[s_1]\,)|s_1\dots s_n\rangle \\ &= sign[\mathcal{P}]^n \sum_{\{s_j\}} \mathrm{tr}\,(\,A^T[s_n]\dots A^T[s_1]\,)|s_1\dots s_n\rangle \\ &= sign[\mathcal{P}]^n \sum_{\{s_j\}} \mathrm{tr}\,(\,A[s_1]\dots A[s_n]\,)|s_1\dots s_n\rangle \\ &= sign[\mathcal{P}]^n|1\rangle\;.\end{aligned}$$

Thus, for the ground state to have definite parity, it is sufficient that such a $Q_\mathcal{P}$ exists. How do we find this matrix, if it exists? We multiply both sides of the defining relation (39) by $A^\dagger[s]$ and sum over s. Using (5) we find that

$$\begin{aligned}Q_\mathcal{P}^{\alpha,\beta} &= sign[\mathcal{P}]\sum_s (A^T[s])^{\alpha,\tau}Q_\mathcal{P}^{\tau,\nu}(A^T[s])^{\nu,\beta} \\ &= sign[\mathcal{P}]\left(\sum_s (A^T[s])^{\alpha,\tau}A^{\beta,\nu}[s]\right)Q_\mathcal{P}^{\tau,\nu} \\ &= sign[\mathcal{P}]\sum_s (A^T[s]\otimes A[s])^{(\alpha,\beta),(\tau,\nu)}Q_\mathcal{P}^{\tau,\nu}\;.\end{aligned}$$

Thus, $Q_{\mathcal{P}}$, if it exists, is the eigenvector of the matrix $\sum_s (A^T[s] \otimes A[s])$ with eigenvalue ± 1. The parity operator of (39) can numerically be computed and it is found that the ground state has parity $(-1)^n$, where n is the number of sites.

B The General State $|Q)$

In this appendix we show in some detail how to rewrite matrix elements between the MP states $|Q)_n$ defined in (4) into ordinary matrix-vector products. The Bloch states $|Q, k)_n$ can be treated in an analogous way but the calculations will be more tedious.

To calculate the norm we use the same trace and tensor product identities as when deriving (15). We find that

$$\begin{aligned} (Q'|Q)_n &= \sum_{\{s_j\}} \mathrm{tr}\left((Q')^* A^*[s_n] \ldots A^*[s_1] \right) \mathrm{tr}\left(Q A[s_n] \ldots A[s_1] \right) \\ &= \mathrm{tr}\left((Q'^* \otimes Q)\, \hat{1}^n \right) . \end{aligned} \tag{40}$$

We can rewrite this trace as ordinary matrix products. To do this we first define the generalized transpose $M^{T_{p_1,p_2,p_3,p_4}}$ of a matrix M by

$$(M^{T_{p_1,p_2,p_3,p_4}})^{(\alpha_1,\alpha_2),(\alpha_3,\alpha_4)} = M^{(\alpha_{p_1},\alpha_{p_2}),(\alpha_{p_3},\alpha_{p_4})} , \tag{41}$$

where $\{p_1, p_2, p_3, p_4\}$ is a permutation of $\{1, 2, 3, 4\}$. We also define a tilde operator $\widetilde{M}$ by the formula

$$\mathrm{tr}\left((Q' \otimes Q) M \right) = \sum_{\alpha',\beta',\alpha,\beta} (Q')^{\alpha',\beta'} (\widetilde{M})^{(\alpha',\beta'),(\alpha,\beta)} Q^{\alpha,\beta} , \tag{42}$$

so that the tilde operator effectively generates the matrix corresponding to the inner product of Q' and Q with M. One finds by writing out (42) in components that

$$\widetilde{M} = M^{T_{3142}} .$$

Hence

$$((Q')^{\alpha',\beta'}|Q^{\alpha,\beta})_n = Q'^{(\alpha',\beta')} G(n)^{(\alpha',\beta'),(\alpha,\beta)} Q^{(\alpha,\beta)} , \tag{43}$$

with

$$G(n) = \widetilde{(\hat{1}^n)} . \tag{44}$$

We have now effectively turned the computation of the trace in (40) for all Q and Q' into a matrix inner product between Q, Q' and a single $m^2 \times m^2$ matrix G, independent of Q and Q'. Note that on the right side in (43) we write Q and Q' as vectors of length m^2.

Similarly we can compute the expectation value of the Heisenberg Hamiltonian defined in (1) with $\beta = 0$ as

$$(Q'|H|Q)_n = \sum_{i=0}^{n-2} \mathrm{tr}\left((Q' \otimes Q)\hat{1}^i \hat{S}\hat{S}\hat{1}^{n-2-i} \right), \tag{45}$$

where $\hat{S}$ denotes the hat mapping in (16) of the spin-1 matrices. Similar to (43) we can define an $m^2 \times m^2$ matrix $H(n)$ by

$$(Q'|H|Q)_n = Q'H(n)Q\ , \tag{46}$$

where $H(n)$ is independent of Q' and Q. Note that the left hand side of (46) is again an ordinary vector-matrix-vector product where Q' and Q are regarded as m^2-dimensional vectors. Other operators like the z-component of total spin can be determined in a similar way.

By rewriting the traces over matrix products as ordinary vector-matrix-vector products we can diagonalize matrices like $H(n)$ in (46) and thereby find the eigenstates in the subspace spanned by all ansatz vectors $|Q)_n$. The energy of a state $|Q)_n$ is given by

$$E_Q = \frac{(Q|H|Q)_n}{(Q|Q)_n} = \frac{QH(n)Q}{QG(n)Q}\ . \tag{47}$$

The eigenvalue equation to be solved is thus (C.f. (37))

$$H(n)Q = E_Q G(n)Q\ . \tag{48}$$

References

1. S. Östlund and S. Rommer, Phys. Rev. Lett. **75**, 3537 (1995)
2. S. Rommer and S. Östlund, Phys. Rev. B **55**, 2164 (1997)
3. S.R. White, Phys. Rev. Lett. **69**, 2863 (1992)
4. S.R. White, Phys. Rev. B **48**, 10345 (1993)
5. L. Accardi, Phys. Rep. **77**, 169 (1981)
6. M. Fannes, B. Nachtergaele and R.F. Werner, Europhys. Lett. **10**, 633 (1989); J. Phys. A **24**, L185 (1991); Commun. Math. Phys. **144**, 443 (1992)
7. A. Klümper, A. Schadschneider and J. Zittartz, Europhys. Lett. **24**, 293 (1993); Z. Phys. B **87**, 281 (1992); C. Lange, A. Klümper and J. Zittartz, Z. Phys. B **96**, 267 (1994)
8. A.K. Kolezhuk and H.-J. Mikeska, Phys. Rev. B **56**, R11380 (1997); A.K. Kolezhuk, J. Phys.: Cond. Matt. **10** 6795 (1998)
9. G. Su, Phys. Lett. A **213**, 93 (1996)
10. A. Schadschneider and J. Zittartz, Ann. der Phys. **4**, 157 (1995)
11. A.K. Kolezhuk, H.J. Mikeska and S. Yamamoto, Phys. Rev. B **55**, R3336 (1997)
12. S. Brehmer, H.J. Mikeska and U. Neugebauer, J. Phys. C: Solid State Phys. **8**, 7161 (1996)
13. J.M. Román, G. Sierra, J. Dukelsky and M.A. Martín-Delgado, J. Phys. A: Math. Gen. **31**, 9729 (1998)

14. S.R. White and D.A. Huse, Phys. Rev. B **48**, 3844 (1993)
15. J. Solyom, Phys. Rev. B **36**, 8642 (1987)
16. G. Fáth and J. Sólyom, Phys. Rev. B **47**, 872 (1993)
17. Y. Xian, J. Phys.: Cond. Matt. **5**, 7489 (1993)
18. R.F. Bishop, J.B. Parkinson and Y. Xian, J. Phys.: Cond. Matt. **5**, 9169 (1993)
19. J. Oitmaa, J.B. Parkinson and J.C. Bonner, J. Phys. C: Solid State Phys. **19**, L595 (1986)
20. U. Schollwöck, T. Jolicœr and T. Garel, Phys. Rev. B **53**, 3304 (1996)
21. T. Xiang and G.A. Gehring, Phys. Rev. B **48**, 303 (1993)
22. R.J. Bursill, T. Xiang and G.A. Gehring, J. Phys. A **28**, 2109 (1995)
23. F.D.M. Haldane, Phys. Lett. A **93**, 464 (1983); Phys. Rev. Lett. **50**, 1153 (1983); J. Appl. Phys. **57**, 3359 (1985)
24. E.S. Sørensen and I. Affleck, Phys. Rev. Lett. **71**, 1633 (1993)
25. T. Sakai and M. Takahashi, Phys. Rev. B **42**, 1090 (1990)
26. L.A. Takhtajan, Phys. Lett. A **87**, 479 (1982)
27. H. Babujian, Phys. Lett. A **90**, 479 (1982)
28. I. Affleck, Phys. Rev. Lett. **55**, 1355 (1985); I. Affleck and F.D.M. Haldane, Phys. Rev. B **36**, 5291 (1987)
29. G.V. Uimin, JEPT Lett. **12**, 225 (1970)
30. C.K. Lai, J. Math. Phys. **15**, 1675 (1974)
31. B. Sutherland, Phys. Rev. B **12**, 3795 (1975)
32. I. Affleck, T. Kennedy, E.H. Lieb and T. Tasaki, Phys. Rev. Lett. **59**, 799 (1987); Commun. Math. Phys. **115**, 477 (1988)
33. See any textbook on Quantum Mechanics, e.g. J.J. Sakurai, Modern Quantum Mechanics, Addison-Wesley (1985)
34. S. Qin, T.K. Ng and Z.B. Su, Phys. Rev. B **52**, 12844 (1995)
35. W. Marshall, Proc. Roy. Soc (London) Ser. A **232**, 48 (1955); E. Lieb and D. Mattis, J. Math. Phys. **3**, 749 (1962)
36. J. Dukelsky, M.A. Martín-Delgado, T. Nishino and G. Sierra, Europhys. Lett. **43**, 457 (1998)
37. W.H. Press, S.A. Teukolsky, W.T. Vetterling and B.P. Flannery, Numerical Recipes, Cambridge University Press (1992)
38. E. S. Sørensen and I. Affleck, Phys. Rev. B **49**, 15771 (1994)
39. Ö. Legeza and G. Fáth, Phys. Rev. B **53**, 14349 (1996)
40. S.R. White and I. Affleck, Phys. Rev. B **54**, 9862 (1996)
41. A. Drzewiński and J.M.J. van Leeuwen, Phys. Rev. B **49**, 403 (1994)
42. M. Andersson, M. Boman and S. Östlund, preprint cond-mat/9810093

4 A Recurrent Variational Approach

Miguel A. Martín-Delgado[1] and Germán Sierra[2]

[1] Departamento de Física Teórica I, Universidad Complutense
E-28040 Madrid, Spain.
[2] Instituto de Matemáticas y Física Fundamental, C.S.I.C.
E-28006 Madrid, Spain.

The introduction of the Density Matrix Renormalization Group (DMRG) by S.R. White [1,2] has changed in many aspects the field of strongly correlated electrons. Originally intended as a new and powerful numerical tool [1–6], the DMRG has also spurred the interest of theorists [7–16] in the search for new analytic albeit non-perturbative methods to deal with the difficulties posed by strongly correlated systems. Introductions can be found in [17–20], and Chap. 2(I).

This chapter is a review of recent work on the Recurrent Variational Approach (RVA) to the DMRG. The RVA is a sort of reduced version of the original DMRG, in which one retains a single state as the best candidate for the ground state of the system. This reduction of degrees of freedom is initially done in order to keep the method manageable analytically. Thus, the RVA falls initially in the category of analytic approaches to the DMRG. The aim of this analytical approach is to try to understand the relevant physical degrees of freedom so that we can figure out what the underlying physics is in a strongly correlated system. This initial analytical goal has been also developed in order to later acquire more numerical precision. To do this, the method becomes more numerical and somehow stands in between an analytical formulation of the DMRG and a numerical one.

The chapter is based on a series of papers [21–25] and several seminars and courses. We focus our attention on the practical purpose of the RVA rather than on the purely theoretical aspects and thus we illustrate the application of the method with examples. Due to the great interest acquired by the subject of ladder systems (see [26] for a review), we have taken these systems as the testgrounds to set up the fundamentals of the method.

We present in Table 1 a sketch of the relationship and differences between the RVA and the DMRG methods. Both in the RVA and the DMRG there is a truncation procedure. In the RVA there is a single state ($m = 1$) which depends on a collection of variational parameters which serve to explore the ground-state manifold. In the DMRG one keeps $m > 1$ states to build the ground state of the system. Likewise, in the RVA the system (chain, ladder, etc) is constructed step by step adding a certain unit (site, rung, etc) and this leads to recursion relations satisfied by the RVA-state (usually the ground state). This is the analogue of the Wilsonian onion scheme used by the DMRG. Finally, in both methods there is an optimization process to select the states. In the RVA it is done by minimizing the ground-state energy while in DMRG it is done by using the density matrix to project onto the most probable states.

Table 1. Schematic comparison between the RVA and the DMRG (RR stands for Recurrence Relation)

Comparison	RVA	DMRG
# States kept	$m = 1$	$m > 1$
Order RR's	$1^{\rm st}, 2^{\rm nd}, \ldots$	$1^{\rm st}$
Optimization	Lowest energy state	Most probable states

Originally, the variational RVA was introduced to solve the RVB (Resonating Valence Bond) for even-legged ladders [21] (see next section). Then it was realized that this recursive method was reminescent of the Wilsonian RG method and contact was made with the matrix-product approach and the thermodynamic limit of DMRG (Östlund and Rommer [15], Chap. 3(I)) There are however some subtle differences between the RVA and the matrix-product (MP) method as presented in the latter references. The ansätze of [15] satisfy periodic boundary conditions, are uniform and translationally invariant. The RVA states satisfy open boundary conditions and can in fact be non uniform. In this respect the RVA is more faithful to the DMRG which can be considered as a highly performant non uniform variational matrix-product method.

On the other hand the RVA allows a simple treatment of systems with holes with and without the Gutzwiller constraint of non double occupancy. In these cases the recurrence relations (RR) are both for the size and the number of holes. We can easily generate doubly projected BCS-like states. The projections being on the number of electrons and the no double occupancy.

What seems to be lacking in the RVA is the concept of a density matrix. As we shall show in Sect. 4 the RVA and more generally the MP methods give rise, in the thermodynamic limit, to a density matrix which emerges as the left eigenvector of a non-symmetric transfer matrix. The latter transfer matrix is constructed out of the MP variational parameters.

The RVA, MP and DMRG methods deal very well with systems with a finite correlation length [37]. However, for gapless systems the DMRG requires a number of states kept m increasing with the size of the system [29].

The organization of this chapter is as follows. In Sects. 2 and 3 we introduce the RVA to solve analytically the resonating valence bond problem in a two-leg Heisenberg ladder. In Sect. 4 we present another extension of the RVA based on first-order recurrence relations, which leads to a matrix-product representation of the ansatz. We apply this to the spin-one Heisenberg chain. It is shown how a density matrix emerges in this formalism and allows us to make contact with the DMRG method. In Sect. 5 we extend the RVA to deal with doped square ladders, taking into account both spin and charge degrees of freedom. In Sect. 6 the RVA is applied to study Heisenberg and t-J models on diagonal ladders. These types of ladders serve as new settings to make extensions of the RVA for different basis of states, such as S_z-states, and for non-uniform ansatz which are those having variational parameters depending locally on the position along the diagonal ladder. This corresponds to RVA applications not in the thermodynamic limit. We close this Chapter in Sect. 7 by discussing new developments and future problems.

1 Two-Legged Spin Ladders

The Recurrent Variational Approach was originally introduced in [21] to solve the computation of RVB states [30] in Heisenberg ladders with an even number of legs. The reason for this restriction comes from two facts. On one hand, theoretical studies have found that antiferromagnetic spin ladders are gapped (gapless) depending on whether the number of legs n_l is an even (odd) number. This prediction has been confirmed experimentally in several compounds [26]. On the other hand, we know that RVB states are gapped. Intuitively, the gap comes from the breaking of a valence bond (singlet) of the covered lattice and thereby forming a triplet state at that bond. This singlet-triplet excitation costs an energy of J which signals the presence of a spin gap in the spectrum of the RVB states. Thus, RVB states are a good basis for the description of even-legged ladders. Moreover, to be more concrete, we choose the simplest even ladder possible, namely, the two-leg ladder with an AF-Heisenberg model described by two coupling constants: J for the links in the legs, and J' for the vertical rungs, see Fig. 1.

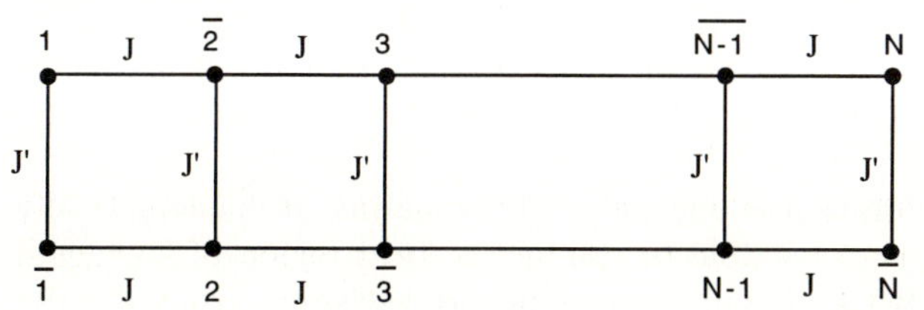

Fig. 1. The two-leg ladder with an AF-Heisenberg model described by two coupling constants: J for the links in the legs, and J' for the vertical rungs

The Hamiltonian reads

$$H = J \sum_{k=1}^{N-1} (\mathbf{S}_k \cdot \mathbf{S}_{\overline{k+1}} + \mathbf{S}_{k+1} \cdot \mathbf{S}_{\overline{k}}) + J' \sum_{k=1}^{N} \mathbf{S}_k \cdot \mathbf{S}_{\overline{k}} \tag{1}$$

where N denotes the number of vertical rungs of the ladder and open boundary conditions have been assumed. As the two-leg ladder is a bipartite lattice, we label the sites with and without bars to distinguish the two sublattices in which the ladder can be decomposed.

What if the number of legs is odd? White, Noack and Scalapino [31] came up with a nice picture to qualitatively distinguish the different behaviour of even-legged vs. odd-legged spin ladders. Their basic idea is to see when the resonance mechanism of valence bonds can be better achieved depending on the number of legs in a spin ladder. For a simple 2×2 plaquette lattice this resonating mechanism is shown in Fig. 2. In this simple lattice, the ground state (g.s.) of the AF-Heisenberg model is precisely the superposition of the two configurations of pair of bonds (vertical and horizontal) depicted in Fig. 2. Recall that a bond between two sites is a singlet state. Thus, the RVB state is exact and it has lower energy than either VB configurations (vertical or horizontal) that form it. We thus say that the *resonance of the VB states lowers the energy of the variational state.*

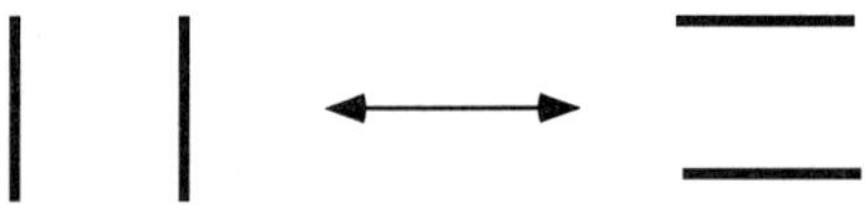

Fig. 2. The basic bond resonance mechanism between horizontal and vertical bonds in an elementary plaquette.

With this resonating mechanism in mind, let us see what happens on spin ladders starting first with the simplest two-leg ladder and then adding more legs. For a two-leg ladder, there are two types of VB configurations, namely,

Resonating: (Fig. 3 (a)) This consists of vertical and horizontal VB states arranged such that the basic resonance mechanism for a single plaquette (Fig. 2) can take place. In particular, this means that whenever there is a horizontal VB state in one leg, there must be another one located opposite in the other leg.

Staggered: (Fig. 3 (b)) This consists of only horizontal VB states arranged such that the basic resonance mechanism for a single plaquette (Fig. 2) cannot take place. The horizontal VB states are shifted by one link from one another.

As the staggered configuration is uncapable of resonating, it is energetically unfavorable. However, it is possible to produce a local region of staggered VB states by introducing two spin defects (uncoupled spins in non-singlet states) at the borders of that region (Fig. 3 (c)). Again, within the short range RVB scenario, we expect that the two spin defects will get confined into a pair on the same rung (Fig. 3 (d)). The reason for this is that in this way we are reducing the staggering regions which are costly in energy. This qualitative picture for the confinement of spin defects is readily extrapolable to any even-legged ladder.

Now let us turn to a three-leg ladder and we immediately see that this setting is qualitatively different. First, in a three-leg ladder we only have one type of order (Fig. 3 (d)), unlike the two possibilities (resonating vs. staggered) in a two-leg ladder. Thus, when a spin defect is introduced there

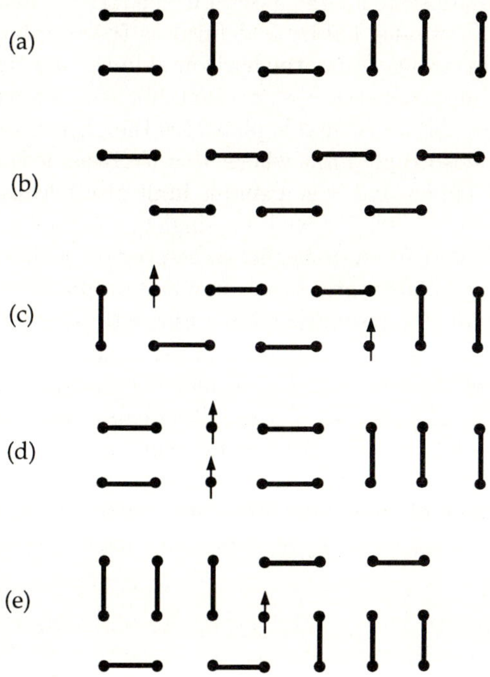

Fig. 3. Different Valence Bond configurations in two-leg and three-leg ladders. (**a**) Resonating. (**b**) Staggered. (**c**) Spin defects bordering a staggered local configuration. (**d**) Confinement of spin defects. (**e**) Non-confinement of spin defects in a three-leg ladder.

is no change in the type of order (Fig. 3 (e)) for it only amounts to a mere shift of configurations. This shift does not cost energy. The final outcome is that spin defects do not come in pairs in a three-leg ladder.

Thus, we arrive at the following qualitative picture distinguishing even-legged and odd-legged ladders:

Even-legged Ladders: They exhibit confinement of spin defects.
Odd-legged Ladders: They do not exhibit confinement of spin defects.

These qualitative results were confirmed with the DMRG method [31].

After this brief discussion of RVB states in general ladders we come to the point of the RVA for two-leg spin ladders.

Resonon states: In order to construct a variational RVB state, we need to classify all the possible short-range VB configurations (dimers) in a two-leg ladder. The guiding idea to achieve this is again the resonating mechanism of valence bonds in the elementary 2×2 plaquettes forming the ladder as explained above. We shall not consider staggered configurations for they raise the energy. To perform this classification, we shall choose a reference state formed by all bonds in the vertical possitions for every rung out of N forming the ladder (Fig. 4). This state we denote as $|0\rangle$. There is a physical reason underlying this state. It is the ground state for the J-term in the Hamiltonian (1). The strong-coupling limit of a two-leg ladder is defined as the coupling region where the vertical rungs are stronger than the horizontal links, namely, $J' \gg J$. In this limit, the most important state is precisely $|0\rangle$. It is known that the strong-coupling limit provides a reasonable picture of the physcis for a two-leg ladder even down to the isotropic limit where $J' = J$. Thus, it is a convenient starting point for the ladder and it is a simple limit also because the diagonalization of the rungs states is trivial: they are singlets and triplets. Once the horizontal coupling J is turned on, it begins to be possible to have a pair of vertical bonds resonating to horizontal bonds in a certain plaquette of the ladder. Let us call a *resonon* to a resonating 2×2 plaquette (Fig. 5). We can place one resonon on any of the $N-1$ plaquettes of the ladder. In this situation we say that the effect of the coupling J is to make the plaquettes start resonating. If we increase the value of J for J' fixed, it will also probably have two resonons present in the ladder (Fig. 6), and eventually, there will be $[N/2]$ plaquettes ($[\ldots]$ meaning integer part) resonating at the same time, see Fig. 7. In summary, the problem of classifying VB states reduces to first decomposing the Hilbert space of dimer states by sectors with a fixed number of resonons, as follows:

Zero-resonon sector: $|0\rangle$. (Fig. 4)

Fig. 4. The zero-resonon state $|0\rangle$ in a two-leg ladder. It is made of all vertical rung singlets. It is the reference state in the strong coupling limit $J' \gg J$.

One-resonon sector: $|x_1\rangle$. (Fig. 5)

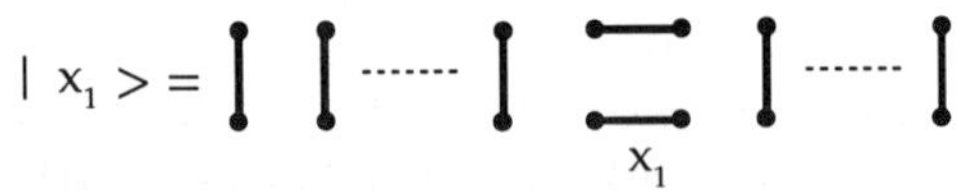

Fig. 5. A generic one-resonon state $|x_1\rangle$ in a two-leg ladder.

Two-resonon sector: $|x_1, x_2\rangle$. (Fig. 6)

Fig. 6. A generic two-resonon state $|x_1, x_2\rangle$ in a two-leg ladder.

$\frac{N}{2}$*-resonon sector:* $|x_1, x_2, \ldots, x_M\rangle$. (Fig. 7)

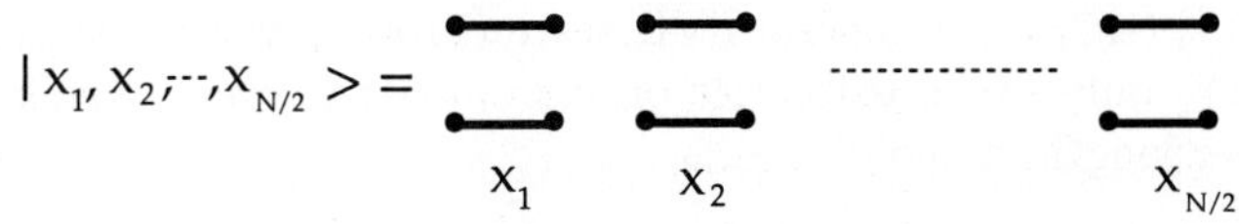

Fig. 7. The all-resonon state $|x_1, x_2, \ldots, x_{\frac{N}{2}}\rangle$ in a two-leg ladder.

With this classification of states we may write the following Generalized Dimer-RVB state for a two-leg spin ladder,

$$|\psi(u)\rangle_{RVB} = \sum_{M=0}^{[N/2]} u^M \sum_{\{x_1, x_2, \ldots, x_M\}} |x_1, x_2, \ldots, x_M\rangle \tag{2}$$

where the sum is taken over all sectors with M resonons and u is the amplitude for a resonon. Here u is a variational parameter to be determined upon minimization of the ground-state energy. As a result, it is a function of the ratio of couplings, namely, $u = u(J/J')$.

Second, it is an interesting combinatorial problem to compute the number $D(N, M)$ of dimer states with N rungs and M resonons which is given by,

$$D(N, M) = \binom{N - M}{M} \tag{3}$$

Thus, the dimension, denoted by F_N, of the whole Hilbert space of dimer RVB states in a two-leg ladder of length N is,

$$F_N = \sum_{M=0}^{[N/2]} \binom{N - M}{M} \tag{4}$$

We have arrived at a well-known formula in the theory of partitions [32] and it corresponds to a combinatorial representation of the Fibonacci number F_N.

Recurrence relations for wave functions: The key point about the result (4) for the dimension of the Hilbert space of dimer RVB states is the fact that Fibonacci numbers are well-kown to satisfy recursion relations which allow us to construct succesive members of the series from previous members, namely,

$$F_{N+2} = F_{N+1} + F_N \quad \text{with initial values,} \quad F_0 = F_1 = 1 \tag{5}$$

In fact, using this recursive idea we could have equally arrived to the generation of all possible dimer-RVB configurations present in the the classification by resonon sectors outlined above. To see this, let us use the Fibonacci

series (5) to generate all possible states: for $N = 1$ there is only one state $F_1 = 1$; for $N = 2$ there are $F_2 = 2$; for $N = 3$ there are $F_3 = 3$; for $N = 4$ there are $F_4 = 5$ and so on and so forth.
Denoting by $|N\rangle \equiv |\psi(u)\rangle_{RVB}$ a generic RVB state (2) for a two-leg ladder of N rungs, there are only two possibilities or *movements* to create generic RVB states of higher length, namely,

- Addition of one vertical rung to create the state $|N+1\rangle$.
- Addition of one pair of horizontal bonds (resonon) to create $|N+2\rangle$.

From these arguments we can establish that the dimer RVB states (2) satisfy a *recursion relation* given by,

$$|N+2\rangle = |N+1\rangle \otimes |\phi_1\rangle_{N+2} + u\,|N\rangle \otimes |\phi_2\rangle_{N+1,N+2} \tag{6}$$

where the state denoted by $|\phi_1\rangle_{N+2}$ is a vertical rung at position $N+2$ and $|\phi_2\rangle_{N+1,N+2}$ is made up of a pair of horizontal bonds located between the rungs at $(N+1, N+2)$, i.e.,

$$|\phi_2\rangle_{N+1,N+2} = (N+1, \overline{N+2})(N+2, \overline{N+1}) \tag{7}$$

where $(N, \overline{M})$ denotes a singlet between the sites N and $\overline{M}$. The recursion relation (RR) (6) is symbolically represented in Fig. 8.

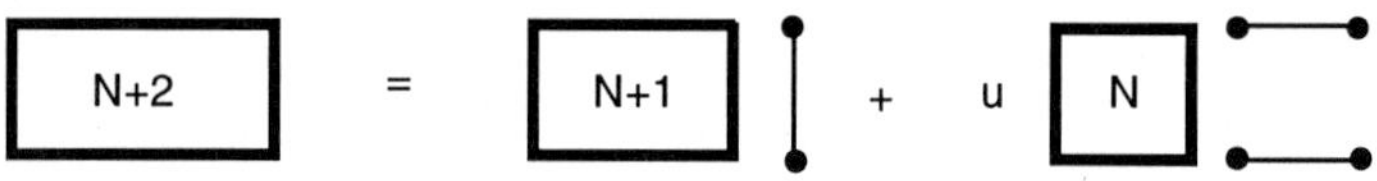

Fig. 8. Diagrammatic representation of the second order recurrence relation (RR).

Using (6) one can generate recursively the dimer RVB state $|N\rangle$ from previous states with lower length and estimate variationally the ground-state energy as,

$$e_N^{(g.s.)} \equiv \frac{\langle N|H_{AFM}^{two-leg}|N\rangle}{(2N)\langle N|N\rangle} = e(N, J'/J; u) \tag{8}$$

The value of u is fixed by minimization of (8). The RVA allows us to find analytically quantities such as, $\langle\psi|\psi\rangle(u)$, $\langle\psi|H_{AFM}|\psi\rangle(u)$, $\langle\psi|\mathbf{S}_i \cdot \mathbf{S}_j|\psi\rangle(u)$, etc ... without going through the tedious task of computing overlappings with the Sutherland rules [33].

It is worthwhile to mention that this recursive method for constructing the states of the ladder bears a great deal of similarity with the iterative method employed by the DMRG for constructing matrix elements of observables in chains, ladders etc..., see Chap. 2(I). In this sense we may consider the RVA as an analytical formulation of the DMRG.

Norms of states: To compute the norms of the states (1) let us define

$$Z_N = \langle N|N\rangle \tag{9}$$

It is also convenient to define an auxiliary function Y_N as follows,

$$Y_N = \left({}_N\langle\phi_1| \otimes \langle N-1|\right)|N\rangle \tag{10}$$

The second order RR's for the states leads (6) to a closed set of RR's for the overlaps Z_N, Y_N , namely,

$$\begin{aligned} Z_{N+2} &= Z_{N+1} + uY_{N+1} + u^2 Z_N \\ Y_{N+2} &= Z_{N+1} + \tfrac{u}{2} Y_{N+1} \end{aligned} \tag{11}$$

where we have made use of the result,

$${}_{N+1}\langle\phi_1|\phi_2\rangle_{N,N+1} = \frac{1}{2}|\phi_1\rangle_N \tag{12}$$

The RR's (11) together with the initial conditions,

$$Z_0 = Z_1 = 1 \ , Y_0 = 0, \ Y_1 = 1 \tag{13}$$

determine Z_N and Y_N for arbitrary values of N as functions of u.

Expectation values: The expectation values of the Hamiltonian can be determined following the same steps outlined above. To this end, we introduce the quantities,

$$E_N = \langle N|H_N|N\rangle \ , D_N = \left({}_N\langle\phi_1| \otimes \langle N-1|\right) H_N |N\rangle \tag{14}$$

To obtain the RR's satisfied by E_N, D_N, one splits the Hamiltonian H_N of a two-leg ladder of length N (1) into two pieces:

$$H_N = H_{N-1} + H_{N-1,N} \tag{15}$$

where H_{N-1} is the Hamiltonian of length $N-1$ and $H_{N-1,N}$ is the rest of the whole H_N Hamiltonian, $H_{N-1,N} \equiv H_N - H_{N-1}$, which is made of one vertical rung and two horizontal links. With this splitting and using (6) and (9), we obtain,

$$\begin{aligned} E_{N+2} &= E_{N+1} + J'\epsilon_0 Z_{N+1} + u(D_{N+1} + (2J+J')\epsilon_0 Y_{N+1}) \\ &\quad + u^2(E_N + 2J\epsilon_0 Z_N) \\ D_{N+2} &= E_{N+1} + J'\epsilon_0 Z_{N+1} + \tfrac{u}{2}(D_{N+1} + (2J+J')\epsilon_0 Y_{N+1}) \end{aligned} \tag{16}$$

where $\epsilon_0 = -3/4$ is the lowest eigenvalue of the operator $\mathbf{S}_1 \cdot \mathbf{S}_2$.

The initial conditions for E_N and D_N are

$$E_0 = 0, \ E_1 = J'\epsilon_0 \ D_0 = 0, \ D_1 = J'\epsilon_0. \tag{17}$$

The RR's for the energies involve the norms of the states and depend both on u and the coupling constants J', J.

2 Results for the Variational Ground-State Energy

The ground-state energy for a ladder of length N is estimated using equation (9). In the thermodynamic limit $N \to \infty$ one can find a closed expression for the density energy per site [21],

$$e_\infty = \lim_{N\to\infty} \frac{1}{2N}\frac{E_N}{Z_N} = \frac{R(\alpha)}{2\alpha Q'(\alpha)P(\alpha)} \tag{18}$$

in terms of three polinomials P, Q, R evaluated at the biggest root α of the cubic polynomial $Q(y)$. Finally one looks for the absolute minimum of e_∞ by varying the parameter u. In Table 2 we show the ground-state energies per site for different values of the coupling constant ratio J/J', varying through strong, intermediate and weak-coupling regimes.

Table 2. The values $-e_\infty^{\rm MF}/J'$ are mean field (MF) values taken from [34], while $-e_\infty^{\rm Lan}/J'$ are Lanczos values taken from [35].

J/J'	u	$-e_\infty^{\rm RVA}/J'$	$-e_\infty^{\rm MF}/J'$	$-e_\infty^{\rm Lan}/J'$
0	0	0.375	0.375000	
0.2	0.128521	0.383114	0.382548	
0.4	0.323211	0.40835	0.405430	
0.6	0.578928	0.44853	0.442424	
0.8	0.87441	0.499295	0.489552	
1	1.18798	0.556958	0.542848	0.578
1.25	1.58519	0.63518	0.614473	0.6687
1.66	2.21853	0.772172	0.738360	0.8333
2.5	3.39153	1.06915	1.002856	1.18
5	5.9777	1.99285		2.265

In the strong coupling regime $J/J' < 1$ the RVA states give a slightly better ground-state energy than the mean field result. This latter state produces rather unphysical results for $J/J' > 1$, which does not occur in our case. The RVA results compare also well with the exact results for $J' \sim J$.

The RR's can also be used to compute the spin correlator $\langle \mathbf{S}_i \cdot \mathbf{S}_j \rangle$ which has an exponential decay behaviour $\exp(-|i-j|/\xi)$, with ξ the spin correlation length which satisfies the exact equation

$$u^3\mathcal{L}^3 - (2+u)u^2\mathcal{L}^2 - (2+4u)u^2\mathcal{L} + 4u^3 = 0 \tag{19}$$

where $\mathcal{L} = \mathrm{e}^{1/\xi}$. For $J = J'$ one finds $\xi = 0.737$ [21], which can be compared with its exact value given by 3.2 [31]. The latter results imply that at the isotropic point the NNN bonds are quite important. The simplest improvement of the dimer-RVB state is to add a bond of length $\sqrt{5}$ (analogue to a knight move [36]), which leads to a third order RR. The g.s. energy per site of this state is given by -0.5713 while the spin correlation length is 0.959 [21].

Hence there is an improvement in both quantities but still one needs longer bonds.

The RVA as well as the matrix-product method work well when the correlation length of the system is small. In the DMRG context this means that the spectrum of the density matrix is dominated by a few eigenvalues [27,28]. However for systems with a large correlation length one has to consider matrix products or DMRG's with high values of m. As shown recently in [29] the correlation length ξ scales as m^β with some exponent β.

3 Matrix-Product States for Integer-Spin Chains

In this section we consider another extension of the RVA which consists in setting up a recursion relation of first order instead of second order and higher RR's that we used before.

Simple examples of RR's of first order were used as a guide to construct the RVA recursion relations in Heisenberg and t-J ladders [21–23]. In those first order RR's the variational ansatz was too simple so as to capture the main features of the systems. Thus we had to resort to higher order RR's. Another alternative that we briefly examine in the following is to keep the RR's to first order at the expense of retaining all possible states available in the problem at that stage.

We shall consider a spin chain with spin S at each site, where S is an integer with a AF-Heisenberg interation. We shall set up the recursion relations as an ansatz for the finite-size construction of the ground state of the spin chain. Then, we shall show how to achieve the thermodynamic limit $N \to \infty$ analytically and then compare the structure of this solution with the DMRG approach revealing a close relationship [24].

Let us denote the basis states as $|a, JM\rangle_N$, where N is the length of the chain, $a = 1, \cdots, d_J$ denotes the multiplicity of the total spin J of the state and M is the third component of the spin. Counting the number of multiplets we have $m = \sum_J \; d_J$, which corresponds to a number $m_W = \sum_J (2J+1) \; d_J$ of states in the standard DMRG. Thus, let us write down the following variational ansatz for a finite size N spin chain:

$$\begin{aligned} |a_1, J_1 M_1\rangle_N &= \textstyle\sum_{a_2 J_2, M_2, M} \; A_{a_1 J_1, a_2 J_2} \\ &\times |SM\rangle_N \; \otimes |a_2, J_2 M_2\rangle_{N-1} \; \langle SM, J_2 M_2 \, | J_1 M_1 \rangle \end{aligned} \tag{20}$$

where $\langle SM, J_2 M_2 \, | J_1, M_1 \rangle$ are Clebsch-Gordan (CG) coefficients which show up in this a rotationally invariant treatment of the spin chain. $A_{a_1 J_1, a_2 J_2}$ are the variational parameters of the ansatz. In (20) we keep not only singlets (valence bonds) but also higher spin states (triplet, etc...) which will improve the numerical results of this type of ansätze.

The ansatz (20) is assumed to be normalized unlike previous RVA treatments. This implies that the variational parameters $A_{a_1 J_1, a_2 J_2}$ are subject to the following orthonormalization conditions,

$$A_{a_1 J_1, a_2 J_2} = 0 \;\text{ unless }\; |J_2 - S| \leq J_1 \leq J_2 + S \tag{21}$$

$$\textstyle\sum_{a_2, J_2} A^*_{a_1 J_1, a_2 J_2} \, A_{a'_1 J_1, a_2 J_2} = \delta_{a_1 a'_1} \tag{22}$$

Equation (21) follows from the CG decomposition $S \otimes J_2 \rightarrow J_1$ in (20).

The initial data of the recurrence relation (20) is given by choosing a spin $S/2$ irrep at the end of the chain. This choice eliminates the multiplicity associated to the effective spins $S/2$ at the ends of the chain. The sum in $J's$ in (20) is of course restricted to a finite set of spins.

After many iterations the DMRG reaches a fixed point, and the ground state exhibits a similar structure as in the ansatz (20) which is called a matrix-product structure (MP) [15]. The parameters $A_{a_1 J_1, a_2 J_2}$ are determined as usual by minimizing the energy of the states $|a, JM\rangle_N$ in the limit where $N \rightarrow \infty$. For this purpose let us define the following quantity,

$$E^{(N)}_{aa'J} = {}_N \langle a, JM| \, H_N \, |a', JM\rangle_N \tag{23}$$

where H_N is the Hamiltonian acting on the chain with N sites. These matrix elements are analogous to the energy matrix elements in (14), etc... From (20) one can derive a recurrence relation for $E^{(N)}_{aa'J}$ similar to RR's in (16),

$$E^{(N)}_{aa'J} = V_{aa'J} + \sum_{bb'J'} T_{aa'J, bb'J'} \, E^{(N-1)}_{bb'J'}, \quad (N \geq 3) \tag{24}$$

where $\mathbf{T}$ is a matrix with entries (we assume from now on that $A_{a_1 J_1, a_2 J_2}$ is real),

$$T_{aa'J, bb'J'} = A_{aJ, bJ'} \, A_{a'J, b'J'} \tag{25}$$

and $V_{aa'J}$ is the matrix element of the part of the Hamiltonian which couples the sites N and $N-1$, which does not dependent on N,

$$V_{aa'J} = {}_N \langle a, JM| \, H_{N-1,N} \, |a', JM\rangle_N \tag{26}$$

The differences between RR's in (24) and (16) have a very definite origin:

1. The absence of auxiliary functions such as D_N (14) in (24) is because the latter are first order RR's instead of second order as in (16).
2. The absence of norm terms such as Z_N (9) in RR's (24) is because the states in the ansatz (20) are normalized.

For the Heisenberg model, $H_{N-1,N} = \mathbf{S}_{N-1} \cdot \mathbf{S}_N$, and applying the Wigner-Eckart theorem, we find the following expression for V in terms of 6-j symbols,

$$V_{a_1 a_2 J_1} = \sum_{a_3 J_2, a_4 J_3, a_5 J_4} \mathcal{H}_{J_1 J_2 J_3 J_4} \times \tag{27}$$
$$A_{a_1 J_1, a_3 J_2} A_{a_2 J_1, a_4 J_3} \, A_{a_3 J_2, a_5 J_4} A_{a_4 J_3, a_5 J_4},$$

$$\mathcal{H}_{J_1,J_2,J_3,J_4} = (-)^{2S+J_1+J_2+J_3+J_4+1}\, S\,(S+1)\,(2S+1) \\ \times \sqrt{(2J_2+1)\,(2J_3+1)} \begin{Bmatrix} 1 & S & S \\ J_1 & J_2 & J_3 \end{Bmatrix} \begin{Bmatrix} 1 & S & S \\ J_4 & J_2 & J_3 \end{Bmatrix} \tag{28}$$

The solution of (24) can be expressed in matrix notation as,

$$|E^{(N)}\rangle = \left(1 + \mathbf{T} + \mathbf{T}^2 + \ldots + \mathbf{T}^{N-3}\right)\, |V\rangle + \mathbf{T}^{N-2}\, |E^{(2)}\rangle \tag{29}$$

where $|E^{(N)}\rangle$ is regarded in (29) as a vector whose components are labeled by $(aa'J)$. The entries of $\mathbf{T}$ are given by (25).

In the limit $N \to \infty$ the contribution from $|E^{(2)}\rangle$ drops off and, as we shall show below, $E^{(N)}_{aa'J}$ behaves as,

$$\lim_{N\to\infty} \frac{1}{N}\, E^{(N)}_{aa'J} = \delta_{aa'}\; e_\infty \tag{30}$$

where e_∞ can be identified with the ground-state energy density and it reads,

$$e_\infty = \sum_{aa'J} \rho_{aa'J}\, V_{aa'J} \tag{31}$$

In (31) $\rho_{aa'J}$ is the left eigenvector of the matrix $\mathbf{T}$ with maximal eigenvalue 1, and plays the role of a density matrix in this formalism. The proof of (30) and (31) follows from the existence of an eigenvalue of the matrix $\mathbf{T}$ equal to 1 [37,15]. This property can be deduced from the normalization condition (22). Let us call $|v\rangle$ and $\langle\rho|$ the right and left eigenvectors associated to the eigenvalue 1 of $\mathbf{T}$, which we shall assume to be unique,

$$\mathbf{T}\, |v\rangle = |v\rangle, \quad \langle\rho|\, \mathbf{T} = \langle\rho| \tag{32}$$

Table 3. Here m is total the number of multiplets, N_A is the number of independent variational parameters, d_J is the number of multiplets with spin J, $e_\infty^{\mathrm{MP,DMRG}}$ is the g.s. energy density of the matrix-product state (DMRG), $1 - P_m$ is the probability of the states truncated out in the DMRG and ξ^{MP} is the spin correlation length of the MP state. The exact results are given by $e_\infty = 1.4014845$ and $\xi = 6.03$ [4].

m	N_A	$d_{1/2}$	$d_{3/2}$	$d_{5/2}$	$-e_\infty^{\mathrm{MP}}$	$-e_\infty^{\mathrm{DMRG}}$	$1-P_m$	ξ^{MP}
1	0	1	0	0	1.333333	1.333333	1.6×10^{-2}	0.910
2	2	1	1	0	1.399659	1.369077	1.4×10^{-3}	2.600
3	4	2	1	0	1.401093	1.392515	1.3×10^{-5}	3.338
4	7	2	2	0	1.401380	1.401380	1.6×10^{-5}	3.937
5	10	2	2	1	1.401443	1.401436	7.6×10^{-6}	4.085
6	13	2	3	1	1.401474	1.401468	1.3×10^{-6}	4.453

Then (22) implies that $|v\rangle$ is given in components by $v_{aa'J} = \delta_{aa'}$. On the other hand the quantities $\rho_{aa'J}$ that appear in (31) are nothing but the components of $\langle\rho|$, and they are found by solving the eigenvalue problem (32). In (31) we have normalized $\langle\rho|$ according to,

$$\langle\rho|v\rangle = 1 \rightarrow \sum_{aJ} \rho_{aaJ} = 1 \tag{33}$$

If one views $\rho_{aa'J}$ as a matrix which consists of blocks labelled by J, it has the properties of a density matrix, and it corresponds precisely to the reduced density matrix of the blocks in the DMRG formalism (see below) [24].

The results of the minimization (with constraints) procedure are summarized in Table 3 and compare with DMRG results [24].

To analyze in more detail the relation between the variational approach and the DMRG we give in Table 2 the eigenvalues of the matrix $\rho_{aa'J}$, and those of the DMRG reduced density matrix in the case where $m = 6$. The latter matrix has dimension $3m$ and the truncation DMRG method consists in choosing m states with highest eigenvalues w^2_{DMRG}, which add up to $1-P_m$ (see Table 3). For this reason we have to scale the DMRG weights of the states kept in order that they sum up to 1.

$$\overline{w}^2_{\text{DMRG}} = w^2_{\text{DMRG}}/P_m \tag{34}$$

Table 4. a and J are the labels of the irrep, w^2_{MP}, are the eigenvalues of the MP density matrix, and $\overline{w}^2_{DMRG}$ are the corresponding DMRG eigenvalues kept in the RG process and normalized to 1. The data corresponds to $m = 6$

a	J	w^2_{MP}	$\overline{w}^2_{DMRG}$
1	1/2	.9695581	.9696232
2	1/2	.0007662	.0007599
1	3/2	.0295443	.0294877
2	3/2	.0001119	.0001089
3	3/2	.0000078	.0000085
1	5/2	.0000118	.0000118

Table 3 and Table 4 suggest that the predictions made by the matrix-product ansatz and the DMRG should become identical for large values of m.

On the other hand, the variational approach introduced above leads to a superblock configuration [B] • [B^r], [24,38] while in the standard DMRG it is [B] • • [B^r]. This is not a major difference though, and as a matter of fact, the superblock scheme [B] • [B^r] is employed in the momentum-space approach to DMRG by T. Xiang, Chap. 5(I). The extensions of the results presented in this section to the two-leg S=1/2 ladder can be found in [25].

4 Two-Legged t-J Ladders

So far we have been dealing only with models involving spin degrees of freedom. Now, we shall present the extension of the RVA when there are also charge degrees of freedom. We shall do this by studying the t-J model in two types of ladders: rectangular and diagonal ladders. It is also possible to extend the method to the Hubbard model.

Together with the Hubbard model, the t-J model is the most studied model in connection to High-T_c Superconductivity. The t-J model is the strong coupling version of the Hubbard model and reduces to the Antiferromagnetic Heisenberg model (AFH) at half-filling. As a matter of fact, we have the following sequence of models,

$$\text{Hubbard Model} \overset{U/t\gg 1}{\Longrightarrow} \text{t-J Model} \overset{\text{half-filling}}{\Longrightarrow} \text{AF Heisenberg Model}$$

Heuristically, we can also motivate the study of the t-J model by simply starting from a AF-Heisenberg model to simulate undoped spin systems and then introducing movable holes with a t-kinetic term added to the Heisenberg model. By doping we always mean the deviation from half-filling.

It is convenient to restrict the study of a complicated model such as the t-J model to ladder lattices and especially we shall be dealing with the simplest case corresponding to the two-leg ladder. Then, the Hamiltonian we are interested in reads as follows,

$$\begin{aligned} H &= H_S + H_K \qquad (35)\\ H_S &= J' \sum_{i=1}^{N} (\mathbf{S}_{i,1}\cdot\mathbf{S}_{i,2} - \frac{1}{4} n_{i,1} n_{i,2}) \\ &\quad + J \sum_{a=1}^{2} \sum_{i=1}^{N-1} (\mathbf{S}_{i,a}\cdot\mathbf{S}_{i+1,a} - \frac{1}{4} n_{i,a} n_{i+1,a}) \\ H_K &= -\, P_G \,[\sum_{a=1}^{2} \sum_{i=1;s}^{N-1} t c^{\dagger}_{i,a;s} c_{i+1,a;s} + \sum_{i=1;s}^{N} t' c^{\dagger}_{i,1;s} c_{i,2;s}\,] P_G + h.c. \end{aligned}$$

where the $c_{i,s}$ ($c^{\dagger}_{i,s}$) is the electron destruction (creation) operator for site i and spin s, n_i is the occupation-number operator, and $\mathcal{P}_G$ is the Gutzwiller projection operator which forbids doubly occupied sites. Here J and t are Heisenberg and hopping couplings, respectively, along the horizontal links of the legs while J' and t' are similar couplings along the vertical rungs of the ladder. We have made this distinction among couplings anticipating the study of the strong-coupling limit we shall carry out, as we did in previous sections with the AF-Heisenberg ladders. The isotropic point is defined as $J' = J$, $t' = t$. The label $a = 1, 2$ stands for the two legs of the ladder.

The big difficulty working with the t-J model is precisely this constraint of no double occupancy. Analytic approaches usually fail to implement the whole physics underlying this constraint or are subject to unproven assumptions. In the RVA treatment of the t-J model, the double occupancy constraint is *built in* from the very beginning

The problem we face is to generalize the RVA of the AF-Heisenberg ladder to the case of doping. Being the RVA a variational method, there is no mechanical way to proceed, unlike the perturbative methods. As usual with variational methods, we need a hint from the physics of the problem in order to set up the methodology. In the case of the two-leg t-J ladder there are some basic features coming from the strong-coupling analysis and other means which are helpful.

We know by now that there is a spin gap in the spectrum of the two-leg ladder when it is undoped. The question is what happens when we dope the system. For low doping, it is expectable that the spin gap remains, but there appear charge gapless excitations associated to the collective motion of pairs of holes which form a sort of superfluid. These facts are reflected in the low-energy physics of this model which falls into the universality class of the Luther-Emery (LE) liquid [43], unlike the one-dimensional version of the t-J model which is known to be a Luttinger Liquid (LL) [44]. In a LE liquid, the spin modes are gapped while the charge degrees of freedom are gapless.

Recurrent relations for doped states

In order to set up recursion relations for doped states in the t-J model it is convenient to recall the elementary recipies we have used so far:

- Set up the strong-coupling limit of the model.
- Select the basic local configurations in this regime.
- Solve the model in a small cluster, e.g. a 2×2 plaquette.

To understand this basic physics, let us take the undoped case $x = 0$ as a reference. In this case the second order RR is given in (6). This corresponds to having no pair of holes $P = 0$ and the state is simply $|N, P = 0\rangle$. It is worthwhile to point out that if we set $N = 2$ in (6) we precisely recover the *exact* ground state for the simple 2×2 plaquette, namely,

$$|N = 2, P = 0\rangle = |\phi_{1,0}\rangle_1 |\phi_{1,0}\rangle_2 + |\phi_{2,0}\rangle_{1,2} \tag{36}$$

where $|\phi_{2,0}\rangle_{1,2}$ is the state defined in (7) times u (see (41)).

We can apply this same reasoning for the construction of the RR for states $|N, P\rangle$ for the case $N = 2$ and $P = 1$. Let us label the sites of the 2×2 lattice as $1, 2, 3, 4$ in a clockwise fashion. To solve for the state $|2, 1\rangle$ we use symmetry arguments to write down this state as (Fig. 9),

$$|2, 1\rangle = a\,|\text{edge}\rangle + b\,|\text{diagonal}\rangle \tag{37}$$

$$|\text{edge}\rangle = (12) + (34) + (23) + (14), \; |\text{diagonal}\rangle = (24) + (13) \tag{38}$$

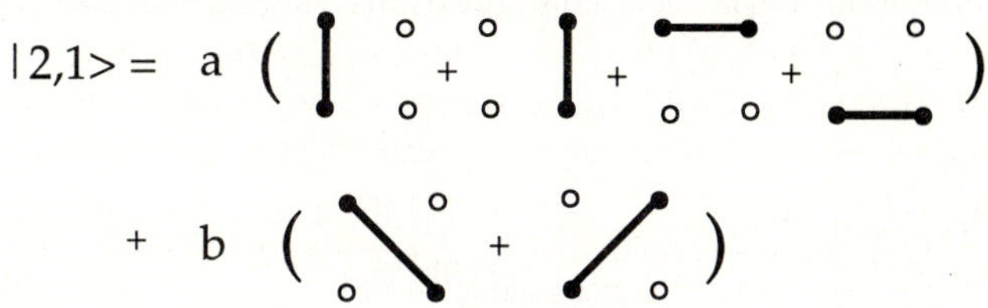

Fig. 9. The exact ground state for a single plaquette with two holes (case N=2 and P=1).

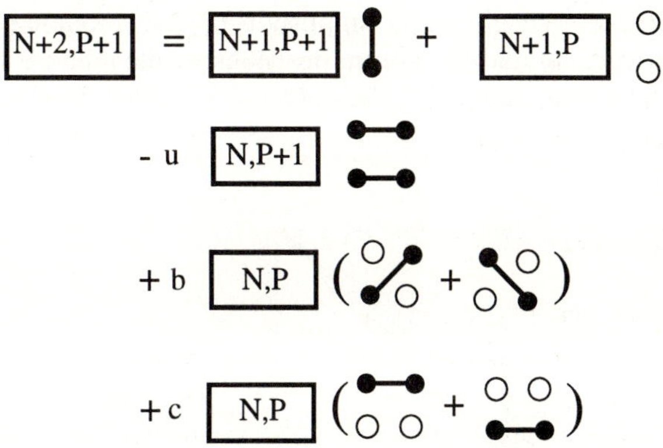

Fig. 10. A pictorical representation of the second-order recurrence relation (40) for two-leg t-J ladder.

where (i, j) symbolically denotes a bond between sites i and j. The ratio of the probability amplitudes a and b and the ground-state energy are given by,

$$\frac{a}{b} = \frac{1}{[2 + (J/4t)^2]^{1/2} - J/4t}\,, E(2,1) = -\frac{1}{2}[J + \sqrt{J^2 + 32t^2}] \tag{39}$$

With the help of these configurations we can set up the RR that generates the dimer-hole RVB states as follows, (Fig. 10)

$$\begin{aligned} |N+2, P+1\rangle = |N+1, P+1\rangle\ |\phi_{1,0}\rangle_{N+2} + |N+1, P\rangle\ |\phi_{1,1}\rangle_{N+2} \\ + |N, P+1\rangle\ |\phi_{2,0}\rangle_{N+1,N+2} + |N, P\rangle\ |\phi_{2,1}\rangle_{N+1,N+2} \end{aligned} \tag{40}$$

where we have gathered the basic local configurations in terms of pair field operators $\Delta^\dagger_{i,j} = \frac{1}{\sqrt{2}}(c^\dagger_{i,\uparrow}c^\dagger_{j,\downarrow} - c^\dagger_{i,\downarrow}c^\dagger_{j,\uparrow})$,

$$\begin{aligned}
|\phi_{1,1}\rangle_x &= |0\rangle_x \\
|\phi_{1,0}\rangle_x &= \Delta^\dagger_{(x,1)(x,2)}\ |0\rangle_x \\
|\phi_{2,0}\rangle_{x,x+1} &= -u\ \Delta^\dagger_{(x,1)(x+1,1)}\ \Delta^\dagger_{(x,2)(x+1,2)}\ |0\rangle_{x,x+1} \\
|\phi_{2,1}\rangle_{x,x+1} &= [b\ (\Delta^\dagger_{(x,1)(x+1,2)} + \Delta^\dagger_{(x,2)(x+1,1)}) \\
&\quad +c\ (\Delta^\dagger_{(x,1)(x+1,1)} + \Delta^\dagger_{(x,2)(x+1,2)})]\ |0\rangle_{x,x+1}
\end{aligned} \tag{41}$$

where $|0\rangle_x$ is the vacuum associated with the rung labelled by the coordinate x. The states $|\phi_{n,p}\rangle$, involve $n = 1, 2$ rungs and $p = 0, 1$ pairs of holes. The variational parameter u gives the amplitude of the resonance of a pair of bonds between vertical and horizontal positions, while b and c are the variational parameters associated with the diagonal and horizontal configurations of two holes, respectively [22].

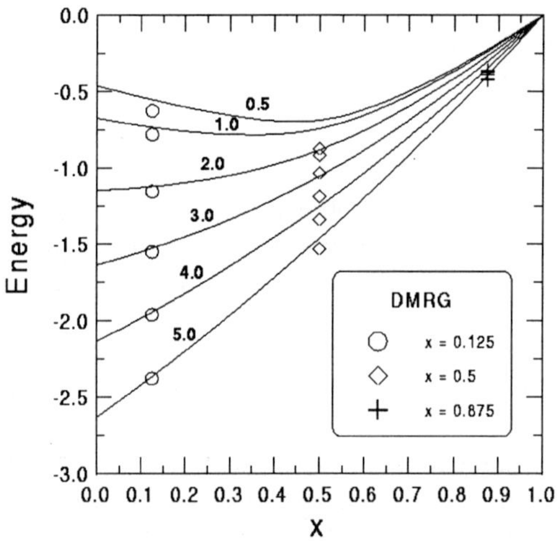

Fig. 11. Ground-state energy per site of the 2 x 32 ladder with $J = 0.5, t = t' = 1$ and $J' = 0.5, 1, 2, 3, 4, 5$. The continuum curves are obtained with the RVA, while the special symbols are the DMRG data corresponding to $x = 1/8, 1/2$ and $7/8$ respectively.

Norms and energies: The parameters u, b, c are found by the standard minimization of the mean value of the energy $\langle N, P|H_N|N, P\rangle/\langle N, P|N, P\rangle$. As usual, (40) implies recursion relations for the energy overlaps. Let us define the following norms $Z_{N,P}$, mean values $E_{N,P}$ and auxiliary functions $Y_{N,P}$, $D_{N,P}$ and $W_{N,P}$

$$\begin{aligned}
Z_{N,P} &= \langle N, P|N, P\rangle & Y_{N,P} &= {}_N\langle\phi_{1,0}|\langle N-1, P|N, P\rangle \\
E_{N,P} &= \langle N, P|H_N|N, P\rangle & D_{N,P} &= {}_N\langle\phi_{1,0}|\langle N-1, P|H_N|N, P\rangle \\
W_{N,P} &= \langle N, P|n_N|N, P\rangle
\end{aligned} \tag{42}$$

where n_N is the number operator acting on the rung N. The off-diagonal overlaps arise from the cross terms when applying (40) to the ket and the bras in $\langle N+2,P+1|N+2,P+1\rangle$ and $\langle N+2,P+1|H_{N+2}|N+2,P+1\rangle$.

The recursion relations satisfied by (42) are given by,

$$\begin{aligned} Z_{N+2,P+1} &= Z_{N+1,P+1} + u^2\, Z_{N,P+1} + u\, Y_{N+1,P+1} \\ &\quad + Z_{N+1,P} + 2(b^2+c^2)\, Z_{N,P} \\ Y_{N+2,P+1} &= Z_{N+1,P+1} + u/2\, Y_{N+1,P+1} \\ E_{N+2,P+1} &= E_{N+1,P+1} - J'\, Z_{N+1,P+1} + u^2\, E_{N,P+1} \\ &\quad -(2J+J'/2)u^2\, Z_{N,P+1} + E_{N+1,P} \\ &\quad +2(b^2+c^2)E_{N,P} - (2Jc^2 + 4bt + 8bct')\, Z_{N,P} \\ &\quad +uD_{N+1,P+1} - 2u(J+J'/2)Y_{N+1,P+1} - 4tbY_{N+1,P} \\ &\quad -\tfrac{1}{4}J\, W_{N+1,P+1} - \tfrac{1}{4}Ju^2W_{N,P+1} - \tfrac{1}{4}J(b^2+c^2)W_{N,P} \\ D_{N+2,P+1} &= E_{N+1,P+1} - J'Z_{N+1,P+1} + u/2D_{N+1,P+1} \\ &\quad -u(J+J'/2)Y_{N+1,P+1} - 2tbZ_{N,P} - \tfrac{1}{4}J\, W_{N+1,P+1} \\ W_{N+2,P+1} &= 2Z_{N+1,P+1} + 2u^2\, Z_{N,P+1} + 2(b^2+c^2)Z_{N,P} \\ &\quad +2uY_{N+1,P+1} \end{aligned} \tag{43}$$

The initial conditions for these equations read as follows,

$$\begin{aligned} &Z_{0,0} = 1, \quad Y_{0,0} = E_{0,0} = D_{0,0} = W_{0,0} = 0 \\ &X_{N,P} = 0, \text{ for } \; N < P \; \text{ and } X = Z, Y, E, D, W \end{aligned} \tag{44}$$

Ground-state properties: It is possible to retrieve now many ground-state properties from the RR's in (43) performing the minimization process with respect to the variational parameters u, b, c for a fixed value of the coupling constants J, J', t, t'. In this case we must also take the finite density of holes limit to get meaningful variable-doping quantities. It is clear that the RR's in (43) are more complicated to carry out an analytic solution in the limit when $N \longrightarrow \infty, P \longrightarrow \infty, x = P/N$ fixed, and we shall not undertake this method of solution this time. Alternatively, we can keep the number of rungs N at a fixed value, though large enough to eliminate finite-size effects, and then vary the number of pairs of holes P in order to cover the whole range of the doping parameter $0 \leq x \leq 1$. This finite-(N,P) minimization process is done numerically by implementing the recursion equations in (43) by means of an iterative algorithm in a computer.

The computer needs are very modest both in memory space and CPU time and with a standard minimization library like those in [46,47] it is possible to minimize very large two-leg ladders, over $N = 200$ rungs, although here we shall present results for sizes 32×2 in order to compare with available DMRG results [22]. In Fig. 11 we show some results comparing the RVA and the DMRG. Observe the huge agreement for the strong-coupling region.

The parameter u starts from a positive value corresponding to the undoped ladder [21], and it decreases upon doping until a critical value $x_c(J/J')$, where it vanishes. For higher dopings u becomes negative. This behaviour is very interesting and it suggests a continuous crossover between the RVB picture and the BCS picture of pairing in the strong coupling regime. We shall call the region below x_c underdoped and above it, overdoped.

For the undoped ladder the parameter u can be interpreted as the square of the RVB amplitude $h_{\rm RVB}$ for having a bond along the legs [21]. The analogous amplitude for a bond along the rungs has been implicitly normalized to 1. For low doping, i.e. $x < x_c$, since $u(x) > 0$, we can similarly define a doping-dependent amplitude for a leg-bond as follows,

$$u(x) = h^2_{\rm RVB}(x) > 0, \quad (x < x_c) \tag{45}$$

In order to fullfill the Marshall theorem [40] for the undoped ladder one requires the RVB amplitude $h_{\rm RVB}(0)$ to be positive [41], which explains why $u(0)$ is also positive. At $x = 0$, $h_{\rm RVB}(0)$ increases with J/J' due to the resonance between rung and leg singlets, according to the RVB scenario. Upon doping, however, the holes act as a destructive interference which degrades progressively the aforementioned resonance mechanism. This explains why $u(x)$ and $h_{\rm RVB}(x)$ decrease with x. For $x < x_c$ the ground state is dominated by the resonanting valence bonds and the RVB picture remains qualitatively correct.

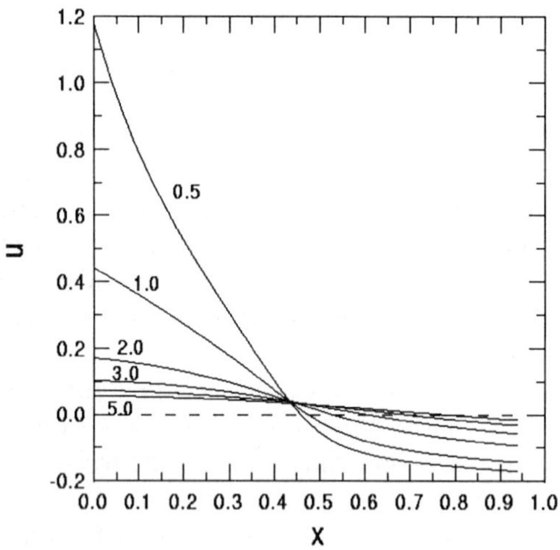

Fig. 12. The variational parameter u as function of the doping for a 2 x 32 ladder with $J = 0.5, t = t' = 1$ and $J' = 0.5, 1, 2, 3, 4, 5$

For $x > x_c$ the parameter u becomes (Fig. 12) negative and the definition (4) makes no longer sense since $h_{\rm RVB}$ is supposed to be positive. The physical interpretation of the overdoped region comes from the solution of the Cooper problem in the t-J two-legged ladder, and its BCS extension. It can be shown analytically that two electrons in the latter system form a bound state only under certain conditions (details will be given elsewhere). For $J = 0.5, t =$

$t' = 1$ one must have $J' > 3.3048$, (the binding of two electrons in the t-J chain requires $J/2t > 1$ [44]). The exact solution for four or more electrons is difficult to construct, but we expect it to be given essentially by a Gutzwiller projected BCS-like wave function with no double occupied sites. A short range version of the latter type of wave function can be generated from the recursion relation (40), with u a negative parameter, which can be written as,

$$u(x) = -h^2_{\mathrm{BCS}}(x) < 0, \quad (x > x_c) \tag{46}$$

where h_{BCS} is the BCS amplitude for two electrons at distance 1 along the legs. Of course the interpretation of u as the square of a BCS amplitude requires it to be negative. As we put more electrons into the ladder the value of h_{BCS} decreases and for electron densities bigger than $1 - x_c$, the Cooper pairs start to resonate and we switch into the RVB regime.

The difference between the underdoped and overdoped regimes can be attributed to two different internal structures of the pairs [22]. In the low doping regime $x < x_c$, holes doped into the spin-liquid RVB state form pairs with an internal $d_{x^2-y^2}$-like structure relative to the undoped system. However for $x > x_c$ one moves into the low-density limit characterized by electrons doped into an internal s-wave like symmetry. The difference between underdoped and overdoped regimes can be attributed to two different symmetries of the order parameter, which may be d-wave or s-wave type.

5 Diagonal t-J Ladders

In this section we shall show how the RVA can also be applied to lattices with shapes different from the standard square-lattice ladders studied so far. More concretely, we shall apply the RVA to a recently introduced class of ladders [23] in the context of strongly coupled electron systems, which are called *diagonal ladders* due to their particular diagonal structure as shown in Fig. 13.

There is a theoretical motivation to study diagonal ladders. They provide another route for going from 1D to 2D, as was one main goal to study ladders in Sec.2. Diagonal ladders are labelled by an integer $n_p = 1, 2, \ldots$ which gives the number of elementary plaquettes needed to generate the entire structure (Fig. 13). The first member of this family, i.e. $n_p = 1$, is also known as the *necklace ladder* and it consists of a collection of N plaquettes joined along a common diagonal. In this chapter we shall focus on the necklace ladder for simplicity, much as we did with the two-leg ladder.

The original motivation for introducing diagonal ladders in the context of the t-J model was to understand the fully doped stripes in the (1,1) direction that have been observed experimentally in materials like $La_{1-x}Sr_xNiO_4$ [49], in Hartree-Fock studies of t-J and Hubbard models, and numerically in DMRG studies of the t-J model [50]. The simplest possible toy model of

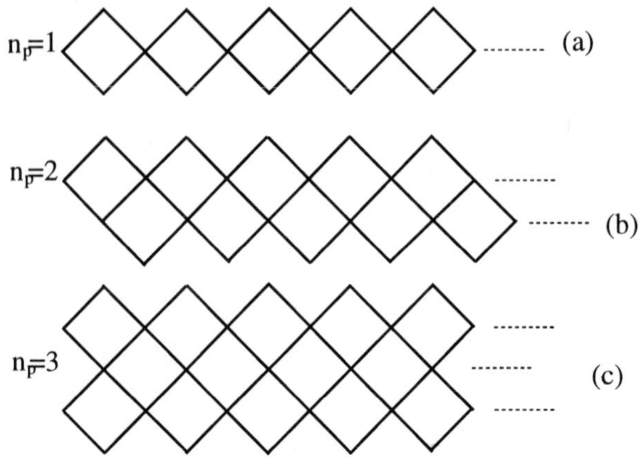

Fig. 13. Examples of diagonal ladders with a number of plaquettes $n_p = 1, 2, 3$ in the unit cell

this type of stripes is provided by a necklace ladder with a hole doping equal to 1/3. As we shall see this doping plays an important role for the necklace ladder but we will be able to deal with doped diagonal ladders much as we did with the two-leg ladder.

Let us also point out that diagonal ladders have recently appeared as constituent parts of some interesting materials like $Sr_{0.4}Ca_{13.6}Cu_{24}O_{41.84}$ known for its superconducting properties at high pressure [39]. In addition there are a variety of materials which contain weakly coupled arrays of ladders [51,52].

From the point of view of the RVA, diagonal ladders are interesting lattices in several regards:

- In the undoped case, they provide a suitable setting to apply RVA ansätze in basis states which need not to be explicitely rotationally invariant. So far we have used basis of states in terms of VB states (singlets). In the diagonal ladder we shall use states which only have a definite quantum number, namely, the S_z-component. These states will be also used in generic doped diagonal ladders.
- Diagonal ladders also provide us with an example in which the RVA ansatz has to be set up in terms of non-uniform configurations, as we shall see later on. This means that we have to work with RR's for a fixed doping x of the ladder, i.e., the RR's do not mix diagonal ladders with different dopings, unlike the square two-leg ladder. The reason for this fact is the existence of a dominant configuration for each doping value which contribute the most to the variational method.

This RVA can be implemented analytically if the variational ansatz is sufficiently simple. Below we shall propose various RVA states for the necklace ladder with dopings $0 \leq x \leq 1/3$.

For a more detailed account on theoretical considerations for diagonal ladders and DMRG results, see [23].

The t-J model on the necklace ladder: It is useful to write the t-J model on the necklace ladder in a similar fashion as the AF-Heisenberg chain, namely, using the diagonal coordinates as shown in Fig. 14,

$$\begin{aligned} H = J \textstyle\sum_{n=1}^{N} & (\mathbf{S}_{1,n} + \mathbf{S}_{2,n}) \cdot (\mathbf{S}_{3,n} + \mathbf{S}_{3,n-1}) \\ & -\mathcal{P}_G \left(t \textstyle\sum_{n=1,s}^{N} \left((c_{1,n;s}^{\dagger} + c_{2,n;s}^{\dagger})\, c_{3,n;s} + h.c. \right) \right) \mathcal{P}_G \end{aligned} \tag{47}$$

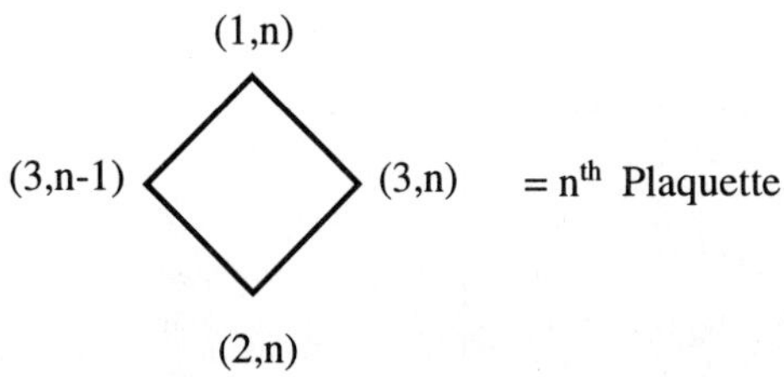

Fig. 14. Diagonal coordinates for a single plaquette

For the necklace t-J ladder, there is a parity plaquette conservation theorem [56]: the t-J Hamiltonian on a necklace ladder commutes with every permutation operator P_n associated with the minor diagonal of the nth plaquette defined by the sites $(1,n)$ and $(2,n)$ in Fig. 14. Here the permutation operator $P_n (n = 1, \ldots, N)$ is defined by its action on the fermionic operators, which is trivial on all the sites except for those on the minor diagonal of the nth plaquette where it acts as

$$P_n\, c_{(1,n),s}\, P_n^{\dagger} = c_{(2,n),s} \quad , P_n\, c_{(2,n),s}\, P_n^{\dagger} = c_{(1,n),s} \tag{48}$$

Of course, the spin and the density-number operators at the sites $(1,n)$ and $(2,n)$ are also interchanged under the action of P_n. The above theorem is the statement that P_n commutes with H, (47), for all n,

$$[H,\ P_n] = 0 \ \text{ for } \ n = 1, \ldots, N \tag{49}$$

and can be easily proved. Equation (49) is not special to the t-J Hamiltonian, since any other lattice Hamiltonian having the permutation symmetry between the two sites on the minor diagonal of every plaquette would share this same property.

The immediate consequence of (49) is that we can simultaneously diagonalize the Hamiltonian $H_{t,J}$ and the whole collection of permutations operators P_n, whose possible eigenvalues are given by $\epsilon_n = \pm 1$. The latter fact is a consequence of the following equation,

$$P_n^2 = 1 \tag{50}$$

Letting ϵ_n denote the parity of the $n^{\rm th}$ plaquette, the 9 possible states associated with the minor diagonal of a plaquette can be classified according to their parity, i.e. $\epsilon_n = 1$ for even parity states and $\epsilon_n = -1$ for odd-parity states (see Table 3).

Table 5. Parity of states in a minor diagonal

	State	ϵ
2 holes	$\|0\rangle$	1
bonding	$(c^\dagger_{1,s} + c^\dagger_{2,s})\|0\rangle$	1
singlet	$\Delta^\dagger_{1,2}\|0\rangle$	1
antibonding	$(c^\dagger_{1,s} - c^\dagger_{2,s})\|0\rangle$	-1
triplet	$(c^\dagger_{1,s}c^\dagger_{2,s'} + s \leftrightarrow s')\|0\rangle$	-1

The Hilbert space $\mathcal{H}_{\rm necklace}$ of the t-J model can be split into a direct sum of subspaces $\mathcal{H}_\epsilon$ classified by the parity of their plaquettes, $\epsilon = \{\epsilon_n\}_{n=1}^N$, namely

$$\mathcal{H}_{\rm necklace} = \oplus_\epsilon \mathcal{H}_\epsilon \tag{51}$$

Thus we can look for the ground state in the sector with N plaquettes, h holes, spin S, and parity ϵ. This seems to be a formidable task. It is more reasonable to look for the the absolute minimum of the g.s. energy, keeping fixed the values of the number of plaquettes N, the number of holes h and the ratio J/t, i.e.

$$E_0(N, h, S_{\rm min}, \epsilon_{\rm min}, J/t) \leq E_0(N, h, S, \epsilon, J/t), \quad \forall S, \epsilon, \tag{52}$$

Even this question is not easy to answer with full generality. However we shall present the most representative cases.

Undoped case ($x = 0$)

A DMRG calculation shows that the g.s. of this spin ladder is ferrimagnetic as shown in the DMRG snapshot of Fig. 15 and that the parity of all plaquettes is $\epsilon = -1$. One can show that the spin necklace ladder is equivalent to an

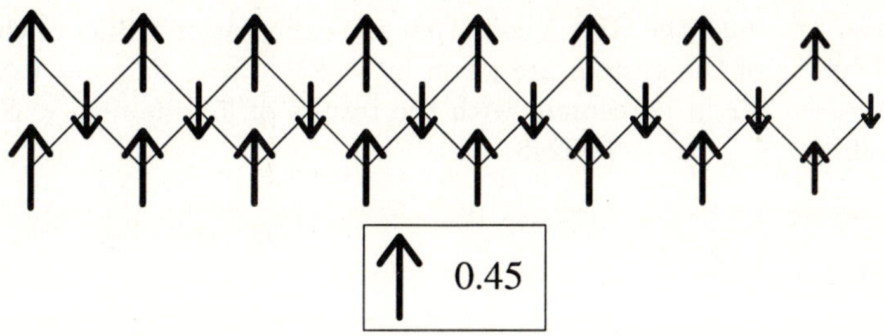

Fig. 15. DMRG results for the spin configuration for the g.s. of an undoped 8 × 3 necklace ladder. The state has total spin $s_G = 4$. The length of the arrow is proportional to $\langle S_z \rangle$, according the scale in the box

alternating spin-1/spin-$\frac{1}{2}$ chain, which is also known to be ferrimagnetic [53–55]. Having $\epsilon = -1$ for every plaquette amounts to having a spin 1 on every minor diagonal (see Table 5).

The most important configuration of the g.s. at $x = 0$ is the Néel state which is slightly perturbed by quantum fluctuations. This can be taken care of by the RVA state generated by the RR's given in Fig. 16.

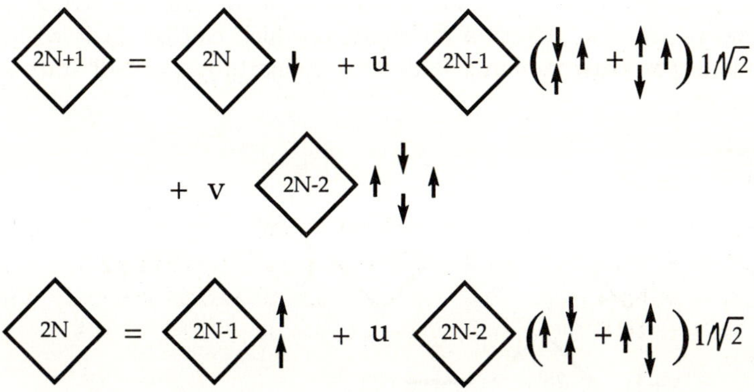

Fig. 16. Diagrammatic representation of the recurrent relations generating the ground state of an undoped necklace ladder using the variational RVA

Minimizing the g.s. energy of this RVA state in the limit $N >> 1$ we find that its value per plaquette is given by -0.4822 J [23], which corresponds to an energy per site of the associated alternating spin chain equal to -0.7233 J. The values of the variational parameters are given by $u = -0.3288$ and $v = 0.1691$. This RVA energy is to be compared with the extrapolated DMRG results $-0.72704J$ or the spin-wave value $-0.718J$ of Pati et. al. [53].

We also find with the RVA ansatz that the expectation values of the spins near the center of the system are given by $< S^z_{1,n} >=< S^z_{2,n} >= 0.396$ and $< S^z_{3,n} >= -0.292$ in agreement with the results of [53] namely $< S^z_{1,n} >= 0.39624$ and $< S^z_{3,n} >= -0.29248$.

The case $x = 1/3$

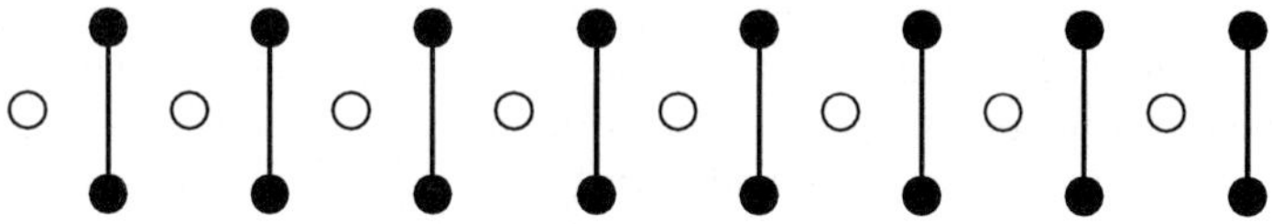

Fig. 17. Pictorical representation of the most probable configuration for doped $x = 1/3$ necklace ladder. Blank circles denote holes and vertical solid lines represent valence bond states

The DMRG shows that for a doping $x = 1/3$ the most important configuration is the one where the holes occupy the main diagonal of the necklace ladder while the minor diagonals are occupied by singlets (Fig. 17). The parity of all the plaquettes is $\epsilon = +1$. This state is a sort of diagonal stripe [23]. Of course, there are also quantum fluctuations which can be dealt with in a RVA fashion. The corresponding RR's that generate a $x = 1/3$ state are given in Fig. 18.

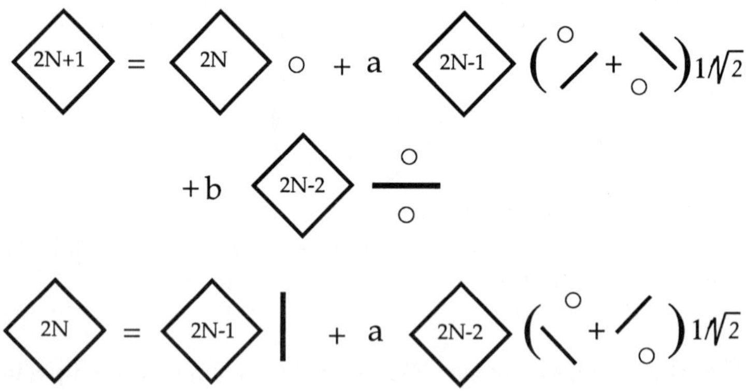

Fig. 18. A pictorical representation of the recurrence relations employed with the RVA to construct variational g.s. states for the doped $x = 1/3$ necklace ladder. The diagonal squares represent bulk states of a given length. Small circles represent holes and solid lines valence bonds

For a necklace with 7 plaquettes ($J = 0.35, t = 1$) the RVA g.s. energy is -16.33996 while the DMRG is -16.554153. We are using a RVA where there 3 variational parameters per plaquette and hence the ansatz is non uniform [23].

We have plotted in Fig. 20 the results for $x = 1/3$ in the necklace ladder and compared it with the DMRG results showing a good agreement.

Intermediate dopings ($0 \le x \le 1/3$)

From Figs. 16 and 18 we observe a similarity in the RR's that generate the RVA states for dopings $x = 0$ and $x = 1/3$. We have seen that for intermediate dopings the g.s. produced by the DMRG contains a dominant state $|\psi_0\rangle$ together with its local fluctuations [23]. The "classical state" $|\psi_0\rangle$ can be generated by a first order RR while its "quantum fluctuations" yield up to a third order RR. The skeleton of these RR's is depicted in Fig. 19 . Observe that there are 3 independent variational parameters a_N, b_N, c_N per plaquette.

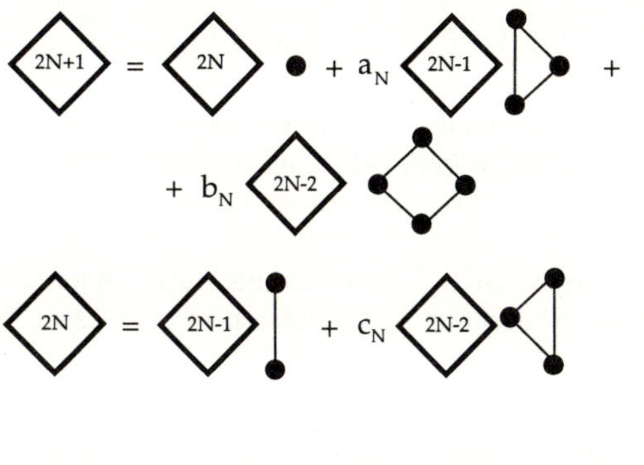

Fig. 19. A pictorical representation of the recursion relations employed to generate RVA states for the underdoped ($0 \le x \le 1/3$) necklace ladder. Here a, b and c are local variational parameters. A square denotes a bulk state on a ladder of a length given by its number inside. The black circles and solid lines represent generic fluctuation states

The procedure to find the ground state of the necklace ladder is summarized as follows,

- Fix the length N of the ladder, the number of holes h and the total S^z.
- Generate all the $|\psi_0\rangle$ configurations with those quantum numbers N, h, S^z. Generically the number of configurations grows exponentially. For example the seven-plaquette case has N=15 and a total of 3×10^4 configurations.
- Compute the energy of the state associated with the zero-order state $|\psi_0\rangle$ using the recursion relations and find the variational parameters which lead to a minimum energy. For the seven-plaquette ladder there are 21 independent variational parameters.
- Extract the state $|\psi_0\rangle$ which has the absolute minimum energy for a given N, h and S^z.

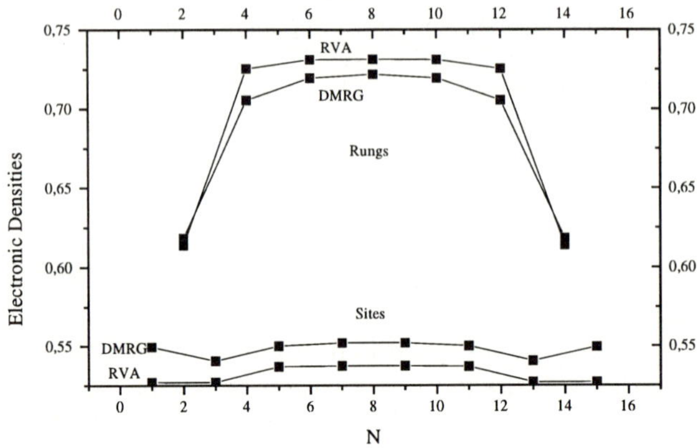

Fig. 20. Electronic densities for a necklace ladder. It corresponds to 7 plaquettes and 8 holes (doping $x = 1/3$). Below are shown the DMRG and RVA values for the sites in the main diagonal, while above are shown the values on the rungs corresponding to subdiagonals

In Table 6 we present some of the results obtained by this RVA and compare with DMRG results [23] finding a very good agreement.

Table 6. DMRG and RVA total energies for a 7-plaquette necklace ladder with h holes and total spin S_z. The string of epsilons is the pattern of parities in the subdiagonals (rungs).

h	S	$(\epsilon_1, \ldots, \epsilon_7)$	E_0^{DMRG}	E_0^{RVA}
8	0	$(+++++++)$	-16.554153	-16.33996
8	1	$(+++-+++)$	-16.284855	-16.00358
7	1/2	$(++++++-)$	-15.489511	-15.18141
6	0	$(-+++++-)$	-14.424805	-14.02286
6	1	$(-+++++-)$	-14.424798	-14.02286

Let us comment on the DMRG results shown in the following figures. In what follows, (h, S) denotes a system with h holes and total spin S as in Table 6.

- *Case (8,1):* The spin excitation of the $x = 1/3$ state is given by a spin-1 magnon strongly localized on an odd-parity plaquette located at the center of the ladder. The other plaquettes remain even and spinless. The value of the spin gap is given by $\Delta_s = 0.27$ (DMRG) and 0.32 (RVA).

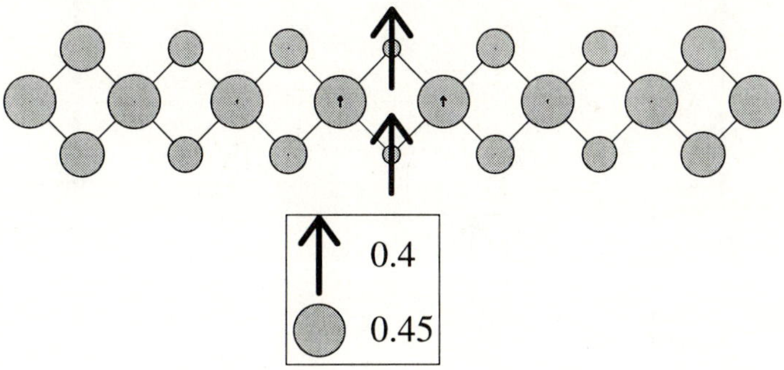

Fig. 21. Results using DMRG showing the necklace state with 8 holes and spin $S_z = 1$. The diameter of the circles are proportional to the hole density, and the length of the arrows are porportional to $\langle S_z \rangle$, according to the scale in the box

- *Case (7,1/2):* This case is obtained by doping the $x = 1/3$ state with an electron. The additional electron goes into anyone of the boundary plaquettes. The corresponding plaquette changes its parity to -1.

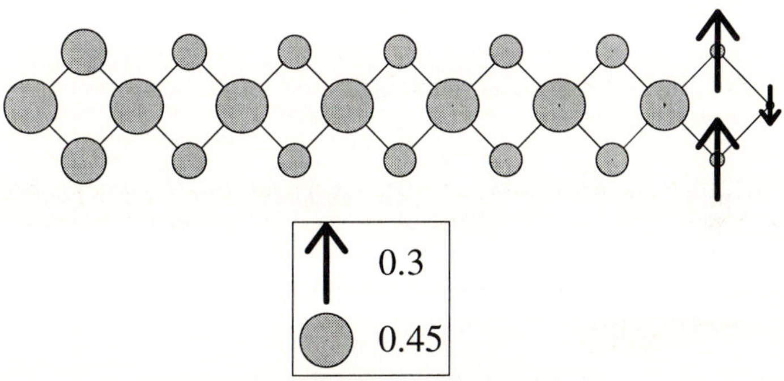

Fig. 22. DMRG results for the hole and spin densities of the necklace state with 7 holes and spin $S_z = 1/2$

- *Cases (6,0) and (6,1)*: The state $x = 1/3$ is now doped with two electrons, which go to the boundary plaquettes, changing their parity. There seems to be a small effective coupling between the two spin 1/2 at the ends of the ladder, which leads to a breaking of the degeneracy between the triplet and the singlet. This is reminiscent of the effective spin 1/2 at the ends of the Haldane and AKLT open spin chains [63]. There also exists a weak effective coupling that breaks the four-fold degeneracy of the open chains.

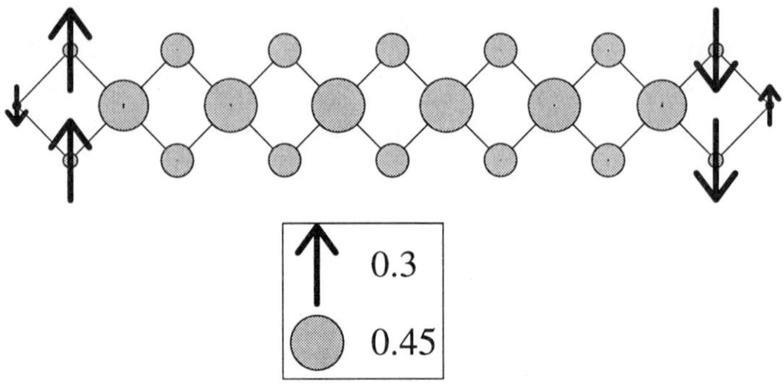

Fig. 23. DMRG results for the hole and spin densities of the necklace state with 6 holes and spin $S_z = 0$

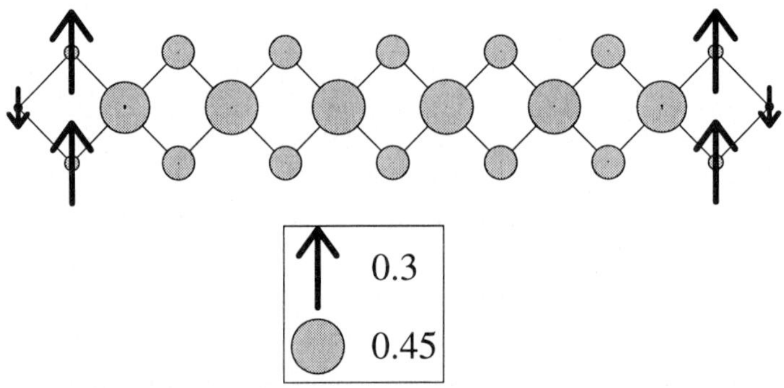

Fig. 24. Results from DMRG showing the necklace state with 6 holes and spin $S_z = 1$

6 Problems and Prospects

In this section we would like to mention some problems that may in principle be studied with the RVA.

Hubbard and polymer models

The application of the RVA goes beyond the specific models such as the Heisenberg and t-J considered in this chapter. There are other models where we can apply RVA techniques. We have already mentioned the study of a two-leg Hubbard ladder in Sect. 5 [57]. Similarly, RVA can be used to study polymers [58] which can be modeled with the Pariser-Parr-Pople (PPP) Hamiltonian [59,60] and Chaps. 2.2(II) and 2.3(II). The two hopping parameters in the PPP Hamiltonian play the role of the J and J' coupling constants in the two-leg Heisenberg ladder.

Other ladders

As explained in Sect. 2 the even-legged ladders have a finite correlation length and a spin gap and hence are good systems to apply the RVA or the MP method. The techniques developed for the 2 leg ladder can be easily extended to the 4 leg ladder. The dimer-RVB state can be generated again by a second-order RR but one needs more than one singlet state associated to the different shapes of the boundary of the four-leg ladder.The number of dimer states approaches from below the exponential behaviour found by Kasteleyn for the 2D dimers.

Stripes

Tranquada et al. have shown experimentally that in certain cuprates the charge degrees of freedom form quasi-1D structures called stripes along the (1,0) direction of the planes [49]. This happens when the doping takes a value $x = 1/8$. Motivated by these results, White and Scalapino have studied large clusters of spins and holes (t-J model) using the DMRG and shown that the (1,0) stripes are essentially two-leg ladders with dopping $x = 1/4$ separated by two-leg undoped spin ladders [50].

This suggests that a simple ansatz for a stripe with doping x could be given as

$$|\text{stripe}\rangle = |A\rangle_1 \, |B\rangle_2 \, |A\rangle_3 \cdots |B\rangle_{N-1} \, |A\rangle_N \tag{53}$$

where $|A\rangle_i$ and $|B\rangle_i$ are two-leg ladder states located at the positions $2i-1$ and $2i$, and dopings x_A and x_B satisfying $x = (x_A + x_B)/2$. In the example of Tranquada $x = 1/8, x_A = 1/4, x_B = 0$. It is clear that the state (53) can be generated by a first-order RR. To go beyond (53) one should consider the quantum fluctuations of the neighbour ladders which can be described by a RVA state generated by the following second order RR

$$\begin{aligned} |2N+1\rangle &= |2N\rangle \, |A\rangle + |2N-1\rangle \, |(BA)\rangle \\ |2N\rangle &= |2N-1\rangle \, |B\rangle + |2N-2\rangle \, |(AB)\rangle \end{aligned} \tag{54}$$

where $|N\rangle$ denotes a state made of N two-leg ladders, $|(AB)\rangle$ is a four-leg ladder state which results from the quantum fluctuations of two parallel two-leg ladders $|A\rangle|B\rangle$ (same for $|(BA)\rangle$).

Equation (54) is similar to the RR that generates the dimer-RVB state of the two-leg ladder with the replacement: rung $\rightarrow$ stripe. The states $|A\rangle$, $|B\rangle$, $|(AB)\rangle$ and $|(BA)\rangle$ can be constructed using again RR's. Hence one can generate the g.s. of the plane in two steps: first building the two-leg and four-leg ladder states and next joining them to get a 2D state. The minimization of the g.s. of the whole system will couple the variational parameters involved in the construction of $|A\rangle, |B\rangle, |(AB)\rangle$ and $|(BA)\rangle$.

2D Matrix-product ansätze

The construction outlined above yields a possible way to build 2D states combining 1D algorithms in the x and y directions of the plane and seems to be appropriate to study stripe phases. However it is not clear that such an algorithm would be able to deal with genuine 2D systems. The problem is related to finding a truly 2D version of the DMRG ([61] and T. Nishino in Chap. 5(I)). The actual DMRG algorithm is essentially one-dimensional and it looses its accuracy for large clusters [62].

The connection between the DMRG and the RVA and MP ansätze may be here of some help. Rather than asking what is the 2D version of the DMRG we may ask whether there are 2D versions of the RVA or the MP. The answer to the latter question is positive since from the work of Affleck et al. we know that there are AKLT states in 2D [63]. The AKLT states can be seen as a simple type of MP states. In fact, Niggemann et al. have constructed a vertex-state model for a spin-3/2 hexagonal lattice which is an example of these 2D MP states [64]. Below we generalize this construction to square lattices.

In 2D there are two types of MP states which can be called, using Statistical-Mechanics terminology, vertex-MP and face-MP ansätze [65]. A vertex-MP ansatz is defined in terms of a set of amplitudes

$$A^{\gamma,\delta}_{\alpha,\beta}[s], \quad (\alpha\,\beta,\ldots = 1,\ldots,m; \quad s = 1,\ldots,m_s) \tag{55}$$

where the labels $\alpha, \beta, \gamma, \delta$ are associated with the links of the square lattice while s labels the quantum state, e.g. spin, associated to the vertex where the 4-links $\alpha, \beta, \gamma, \delta$ meet. $A^{\gamma,\delta}_{\alpha,\beta}[s]$ is a sort of Boltzmann weight of a vertex model. The vertex-MP wave function $\psi(s_1, s_2, \ldots, s_N)$ can be obtained by multiplying all the Boltzmann weights $A^{\gamma_i,\delta_i}_{\alpha_i,\beta_i}[s_i]$ and contracting and summing over the link variables according to the same pattern of a vertex model in Statistical Mechanics [66]. Hence the value of the wave function $\psi(s_1, s_2, \ldots, s_N)$ is given by the partition function of a vertex model where the Boltzmann weights depend on the value of the local states s_i. Moreover the norm $\langle\psi|\psi\rangle$ of the vertex-MP state is given by the partition function of another vertex model whose Boltzmann weights are defined as,

$$R^{\gamma\gamma',\delta\delta'}_{\alpha\alpha',\beta\beta'} = \sum_s A^{\gamma,\delta}_{\alpha,\beta}[s]\, A^{\gamma',\delta'}_{\alpha',\beta'}[s] \tag{56}$$

This R matrix is the 2D version of the T matrix defined in (25).

The face-MP models can be defined in a similar manner by a set of variational parameters as in (55) where now the variables $\alpha, \beta, \ldots$ are now associated to the vertices of the squares while the quantum variable s is associated to the face whose vertices are $\alpha, \beta, \gamma, \delta$. This is similar to the face or Interaction Round a Face models (IRF) in Statistical Mechanics [66].

Hence in 2D there are two generic ways to produce MP ansätze which are in fact the straightforward generalization of the 1D MP ansätze. These two generalizations suggest to use some well-know models such as the 6-vertex model to test some of the ideas presented above.

We would like first of all to thank J. Dukelsky, T. Nishino, S. R. White, D.J. Scalapino and J.M. Román for their collaboration and many discussions on the several aspects related to the common work presented in these lectures.

We would like to thank the organizers of the DMRG98 at the Max-Planck-Institut in Dresden for their kind invitation to lecture in this meeting, for their warm hospitality and for making these lecture notes come out, as well as Prof. P.Fulde for supporting the DMRG98 at the MPI Dresden.

We also acknowledge many interesting discussions during this meeting with S. White, R. Noack, K. Hallberg, T. Nishino, S. Pleutin, X. Wang, I. Peschel, S. Ramasesha, S. Rommer, T. Xiang.

We acknowledge support from the DGES under contract PB96-0906.

References

1. S.R. White, Phys. Rev. Lett. **69**, 2863 (1992)
2. S.R. White, Phys. Rev. B **48**, 10345 (1993)
3. S.R. White and R.M. Noack, Phys. Rev. Lett. **68**, 3487 (1992)
4. S.R. White and D.A. Huse Phys. Rev. B **48**, 3844 (1993)
5. R.M.Noack, S.R. White and D.J. Scalapino Phys. Rev. Lett. **73**, 882 (1994)
6. As a proof of the relevance acquired by this method in the recent years, it is impossible to fully account for all the existing bibliography on the DMRG, and the reader is refered to search the cond-mat archive for a complete documentation on the subject.
7. M.A. Martín-Delgado and G. Sierra, Int. J. Mod. Phys. A **11**, 3145 (1996)
8. M.A. Martín-Delgado and G. Sierra, Phys. Lett. B **364**, 41 (1995)
9. M.A. Martín-Delgado, J. Rodriguez-Laguna and G. Sierra, Nucl. Phys. B **473**, 685 (1996)
10. M.A. Martín-Delgado and G. Sierra, Phys. Rev. Lett. **76**, 1146 (1996)
11. M.A. Martín-Delgado and G. Sierra, in From Field Theory to Quantum Groups, World Scientific Publishers (1996)
12. G. Sierra and T. Nishino, cond-mat/9610221; Nucl. Phys. B **495**, 505 (1997)
13. T. Nishino, J. Phys. Soc. Jpn. **64**, 3958 (1995)
14. T. Nishino and K. Okunishi, J. Phys. Soc. Jpn. **65**, 891-894 (1996); cond-mat/9507087
15. S. Östlund and S. Rommer, Phys. Rev. Lett. **75**, 3537 (1996); Phys. Rev. B **55**, 2164 (1997)
16. T. Nishino and K. Okunishi. J. Phys. Soc. Jpn. **64**, 4084 (1995); cond-mat/9510004
17. S.R. White, Phys. Rep. **301**, 187 (1998)
18. J. González, M.A. Martín-Delgado, G. Sierra and A.H. Vozmediano, Quantum Electron Liquids and High-T_c Superconductivity, Chapter 11, Lecture Notes in Physics, Monographs, Springer **38** (1995)
19. T. Nishino and K. Okunishi, in Strongly Correlated Magnetic and Superconducting Systems, Lecture Notes in Physics, Springer, **478** (1997)

20. M.A. Martín-Delgado, in Strongly Correlated Magnetic and Superconducting Systems, Lecture Notes in Physics, Springer, **478** (1997)
21. G. Sierra and M.A. Martín-Delgado, Phys. Rev. B **56**, 8774 (1997)
22. G. Sierra, M.A. Martín-Delgado, J. Dukelsky, S. R. White and D. J. Scalapino, Phys. Rev. B **57**, 11666 (1998)
23. G. Sierra, M.A. Martín-Delgado, S.R. White, D.J. Scalapino, J. Dukelsky, cond-mat/9806251, to be published in Phys. Rev. B
24. J. Dukelsky, M.A. Martín-Delgado, T. Nishino, G. Sierra. Europhys. Lett. **43**, 457 1998
25. J.M. Román, G. Sierra, J. Dukelsky and M.A. Martín-Delgado: cond-mat/9802150; J. Phys. A: Math. Gen. **31**, 9729 (1998)
26. E. Dagotto and T M. Rice, Science **271**, 619 (1996)
27. I. Peschel, M. Kaulke and Ö. Legeza, Ann. Physik (Leipzig) **8**, 153 (1999); cond-mat/9810174
28. K. Okunishi, Y. Hieida and Y. Akutsu, cond-mat/9810239
29. M. Andersson, M. Boman and S. Östlund, cond-mat/9810093
30. P.W. Anderson, Science **235**, 1196 (1987)
31. S. R. White, R. M. Noack and D. J. Scalapino, Phys. Rev. Lett. **73**, 886 (1994)
32. G. Andrews, The Theory of Partitions, Encyclopedia of Mathematics and its Applications, Addison-Wesley **2** (1976)
33. B. Sutherland, Phys. Rev. B **37**, 3786 (1988)
34. S. Gopalan, T.M. Rice and M. Sigrist, Phys. Rev. B **49**, 8901 (1994)
35. T. Barnes, E. Dagotto, J. Riera and E. S. Swanson, Phys. Rev. B **47**, 3196 (1993)
36. G. Fano, F. Ortolani and L. Ziosi, Phys. Rev. B **54**, 17557 (1996)
37. M. Fannes, B. Nachtergaele and R.F. Werner, Commun. Math. Phys. **144**, 443 (1992)
38. H. Takasaki, T. Hikihara and T. Nishino, cond-mat/9810241
39. M. Uehara , T. Nagata, J. Akimitsu, H. Takahashi, N. Mori and K. Kinoshita, J. Phys. Jpn, **65**, 2764 (1997)
40. W. Marshall, Proc. Roy. Soc. A **232**, 48 (1955)
41. S. Liang, B. Doucot and P.W. Anderson, Phys. Rev. Lett. **50**, 1153 (1988)
42. W. Marshall, Proc. Roy. Soc. A **232**, 48 (1955)
43. M. Troyer, H. Tsunetsugu and T. M. Rice, Phys. Rev. B **53**, 251 (1996)
44. M. Ogata, M.U. Luchini, S. Sorella and F.F. Assaad, Phys. Rev. Lett. **66**, 2388 (1991)
45. S.R. White and D.J. Scalapino, Phys. Rev. B **55**, 6504 (1997)
46. The IMSL Fortran-77 Library. Visual Numerics Inc.
47. The Numerical Recipes in C. W. Press, S.A. Teukolsky, W.T. Vetterling, B.P. Flannery, Cambridge University Press (1993)
48. A. Klümper, A. Schadschneider and J. Zittartz, Europhys. Lett. **24**, 293 (1993)
49. J.M. Tranquada et al. Nature **375**, 561 (1995); Phys. Rev. B **54**, 7489 (1996)
50. S.R. White and D.J Scalapino, Phys. Rev. Lett. **80**, 1272 (1998); cond-mat/9801274
51. S.A. Carter, B. Battlogg, R.J. Cava, J.J. Krajewski, W.F. Peck, Jr. and T.M. Rice, Phys. Rev. Lett. **77**, 1378 (1996)
52. M. Matsuda, K.M. Kojima, Y.J. Uemura, J.L. Zarestky, K. Nakajima, K. Kakurai, T. Yokoo, S.M. Shapiro and G. Shirane, Phys. Rev. B **57**, 11467 (1998)

53. S.K. Pati, S. Ramasesha and D. Sen, Phys. Rev. B **55**, 8894 (1997); cond-mat/9704057
54. A.K. Kolezhuk, H.J. Mikeska and S. Yamamoto, Phys. Rev. B **55**, 3336 (1997)
55. S. Yamamoto, S. Brehmer and H.J. Mikeska, Phys. Rev. B **57**, 13610 (1998); S. Brehmer, H.J. Mikeska and S. Yamamoto, J. Phys. Cond. Matt. **9**, 3921 (1997)
56. A. Giesekus, Phys. Rev. B **52**, 2476 (1995)
57. E. Kim, private communication.
58. S. Pleutin, private communication.
59. S. Pleutin and J.L. Fave, J. of Phys. : Condens. Matter **10**, 3941 (1998)
60. M.A. Garcia-Bach, R. Valenti, D.J. Klein, cond-mat/9705121 and references therein; Phys. Rev. B in press
61. H. Otsuka Phys. Rev. B **53**, 14004 (1996)
62. S. Liang and H. Pang, Phys. Rev. B **49**, 9214 (1994)
63. I. Affleck, T. Kennedy, E.H. Lieb and H. Tasaki, Commun. Math. Phys. **115**, 477 (1988)
64. H. Niggemann, J. Zittarz, Z. Phys. B **101**, 289 (1996)
H. Niggemann, A. Klümper, J. Zittarz, Z. Phys. B **104**, 103 (1997)
65. G. Sierra and M.A. Martín-Delgado, Proceedings of the Workshop on the Recent Developments in Exact RG Methods, Faro, Portugal, (1998)
66. R. J. Baxter, Exactly Solved Models in Statistical Mechanics, Academic Press (1982)

5 Transfer-Matrix Approach to Classical Systems

T. Nishino[1], K. Okunishi[2], Y. Hieida[2], T. Hikihara[1], and H. Takasaki[1]

[1] Department of Physics, Graduate School of Science and Technology Kobe University, Inoshishi-dai, Nada, Kobe 657-8501, Japan
[2] Department of Physics, Graduate School of Science, Osaka University Toyonaka, Osaka 560-0043, Japan

The development of the DMRG method by White [1] is one of the major achievements in computational condensed-matter physics (see Chap. 2(I)). DMRG enables one to calculate ground states of relatively large-scale one-dimensional (1D) quantum systems. Since 1D quantum systems are deeply related to 2D classical systems [2–4], it is natural to import the DMRG method to 2D classical systems. The infinite-system algorithm was first applied to the Ising model by one of the authors [5]. Carlon *et al.* [6–8] applied the finite-system algorithm to the Potts model (Chap. 3.2(II)), and they calculated very accurate density profiles and critical indices. After that, the DMRG formulation for classical systems was applied to 1D quantum system at finite temperature [9,10,12] (Chaps. 6(I) and 4(II)).

In this review we explain the DMRG applied to 2D classical systems ('classical DMRG') by looking at the renormalization group (RG) transformation for the transfer matrix T. For the distinction, we call the DMRG for quantum systems 'quantum DMRG' in the following. We first introduce the infinite-system algorithm (Sects. 1-4). It then follows naturally that the classical DMRG is a kind of variational method that maximizes the partition function using a limited number of degrees of freedom, where the variational state is written as a product of local matrices [15] (Sect. 5). Actually, such a variational formulation has been used for a long time; we give a brief review in Sect. 6. The variational state obtained by the infinite-system algorithm can be improved systematically by use of the finite-system algorithm (Sect. 7). As a variation of the infinite-system algorithm, we introduce the corner transfer matrix (CTM) formulation of the classical DMRG, where the formulation can be generalized to higher-dimensional systems (Sect. 8). We finally discuss some remaining problems in the last section.

1 Transfer Matrix

As an example of a 2D classical system, let us consider the isotropic square-lattice Ising model [17] on a cylinder, which is defined by imposing periodic boundary conditions in the vertical direction and open boundary conditions for the horizontal direction of an $(N \times \ell)$ spin lattice (Fig. 1) [16]. The system

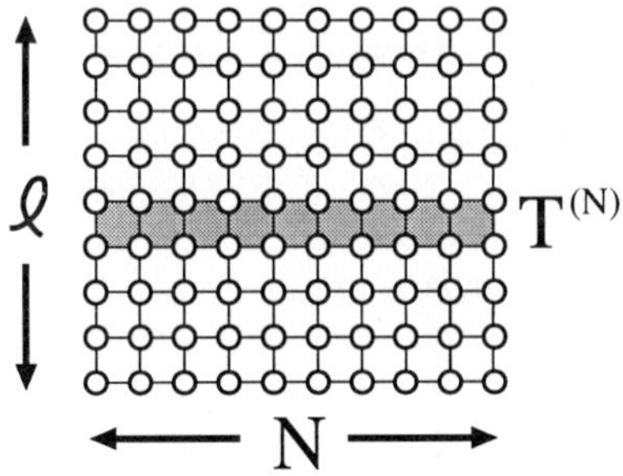

Fig. 1. Square-lattice Ising model on a N by ℓ transfer matrices $T^{(N)}$ defined in (1). Open boundary conditions are imposed in the horizontal direction and periodic ones in the vertical direction.

consists of ℓ rows of width N. We assume $\ell \gg N$, and refer to N as the system size. We label the spins in a row from left to right as $s_1, s_2, \dots, s_{N-1}, s_N$, and occasionally use the vector notation $\boldsymbol{s} = s_{1 \dots} s_N$ for simplicitely. When the Ising interaction is restricted to the nearest neighbor pairs, the transfer matrix is expressed as a 2^N-dimensional matrix (Fig. 2)

$$
\begin{aligned}
T^{(N)}(\boldsymbol{s}'|\boldsymbol{s}) &= \exp\left\{ \frac{K}{2} \sum_{i=1}^{N-1} (s'_i s'_{i+1} + s_i s_{i+1}) + K \sum_{i=1}^{N} s'_i s_i \right\} \qquad (1) \\
&= \exp\left(\frac{K}{2} s_1 s'_1\right) \left\{ \prod_{i=1}^{N-1} W(s'_i s'_{i+1} | s_i s_{i+1}) \right\} \exp\left(\frac{K}{2} s_N s'_N\right) ,
\end{aligned}
$$

where $K = J/k_{\mathrm{B}}T$ is the coupling and

$$
W(s'_i s'_{i+1} | s_i s_{i+1}) = \exp\left\{ \frac{K}{2} \left(s_i s_{i+1} + s_{i+1} s'_{i+1} + s'_{i+1} s'_i + s'_i s_i \right) \right\} \qquad (2)
$$

is the local Boltzmann weight for a square between i and $i+1$. We have symmetrized the transfer matrix in order to simplify the following formulation. It may be helpful for the readers to recall the quantum-classical correspondence $T^{(N)} = \exp(-\Delta H^{(N)})$, where $H^{(N)}$ is the corresponding quantum Hamiltonian. (However, $H^{(N)}$ is normally non-local and the correspondence is more of a formal character.)

For a tutorial purpose, we only consider the symmetric transfer matrix in the following. Extensions to the asymmetric case are straightforward [5,18]; to treat an asymmetric transfer matrix is not rare in classical statistical mechanics. More details can be found in Chaps. 6(I) and 7(I) [10,11], as

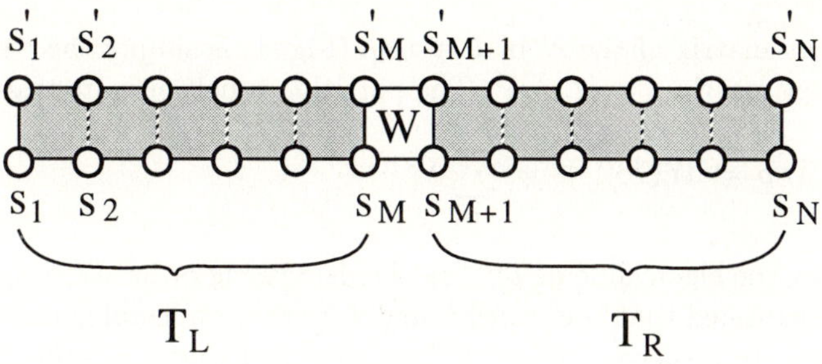

Fig. 2. Division of the transfer matrix $T^{(N)}$ into the half-row transfer matrices T_L and T_R, and a Boltzmann weight W, compare (3).

well as in Chap. 4.1(II) [12,13]. More generally, it is possible to treat various models such as

- The q-state Potts model [19,20] or the n-vector model, that have *discrete spin symmetry and short range interaction.*
- The interaction round a face (IRF) model [18] whose Boltzmann weight $W(s'_i s'_{i+1}|s_i s_{i+1})$ is represented by arbitrary matrix; *the weight can even be negative.* [14] (There is no sign problem.)
- The vertex model [18] that has spin variables in the middle of each bond.

in the framework of DMRG.

Following the convention of the quantum DMRG, let us divide the spin row into the left and the right as $\boldsymbol{s} = \boldsymbol{s}_L \boldsymbol{s}_R$, where $\boldsymbol{s}_L = s_1 ... s_M$ and $\boldsymbol{s}_R = s_{M+1} ... s_N$. For the moment we assume that N is an even integer, and that $M = N/2$. According to the division, we can decompose $T^{(N)}$ into three factors

$$T^{(N)}(\boldsymbol{s}'|\boldsymbol{s}) = T_L(\boldsymbol{s}'_L|\boldsymbol{s}_L) W(s'_M s'_{M+1}|s_M s_{M+1}) T_R(\boldsymbol{s}'_R|\boldsymbol{s}_R) \,, \tag{3}$$

where $T_L(\boldsymbol{s}'_L|\boldsymbol{s}_L)$ and $T_R(\boldsymbol{s}'_R|\boldsymbol{s}_R)$ are the half-row transfer matrices

$$\begin{aligned} T_L(\boldsymbol{s}'_L|\boldsymbol{s}_L) &= \exp\left(\frac{K}{2} s'_1 s_1\right) \prod_{i=1}^{M-1} W(s'_i s'_{i+1}|s_i s_{i+1}) \\ T_R(\boldsymbol{s}'_R|\boldsymbol{s}_R) &= \exp\left(\frac{K}{2} s'_N s_N\right) \prod_{i=M+1}^{N-1} W(s'_i s'_{i+1}|s_i s_{i+1}) \,. \end{aligned} \tag{4}$$

2 Density Submatrix

The density matrix of the N by ℓ system (Fig. 1) is simply the ℓ-th power of the transfer matrix $\rho = (T^{(N)})^\ell$. The partition function is its trace

$$Z = \mathrm{Tr}\,\rho = \mathrm{Tr}\,(T^{(N)})^\ell = \sum_\kappa \lambda_\kappa^{\,\ell}\,, \tag{5}$$

where λ_κ is the eigenvalue of $T^{(N)}$ in the decreasing order $\lambda_1 \geq \lambda_2 \geq \cdots \geq 0$. (We have assumed that the transfer matrix is diagonalizable, and that all the eigenvalues are positive.) Since we have assumed that ℓ is sufficiently larger than N, the partition function can be approximated as $Z \simeq \lambda_1^{\,\ell}$, where the symbol '$\simeq$' denotes that the ratio $\lambda_1^{\,\ell}/Z$ converges to unity in the limit $\ell/N \to \infty$. In this situation, the density matrix can be well approximated as

$$\rho \simeq |v^{(N)}\rangle\, \lambda_1^{\,\ell}\, \langle v^{(N)}| \tag{6}$$

where $|v^{(N)}\rangle$ is the eigenvector of $T^{(N)}$ that corresponds to the largest eigenvalue λ_1. We assume the normalization $\langle v^{(N)}|v^{(N)}\rangle = 1$.

What is called 'density matrix' in the context of DMRG is not the density matrix itself, but the density submatrix (DSM) — also called 'reduced density matrix' — which is obtained by partially tracing out spin indices from ρ. The DSM for the left and the right halves of the system are defined as [21]

$$\begin{aligned} \rho_L(s'_L|s_L) &= \sum_{s_R} \rho(s'_L\, s_R|s_L\, s_R) \\ \rho_R(s'_R|s_R) &= \sum_{s_L} \rho(s_L\, s'_R|s_L\, s_R)\,. \end{aligned} \tag{7}$$

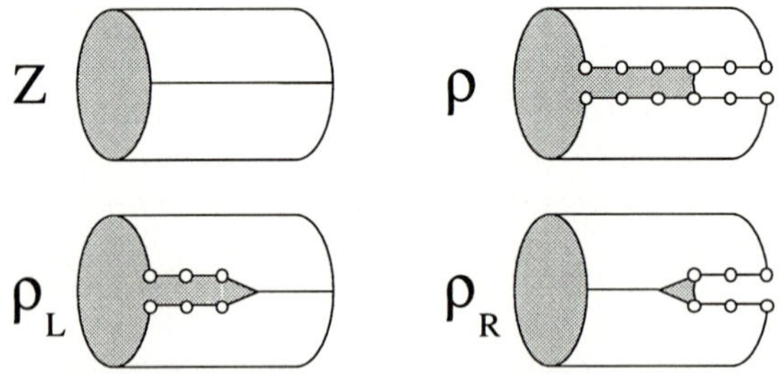

Fig. 3. ρ_L and ρ_R (7). The DSM can be interpreted as a cut in the 2D system shown in Fig. 1.

Figure 3 shows the graphical representation of the DSM, where ρ_L and ρ_R correspond to the cuts on the cylindrical system shown in Fig. 1 [22]. To take the trace of the DSM one has to join the system along the cut and to reconstruct the cylinder: $Z = \mathrm{Tr}\,\rho_L = \mathrm{Tr}\,\rho_R$. Using the eigenvector $v^{(N)}$ in (6), we can express the DSM in the more familiar form

$$\rho_L(s'_L|s_L) \simeq \sum_{s_R} v^{(N)}(s'_L\, s_R)\, \lambda_1{}^{\ell}\, v^{(N)}(s_L\, s_R)$$
$$\rho_R(s'_R|s_R) \simeq \sum_{s_L} v^{(N)}(s_L\, s'_R)\, \lambda_1{}^{\ell}\, v^{(N)}(s_L\, s_R) \tag{8}$$

that has been used for the quantum DMRG.

As for the quantum DMRG, the diagonalization of the DSM is one of the important steps for the classical DMRG. The DSM ρ_L and ρ_R can be diagonalized as

$$\rho_L = \sum_{\xi} \boldsymbol{a}_\xi\, \omega_\xi{}^2\, \boldsymbol{a}_\xi{}^T\,, \qquad \rho_R = \sum_{\zeta} \boldsymbol{b}_\zeta\, \mu_\zeta{}^2\, \boldsymbol{b}_\zeta{}^T\,, \tag{9}$$

where $\boldsymbol{a}_\xi$ and $\boldsymbol{b}_\xi$ are the eigenvectors of ρ_L and ρ_R, respectively. The eigenvectors satisfy the orthogonality relations $\boldsymbol{a}_\xi{}^T\, \boldsymbol{a}_{\xi'} = \delta_{\xi\xi'}$ and $\boldsymbol{b}_\zeta{}^T\, \boldsymbol{b}_{\zeta'} = \delta_{\zeta\zeta'}$, and they are complete

$$I_L = \sum_{\xi} \boldsymbol{a}_\xi\, \boldsymbol{a}_\xi{}^T\,, \qquad I_R = \sum_{\zeta} \boldsymbol{b}_\zeta\, \boldsymbol{b}_\zeta{}^T\,, \tag{10}$$

where I_L and I_R are 2^M- and 2^{N-M}-dimensional unit matrices, respectively; $I_L = I_R$ when $N = 2M$. It is convenient to use the matrix notation $A = (\boldsymbol{a}_1, \boldsymbol{a}_2, \cdots)$ and $B = (\boldsymbol{b}_1, \boldsymbol{b}_2, \cdots)$ when we explicitly show the spin indices. For example, (9) can be written as

$$\rho_L(s'_L|s_L) = \sum_{\xi} A(s'_L|\xi)\, \omega_\xi{}^2\, A(s_L|\xi)$$
$$\rho_R(s'_R|s_R) = \sum_{\zeta} B(s'_R|\zeta)\, \mu_\zeta{}^2\, B(s_R|\zeta)\,. \tag{11}$$

In (9) and (11) we have expressed the eigenvalues of ρ_L and ρ_R as squares of real numbers, because normally they are non-negative [23]. As to the eigenvalues of ρ, we assume decreasing order for both $\omega_\xi{}^2$ and $\mu_\zeta{}^2$

$$\omega_1{}^2 \geq \omega_2{}^2 \geq \cdots \geq 0\,, \qquad \mu_1{}^2 \geq \mu_2{}^2 \geq \cdots \geq 0\,, \tag{12}$$

in the following. In general, $\omega_i = \mu_i$ holds for $1 \leq i \leq \min(2^M, 2^{N-M})$ in the limit $\ell/N \to \infty$.

3 Renormalization-Group Transformation

Unlike the density matrix ρ, not only the largest eigenvalue is dominant in (12). This is because $\boldsymbol{s}'_L$ and $\boldsymbol{s}_L$ in $\rho_L(\boldsymbol{s}'_L|\boldsymbol{s}_L)$ are correlated through the junction s_R (7)-(8). It has been known that the eigenvalues ${\omega_i}^2$ (or ${\mu_i}^2$) in (12) decay nearly exponentially when the correlation length of the system is finite [18,24,25] (Chap. 3.1(II)), and that the decay is rapid when the correlation length is short. Even at the critical temperature, we expect a certain decay in ${\omega_i}^2$, because the system size N is finite and the correlation length is of the order of N. As a result we can say that the partition function can be approximated by the partial sum of the DSM eigenvalues

$$\tilde{Z} = \sum_{i=1}^{m} {\omega_i}^2 , \tag{13}$$

where m is the number of the eigenvalues kept. It is obvious that $Z \geq \tilde{Z}$, and the difference $Z - \tilde{Z}$ is fairly small when m is sufficiently large. We can also express $\tilde{Z}$ as

$$\tilde{Z} = \mathrm{Tr}\,(P_L\, \rho_L) = \mathrm{Tr}\, \tilde{\rho}_L \tag{14}$$

where P_L is a projection operator

$$P_L(s'_L|s_L) = \sum_{\xi=1}^{m} A(s'_L|\xi)A(s_L|\xi) , \tag{15}$$

and therefore $\tilde{\rho}_L$ is an m-dimensional diagonal matrix

$$\tilde{\rho}_L(\xi'|\xi) = \sum_{s'_L s_L} A(s'_L|\xi')\rho_L(s'_L|s_L)A(s_L|\xi) = \delta_{\xi'\xi}\,\omega_\xi^2 \tag{16}$$

where $\xi', \xi \leq m$. (If $m = 2^M$, P_L coincides with I_L in (10).) From (14)-(16), the transformation $\rho_L \rightarrow \tilde{\rho}_L$ by transformation from the row spin s_L to an m-state block spin ξ. In the same manner $B(s_R|\zeta)$ is related to the RG transformation $s_R \rightarrow \zeta$, that performs the mapping $\rho_R \rightarrow \tilde{\rho}_R$. In the following, we put a tilde over the renormalized matrices.

It should be noted that the RG transformation $\rho_L \rightarrow \tilde{\rho}_L$ is performed so that it maximizes the approximate partition function $\tilde{Z}$ within the restricted number of degrees of freedom m. On the other hand, the RG transformation in quantum DMRG minimizes the ground-state energy; this is consistent with the thermodynamic relation $F = -k_{\mathrm{B}}T \log Z = U - TS$ in the limit $T \rightarrow 0$.

4 Infinite-System Algorithm

In quantum DMRG, the infinite-system algorithm consists of the iterative use of the system expansion and the RG transformation for the Hamiltonian. It is easy to introduce the RG process to 2D classical system, using the quantum-classical correspondence $T^{(N)} = \exp(-\Delta H^{(N)})$; as far as the formulation is concerned, the classical DMRG can be obtained just by replacing the Hamiltonian $H^{(N)}$ in quantum DMRG by the transfer matrix $T^{(N)}$. As we show in this section, the infinite-system algorithm constructs renormalized transfer matrices $\tilde{T}^{(6)}, \tilde{T}^{(8)}, \ldots, \tilde{T}^{(N)}$ successively up to arbitrary N.

The infinite-system algorithm starts from the 4-site system, whose transfer matrix is given by

$$T^{(4)}(s_1's_2's_3's_4'|s_1s_2s_3s_4) = T_L(s_1's_2'|s_1s_2)W(s_2's_3'|s_2s_3)T_R(s_3's_4'|s_3s_4)\,. \quad (17)$$

Diagonalizing $T^{(4)}$, we obtain the largest eigenvalue and the corresponding eigenvector $v^{(4)}(s_1s_2s_3s_4)$, that satisfies $T^{(4)}v^{(4)} = \lambda\, v^{(4)}$. We then obtain $\rho_L(s_1's_2'|s_1s_2)$ and $\rho_R(s_3's_4'|s_3s_4)$ using (8) and get the RG transformation matrices $A(s_1s_2|\xi)$ and $B(s_3s_4|\zeta)$ from the diagonalization of ρ_L and ρ_R (11).

The next step is the enlargement of the system from $N = 4$ to $N = 6$ and the RG transformation via the substitution

$$\sum_{s_1's_2's_1s_2} A(s_1's_2'|\xi')T_L(s_1's_2'|s_1s_2)W(s_2's_3'|s_2s_3)A(s_1s_2|\xi) \rightarrow \tilde{T}_L(\xi's_3'|\xi s_3)$$

$$\sum_{s_5's_6's_5s_6} B(s_5's_6'|\zeta')W(s_4's_5'|s_4s_5)T_R(s_5's_6'|s_5s_6)B(s_5s_6|\zeta) \rightarrow \tilde{T}_R(s_4'\zeta'|s_4\zeta)\,, \quad (18)$$

where $\tilde{T}_L$ and $\tilde{T}_R$ are the renormalized half-row transfer matrices. (We have changed the spin indices of T_R according to the enlargement $N = 4$ to $N = 6$.) The greek indices ξ and ζ represent the block variables, that take (at most) m values. This is shown graphically in Fig. 4. Joining $\tilde{T}_L$ and $\tilde{T}_R$, we can construct the renormalized transfer matrix for $N = 6$

$$\tilde{T}^{(6)}(\xi's_3's_4'\zeta'|\xi s_3s_4\zeta) = \tilde{T}_L(\xi's_3'|\xi s_3)W(s_3's_4'|s_3s_4)\tilde{T}_R(s_4'\zeta'|s_4\zeta)\,. \quad (19)$$

At this stage, ξ and ζ represent the 2-spin blocks.

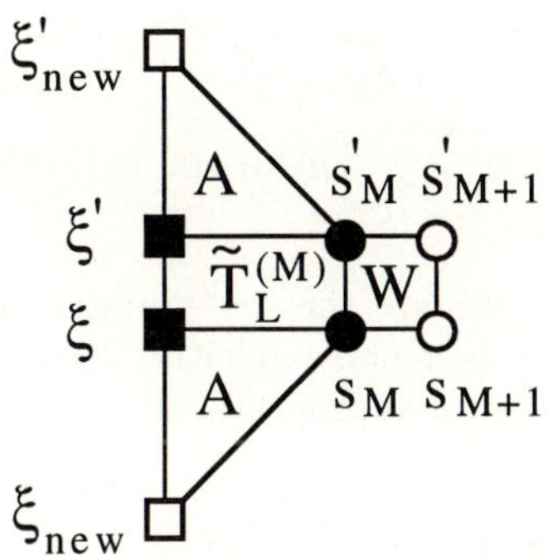

Fig. 4. Enlargement by W and RG transformation through A of the half-row transfer matrix $\tilde{T}_L$ in (18)-(21). The filled circles and squares denote variables which are summed.

It is obvious that we can obtain $\tilde{T}^{(8)}$ in the same manner, i.e., diagonalizing $\tilde{T}^{(6)}$ in (19) to obtain the largest eigenvalue and the corresponding eigenvector $\tilde{v}^{(6)}(\xi s_3 s_4 \zeta)$, and create $\tilde{\rho}_L(\xi' s_3' | \xi s_3)$ and to $\tilde{\rho}_R(s_4' \zeta' | s_4 \zeta)$, using an extension of (8). We obtain the RG transformations $A(\xi s_3 | \xi_{\rm new})$ and $B(s_4 \zeta | \zeta_{\rm new})$ by diagonalizing $\tilde{\rho}_L$ and $\tilde{\rho}_R$, (an extension of (11)) respectively, where $\xi_{\rm new}$ and $\zeta_{\rm new}$ are m-state variables that represent the 3-spin block. The $\tilde{T}^{(8)}$ is then constructed as

$$\begin{aligned}\tilde{T}^{(8)}&(\xi'_{\rm new} s'_4 s'_5 \zeta'_{\rm new} | \xi_{\rm new} s_4 s_5 \zeta_{\rm new}) \\ &= \tilde{T}_L(\xi'_{\rm new} s'_4 | \xi_{\rm new} s_4) W(s'_4 s'_5 | s_4 s_5) \tilde{T}_R(s'_5 \zeta'_{\rm new} | s_5 \zeta_{\rm new}) ,\end{aligned} \tag{20}$$

where $\tilde{T}_L(\xi'_{\rm new} s'_4 | \xi_{\rm new} s_4)$ and $\tilde{T}_R(s'_5 \zeta'_{\rm new} | s_5 \zeta_{\rm new})$ are created as

$$\begin{gathered}\sum_{\xi' s'_3 \xi s_3} A(\xi' s'_3 | \xi'_{\rm new}) \tilde{T}_L(\xi' s'_3 | \xi s_3) W(s'_3 s'_4 | s_3 s_4) A(\xi s_3 | \xi_{\rm new}) \to \tilde{T}_L(\xi'_{\rm new} s'_4 | \xi_{\rm new} s_4) \\ \sum_{s'_6 \zeta' s_6 \zeta} B(s'_6 \zeta' | \zeta'_{\rm new}) W(s'_5 s'_6 | s_5 s_6) \tilde{T}_R(s'_6 \zeta' | s_6 \zeta) B(s_6 \zeta | \zeta_{\rm new}) \to \tilde{T}_R(s'_5 \zeta'_{\rm new} | s_5 \zeta_{\rm new}) .\end{gathered} \tag{21}$$

In this way, we further obtain $\tilde{T}^{(10)}$, $\tilde{T}^{(12)}$, $\tilde{T}^{(14)}$, ..., etc., up to arbitrary system size. This is the outline of the infinite-system algorithm.

Numerical diagonalization for $\tilde{T}^{(N)}$ is normally performed via the Lanczos method [26], that requires multiplication of $\tilde{T}^{(N)}$ with a $4m^2$-dimensional vector $\boldsymbol{x}$. Since $\tilde{T}^{(N)}$ is represented as a product of $\tilde{T}_L$, W, and $\tilde{T}_R$, the numerical multiplication $\boldsymbol{x}''' = \tilde{T}^{(N)} \boldsymbol{x}$ can be done very rapidly via the following three steps (Chap. 2(I))

$$\begin{aligned} x'(\xi' s'_M s_M s_{M+1} \zeta) &= \sum_{\xi} \tilde{T}_L(\xi' s'_M | \xi s_M) x(\xi s_M s_{M+1} \zeta) , \\ x''(\xi' s'_M s'_{M+1} s_{M+1} \zeta) &= \sum_{s_M} W(s'_M s'_{M+1} | s_M s_{M+1}) x'(\xi' s'_M s_M s_{M+1} \zeta) , \\ x'''(\xi' s'_M s'_{M+1} \zeta') &= \sum_{s_{M+1} \zeta} \tilde{T}_R(s'_{M+1} \zeta' | s_{M+1} \zeta) x''(\xi' s'_M s'_{M+1} s_{M+1} \zeta) .\end{aligned} \tag{22}$$

One does not have to prepare the $4m^2$-dimensional transfer matrix $\tilde{T}^{(N)}$, containing $16m^4$ matrix elements, explicitly. The Lanczos diagonalization can be accelerated further by an appropriate choice of the initial vector [44–46].

5 Variational State in DMRG

the classical (and also quantum) DMRG. Suppose that we have $\tilde{T}^{(N)}$ and the corresponding eigenvector $\tilde{v}^{(N)}(\xi s_M s_{M+1} \zeta)$ from the infinite-system algorithm, where $\langle \tilde{v}^{(N)} | \tilde{v}^{(N)} \rangle = 1$ is satisfied. If we form the expectation value

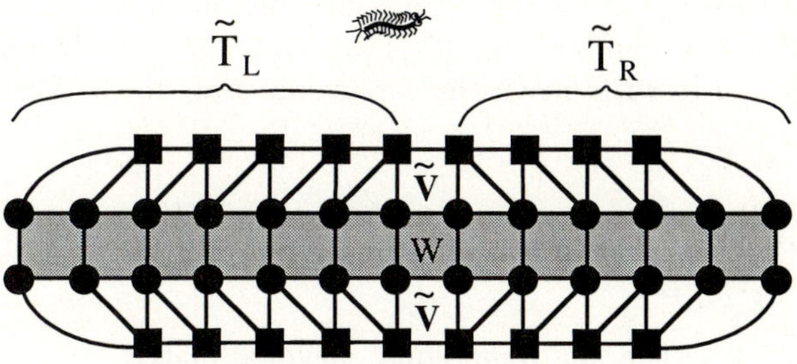

Fig. 5. Graphical representation of the variational eigenvalue of the transfer matrix $T^{(N)}$, see (23). The shaded region represents $T^{(N)}$. The upper and lower parts correspond to the variational state $v^{(N)}$ in (24).

$\tilde{\lambda} = \langle \tilde{v}^{(N)} | \tilde{T}^{(N)} | \tilde{v}^{(N)} \rangle$, this can be represented as shown in Fig. 5; the squares denote ξ, ξ', ζ, and ζ' variables which are defined in the renormalization steps. The left and the right halves of the diagram correspond to $\tilde{T}_L(\xi' s'_M | \xi s_M)$ and $\tilde{T}_R(s'_{M+1} \zeta' | s_{M+1} \zeta)$, respectively. As indicated by the filling, all the spin variables are all summed. If we recall that $\tilde{T}_L$ and $\tilde{T}_R$ are created through the iterative use of (21), it is possible to write

$$\langle \tilde{v}^{(N)} | \tilde{T}^{(N)} | \tilde{v}^{(N)} \rangle = \langle v^{(N)} | T^{(N)} | v^{(N)} \rangle , \tag{23}$$

where $v^{(N)}(s_{1 \ldots} s_N)$ is the variational state for $T^{(N)}$ defined as

$$\sum_{\{\xi\}\{\zeta\}} \prod_{i=2}^{M-1} A(\xi_{i-1} s_i | \xi_i) \, \tilde{v}^{(N)}(\xi_{M-1} s_M s_{M+1} \zeta_{M+2}) \prod_{j=M+2}^{N-1} B(s_j \zeta_{j+1} | \zeta_j) , \tag{24}$$

where $\xi_1 \equiv s_1$ and $\zeta_N \equiv s_N$. This corresponds to the portions in Fig. 5 above and below the original (= unrenormalized) transfer matrix, respectively. We have put the indices to ξ and ζ in order to distinguish the block spin variables. The matrices $A(\xi_{i-1} s_i | \xi_i)$ and $B(s_j \zeta_{j+1} | \zeta_j)$ depend on their position i and j, respectively. Note that $v^{(N)}(s_{1 \ldots} s_N)$ is normalized because of the assumed normalization $\langle \tilde{v}^{(N)} | \tilde{v}^{(N)} \rangle = 1$ and the orthogonal relations

$$\begin{aligned} \sum_{\xi_{i-1} s_i} A(\xi_{i-1} s_i | \xi_i) \, A(\xi_{i-1} s_i | \xi'_i) &= \delta_{\xi_i \, \xi'_i} \\ \sum_{s_j \zeta_{j+1}}^{m} B(s_j \zeta_{j+1} | \zeta_j) B(s_j \zeta_{j+1} | \zeta'_j) &= \delta_{\zeta_j \, \zeta'_j} \, . \end{aligned} \tag{25}$$

Since $T^{(N)}$ is symmetric and positive definite, the largest eigenvalue $\tilde{\lambda}$ of $\tilde{T}^{(N)}$ is a variational lower bound for the largest eigenvalue λ of $T^{(N)}$

$$\tilde{\lambda} = \langle v^{(N)} | T^{(N)} | v^{(N)} \rangle \leq \lambda , \tag{26}$$

and the difference $\epsilon = \lambda - \tilde{\lambda}$ is a decreasing function of m.

It is possible to decompose $\tilde{v}^{(N)}(\xi_{M-1} s_M s_{M+1} \zeta_{M+2})$ further into a matrix product [27], and to write down the variational state in the form

$$v^{(N)}(s_{1 \ldots} s_N) = \sum_{\{\xi\}\{\zeta\}} \prod_{i=2}^{M} A(\xi_{i-1} s_i | \xi_i) \, \Omega(\xi_M | \zeta_{M+1}) \prod_{j=M+1}^{N-1} B(s_j \zeta_{j+1} | \zeta_j) \,, \tag{27}$$

where $\Omega(\xi_M|\zeta_{M+1})$ is a $2m$-dimensional diagonal matrix

$$\Omega(\xi_M|\zeta_{M+1}) = \delta_{\xi_M \zeta_{M+1}} \frac{\omega_{\xi_M}}{\sqrt{\tilde{Z}}} \,. \tag{28}$$

We have imposed the normalization $\Omega^2 = 1$. (Strictly speaking, Ω need not to be diagonal, but $\Omega\Omega^T$ should be.) In the thermodynamic limit $N \to \infty$, the matrices $A(\xi_{i-1} s_i|\xi_i)$ and $B(s_j \zeta_{j+1}|\zeta_j)$ lose the position dependence, and the variational state in (27) coincides with Östlund's matrix-product state (See Chap. 3(I)).

As a result of the infinite-system algorithm, we obtain the variational free energy per site

$$\tilde{f} = -\frac{1}{N} k_{\mathrm{B}} T \log \tilde{\lambda} \tag{29}$$

and other thermodynamic quantities via the numerical derivative of $\tilde{f}$. Since we have the variational state $v^{(N)}(s_{1 \ldots} s_N)$ explicitly ((24) and (27)), we can calculate arbitrary spin correlation functions, such as $\langle s_i \rangle = \langle v^{(N)} | \hat{s}_i | v^{(N)} \rangle$ and $\langle s_i s_j \rangle = \langle v^{(N)} | \hat{s}_i \hat{s}_j | v^{(N)} \rangle$, etc. Generally speaking, the numerical precision of $\langle s_i s_j \rangle$ decreases as $|i - j|$ increases.

When we calculate spin correlation functions, we have to check that the energy scale ϵ_m introduced by the restriction of the degrees of freedom is sufficiently small compared with the excitation gap ϵ_g (of $H^{(N)} \equiv -\log \tilde{T}^{(N)}/\Delta$). If not, we have to increase m to keep the numerical precision. If the system is just at the critical temperature $T = T_{\mathrm{c}}$, the gap ϵ_g is of the order of $1/N$. For this reason, it is not reliable to apply the infinite-system algorithm to a classical system just at T_{c}, and directly analyze the thermodynamic limit $N \to \infty$. It would be better to use the finite-system algorithm, that improves the variational state obtained by the infinite-system algorithm further (see Sect. 7).

Let us see how accurate the infinite-system algorithm is. Figure 6 shows the specific heat $\mathrm{C_v}(T)$ of the square-lattice Ising model, which is obtained by taking the temperature derivative of the nearest-neighbor correlation function $E(T) = \langle s_M s_{M+1} \rangle$. We calculate $E(T)$ for the system up to $N = 2M = 2048$, which is sufficiently larger than the correlation length for each plotted temperature, except at $T = T_{\mathrm{c}}$. The m-dependence for $E(T)$ is negligible when we keep $m = 60$ states. Since we know the exact solution for this

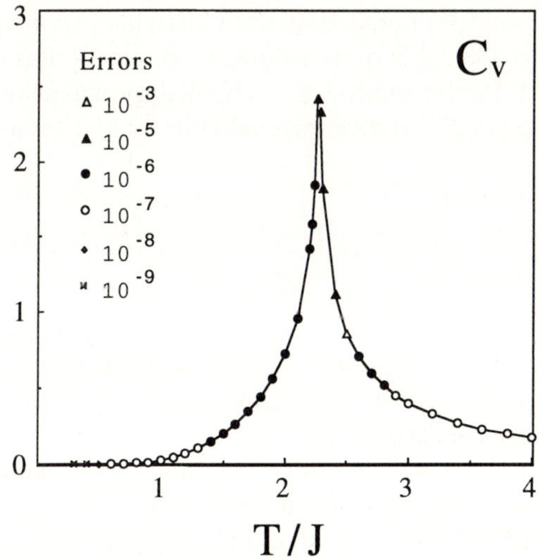

Fig. 6. Numerical results for the specific heat of the 2D Ising model [5].

model [28], we can directly observe the numerical error in $C_v(T)$; in Fig. 6, it is indicated by several marks. The numerical error is not negligible near the critical temperature T_c, partly because $E(T)$ is singular at T_c, and because the numerical derivative

$$\frac{E(T + \Delta T/2) - E(T - \Delta T/2)}{\Delta T} \tag{30}$$

is sensitive to ΔT; typically, we have chosen $\Delta T = 10^{-4}$. The other source of numerical errors is the increase of the cut-off energy scale ϵ_m near T_c, that spoils the numerical precision of the block-spin transformation.

6 History of the Matrix-Product State

Variational states written in the form of matrix products, such as $v^{(N)}$ in (24), have been used for a long time. In this section we review briefly the history of the matrix-product state. In 1941 Kramers and Wannier (K-W) [29,30] investigated the square-lattice Ising model, assuming that the variatonal state can be written as

$$v(\ldots, s_{i-1}, s_i, s_{i+1}, s_{i+2}, \ldots) = \ldots F(s_{i-1}|s_i)F(s_i|s_{i+1})F(s_{i+1}|s_{i+2}) \ldots , \tag{31}$$

where $F(s'|s)$ is a 2-dimensional symmetric matrix. The transition temperature T_c and the specific heat calculated from this variational state are more accurate than those obtained by the molecular-field and the Bethe approximations [31]. It should be noted that the Gutzwiller approximation [32] for the Hubbard model [33,34] is quite similar to the K-W approximation.

Around 1960-70 Baxter improved the K-W approximation by introducing additional freedom [18,35]. His variational state is written as

$$v(\dots, s_{i-1}, s_i, s_{i+1}, s_{i+2}, \dots) = \sum_{\dots,a,b,c,d,\dots} \dots F_{ab}(s_{i-1}|s_i) F_{bc}(s_i|s_{i+1}) F_{cd}(s_{i+1}|s_{i+2}) \dots , \qquad (32)$$

where ...a, b, c, d..., denote the additional m-state variables. Since $F_{ab}(s'|s)$ contains $4m^2$ adjustable parameters, the way of finding out the best $F_{ab}(s'|s)$ is non trivial. He performed the optimization using a self-consistent equation for the corner transfer matrix (CTM) [18].

Applications of the matrix-product formulation to quantum systems began with the investigations of Haldane's conjecture. In 1985 Nightingale and Blöte [36] used the K-W matrix product (31) as the initial vector of their projector Monte Carlo simulation. It is interesting that they commented on Baxter's method as follows "*... This method was formulated by Baxter for classical models in statistical mechanics. The generalization to quantum mechanical system is straightforward.*" (The RVA formulation by Martín-Delgado and Sierra can be seen as a realization of this concept; see Chap. 4(I).) In 1987 Affleck, Lieb, Kennedy, and Tasaki [37] showed that the ground state of a special $S = 1$ spin chain can be exactly expressed as

$$\sum_{\dots,a,b,c,d,e,\dots} \dots M_{ab}(s_{i-1}) M_{bc}(s_i) M_{cd}(s_{i+1}) M_{de}(s_{i+2}) \dots , \qquad (33)$$

where ..., a, b, c, d, e,... are 2-state variables. They also showed that ground states of this kind exist also for a two-dimensional $S = 3/2$ quantum spin system. Fannes *et al.* generalized the above wave function (33) by assigning m degrees of freedom to ..., a, b, c, d, e,.... Their variational state is known as 'finitely correlated state,' since the correlation length is always finite [38,39]. Although (33) does not look like (32), they are essentially the same; they are related via a duality transformation. Such a product state has been considered independently in the field of classical diffusion models [40–43].

The variational states in (31)-(33) are uniform. The advantage of the variational state in DMRG ((24) and (27)) is that it allows for a position dependence of the matrices. Because of this, it is possible to treat finite-size systems in the framework of DMRG.

7 Finite-System Algorithm

The variational state $v^{(N)}(s_1 \dots s_N)$ in (24) obtained by the infinite-system algorithm is a good approximation for the eigenvector of $T^{(N)}$, however, it

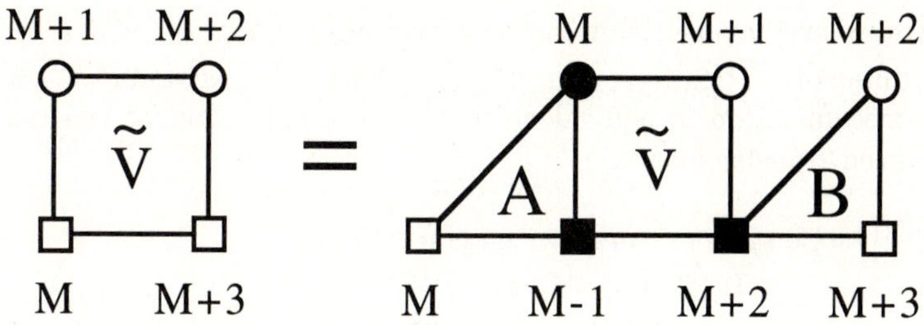

Fig. 7. Position shift of the renormalized state vector $\tilde{v}^{(N)}$ according to (35)-(36).

is not the best one. The finite-system algorithm improves the state, so that $\langle v^{(N)}|T^{(N)}|v^{(N)}\rangle$ is maximized under the constraint $\langle v^{(N)}|v^{(N)}\rangle = 1$, keeping at most m states for each block variable.

The algorithm gradually improves the whole variational state

$$\sum_{\{\xi\}\{\zeta\}} \prod_{i=2}^{M-1} A(\xi_{i-1}s_i|\xi_i)\; \tilde{v}^{(N)}(\xi_{M-1}s_M s_{M+1}\zeta_{M+2}) \prod_{j=M+2}^{N-1} B(s_j\zeta_{j+1}|\zeta_j)\,. \tag{34}$$

by directly improving the vector $\tilde{v}^{(N)}$, and indirectly improving other parts by shifting the position of $\tilde{v}^{(N)}$. Assuming that we have already obtained all the matrices shown in (34), we explain the following numerical procedures of the finite-system algorithm of the classical DMRG; the main point is the position shift for $\tilde{v}^{(N)}$ [44,45]. Let us create the DSM

$$\begin{aligned}&\tilde{\rho}_L(\xi'_{M-1}s'_M|\xi_{M-1}s_M)\\ &\quad = \sum_{s_{M+1}\zeta_{M+2}} \tilde{v}^{(N)}(\xi'_{M-1}s'_M s_{M+1}\zeta_{M+2})\, \tilde{v}^{(N)}(\xi_{M-1}s_M s_{M+1}\zeta_{M+2})\,,\end{aligned} \tag{35}$$

and obtain $A(\xi_{M-1}s_M|\xi_M)$ by diagonalizing it. Then, by shifting the position of $\tilde{v}^{(N)}$ as shown in Fig. 7

$$\begin{aligned}&\tilde{v}^{(N)}(\xi_M s_{M+1}s_{M+2}\zeta_{M+3}) =\\ &\quad \sum_{s'_M\xi'_{M-1}\zeta'_{M+2}} \tilde{A}(\xi'_{M-1}s'_M|\xi_M)\tilde{v}^{(N)}(\xi'_{M-1}s'_M s_{M+1}\zeta'_{M+2})\tilde{B}(s_{M+2}\zeta_{M+3}|\zeta'_{M+2})\,,\end{aligned} \tag{36}$$

we can construct the new variational state

$$\begin{aligned}&v^{(N)}_{\text{new}}(s_1 \ldots s_N) =\\ &\quad \sum_{\{\xi\}\{\zeta\}} \prod_{i=2}^{M} A(\xi_{i-1}s_i|\xi_i)\; \tilde{v}^{(N)}(\xi_M s_{M+1}s_{M+2}\zeta_{M+3}) \prod_{j=M+3}^{N-1} B(s_j\zeta_{j+1}|\zeta_j)\,.\end{aligned} \tag{37}$$

Compare (37) with (34), the position of $\tilde{v}^{(N)}$ is shifted by one.

The finite-system algorithm can be viewed as maximizing $\langle v_{\rm new}^{(N)}|T^{(N)}|v_{\rm new}^{(N)}\rangle$ via the tuning of $\tilde{v}^{(N)}(\xi_M s_{M+1} s_{M+2} \zeta_{M+3})$, under the constraint $\langle v_{\rm new}^{(N)}|v_{\rm new}^{(N)}\rangle = 1$. This maximization is equivalent to the diagonalization of the (shifted) renormalized transfer matrix

$$\begin{aligned}\tilde{T}^{(N)}&(\xi'_M s'_{M+1} s'_{M+2} \zeta'_{M+3}|\xi_M s_{M+1} s_{M+2} \zeta_{M+3})\\ &= \tilde{T}_L(\xi'_M s'_{M+1}|\xi_M s_{M+1}) W(s'_{M+1} s'_{M+2}|s_{M+1} s_{M+2})\\ &\qquad \tilde{T}_R(s'_{M+2} \zeta'_{M+3}|s_{M+2} \zeta_{M+3}) \end{aligned} \tag{38}$$

to obtain its eigenvector $\tilde{v}^{(N)}(\xi_M s_{M+1} s_{M+2} \zeta_{M+3})$, where we already have $\tilde{T}_R(s'_{M+2} \zeta'_{M+3}|s_{M+2} \zeta_{M+3})$. The half-row transfer matrix $\tilde{T}_L(\xi'_M s'_{M+1}|\xi_M s_{M+1})$ can be easily obtained as

$$\begin{aligned}\sum_{\xi'_{M-1} s'_M \xi_{M-1} s_M} &A(\xi'_{M-1} s'_M|\xi'_M) \tilde{T}_L(\xi'_{M-1} s'_M|\xi_{M-1} s_M)\\ &W(s'_M s'_{M+1}|s_M s_{M+1}) A(\xi_{M-1} s_M|\xi_M) \end{aligned} \tag{39}$$

Thus, by choosing $\tilde{v}^{(N)}$ in (36) as the initial vector of the Lanczos diagonalization for $\tilde{T}^{(N)}$ in (38) [44], one can rapidly improve $\tilde{v}^{(N)}(\xi_M s_{M+1} s_{M+2} \zeta_{M+3})$.

In such a way the finite-system algorithm shifts the position of $\tilde{v}^{(N)}$ to an arbitrary place and improves the variational state site by site [47]. The numerical procedures are basically the same as those in the quantum DMRG explained in Chap. 2(I). It is interesting that such a local improvement (or update) is also used in the zero-temperature QMC simulation for fermionic systems [48].

The advantage of the finite-system algorithm, compared to the infinite one, is its high numerical precision in the calculated thermodynamic quantities and spin correlation functions. The precision is high enough to determine the critical exponents and minor corrections to the scaling hypothesis of a classical system with the help of finite-size scaling [49,50]. Examples are shown in Chaps. 3.2(II) and 3.3(II) by Carlon and Drzewinski [6–8].

8 Corner Transfer Matrix Formulation

We have treated the 2D Ising model on a cylinder, and applied the RG transformation to the row transfer matrix $T^{(N)}$. Remember that the block spin transformation is obtained from the diagonalization of the DSM, which corresponds to a cut in the cylinder (Fig. 3). The pictorial image of the DSM suggests that we can define a DSM for any system with arbitrary geometry, just by creating a cut in it. Baxter considered such a construction of the DSM more than 30 years ago [18,35]. He took a square cluster, and expressed the DSM of the system as the fourth power of the so called corner transfer matrix (CTM).

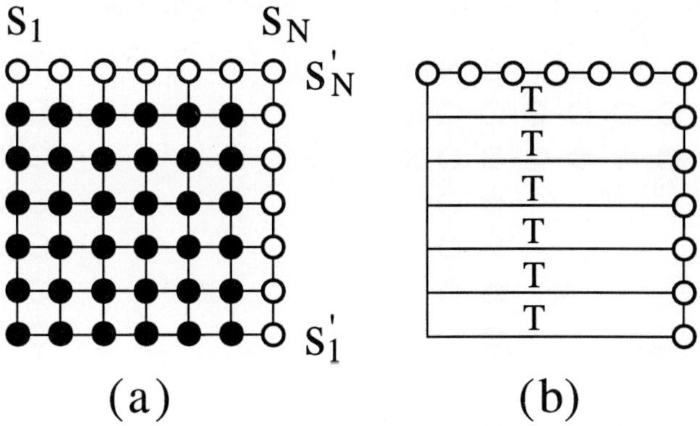

Fig. 8. Construction of the corner transfer matrix $C^{(N)}$, see (40). (**a**) The configuration sum is taken over the spins shown in black. (**b**) It is possible to regard $C^{(N)}$ as a stack of N copies of $T^{(N)}$ with appropriate boundary conditions.

The CTM represents the Boltzmann weight of a square system. Figure 8 (a) shows the corner transfer matrix of the Ising model

$$C^{(N)}(\boldsymbol{s}'|\boldsymbol{s}) = \sum_{\{s\}} \prod_{\langle ijkl \rangle} W(s_i s_j | s_k s_l) , \tag{40}$$

where $\langle ijkl \rangle$ represents the neighboring spins around a plaquette W, and the sum is taken over all the spins shown in black. By definition, $C^{(N)}(\boldsymbol{s}'|\boldsymbol{s})$ is block diagonal, because s'_N in $\boldsymbol{s}' = s'_1 ... s'_N$ and s_N in $\boldsymbol{s} = s_1 ... s_N$ are the same; in other words, $C^{(N)}(\boldsymbol{s}'|\boldsymbol{s})$ for $s'_N \neq s_N$ is always zero. It is also possible to construct $C^{(N)}$ by stacking N transfer matrices: $N \times T^{(N)}$. (See Fig. 8 (b).) The system is then enlarged by joining another row and column to the square. This is done by multiplying horizontal and vertical transfer matrices T and a plaquette W as shown in Fig. 9 (a). Formally this means $C^{(N)}(\boldsymbol{s}'|\boldsymbol{s}) \rightarrow C^{(N+1)}(\boldsymbol{s}'_{\rm new}|\boldsymbol{s}_{\rm new}) = C^{(N+1)}(s'_1 ... s'_{N+1}|s_1 ... s_{N+1})$ with

$$C^{(N+1)}(\boldsymbol{s}'_{\rm new}|\boldsymbol{s}_{\rm new}) = \sum_{\boldsymbol{s}'''\boldsymbol{s}''} \delta_{s'_{N+1} s_{N+1}} W(s_N s_{N+1} | s''_N s'_N) \tag{41}$$
$$T^{(N)}(\boldsymbol{s}'|\boldsymbol{s}''')T^{(N)}(\boldsymbol{s}|\boldsymbol{s}'')C^{(N)}(\boldsymbol{s}'''|\boldsymbol{s}'') .$$

At the same time we extend the length of $T^{(N)}$ by joining a plaquette W as shown in Fig. 9 (b) in order to prepare for the next extension $C^{(N+1)} \rightarrow C^{(N+2)}$. The extension $T^{(N)} \rightarrow T^{(N+1)}$ is essentially the same as (21).

Baxter constructed the DSM as the fourth power of the CTM

$$\rho_{\rm c}(\boldsymbol{s}''''|\boldsymbol{s}) = \sum_{\boldsymbol{s}'''\boldsymbol{s}''\boldsymbol{s}'} C^{(N)}(\boldsymbol{s}''''|\boldsymbol{s}''')C^{(N)}(\boldsymbol{s}'''|\boldsymbol{s}'')C^{(N)}(\boldsymbol{s}''|\boldsymbol{s}')C^{(N)}(\boldsymbol{s}'|\boldsymbol{s}) . \tag{42}$$

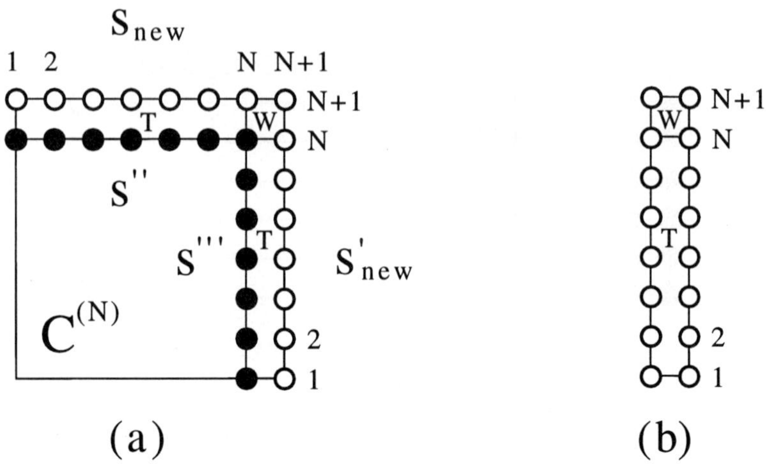

Fig. 9. Enlargement of the CTM: (a) $C^{(N+1)}$ is obtained by joining two $T^{(N)}$ and a W to $C^{(N)}$, see (41). (b) The extension $T^{(N)} \to T^{(N+1)}$, see (21).

This is not far from the conventional definition of DSM in DMRG (8), because $\rho_{\rm c}$ can be written as

$$\rho_{\rm c}(\boldsymbol{s}'_L|\boldsymbol{s}_L) = \sum_{\boldsymbol{s}_R} v(\boldsymbol{s}'_L|\boldsymbol{s}_R) v(\boldsymbol{s}_L|\boldsymbol{s}_R) , \tag{43}$$

where the vector $v(\boldsymbol{s}_L|\boldsymbol{s}_R)$ is given by

$$v(\boldsymbol{s}_L|\boldsymbol{s}_R) = \sum_{\boldsymbol{s}} C^{(N)}(\boldsymbol{s}_L|\boldsymbol{s}) C^{(N)}(\boldsymbol{s}|\boldsymbol{s}_R) . \tag{44}$$

Note that $v(\boldsymbol{s}_L|\boldsymbol{s}_R)$ is block diagonal, as $C^{(N)}(\boldsymbol{s}'|\boldsymbol{s})$ is. Figure 10 shows the graphical representation of $\rho_{\rm c}$, that corresponds to a $2N-1$ by $2N-1$ square with a cut, where the cut extends from an edge to the center. Thus $Z^{(2N-1)} = \mathrm{Tr}\, \rho_{\rm c}$ is the partition function of the $2N-1$ by $2N-1$ square system.

Using the CTM, Baxter calculated the variational free energy (per site) of 2D lattice models in the limit $N \to \infty$ by solving a self-consistent equation. Okunishi pointed out that Baxter's method is basically the same as the infinite-system algorithm of the classical DMRG [51]. Compared to the classical DMRG, Baxter's method has the advantage that the numerical calculation is very fast, since it does not require diagonalizations of large-scale matrices. Nishino and Okunishi introduced this advantage into the DMRG and formulated a numerical RG algorithm, which is called 'corner transfer matrix renormalization group' (CTMRG) [52,53].

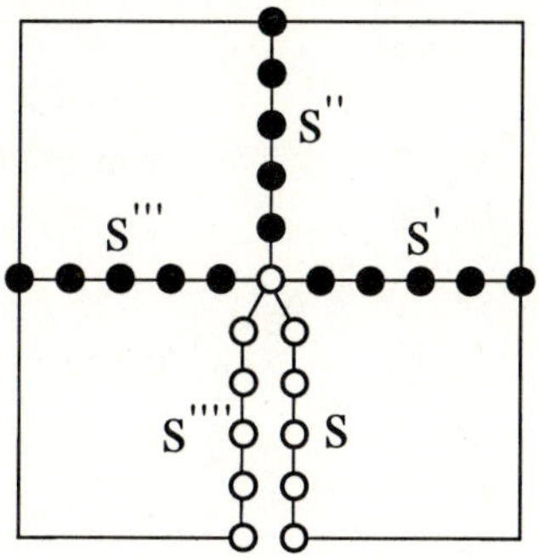

Fig. 10. Baxter's construction of the DSM as the fourth power of a CTM (42).

The outline of the CTMRG is as follows. The diagonalization of the DSM

$$\rho_c(\boldsymbol{s}'|\boldsymbol{s}) = \sum_{\xi} A(\boldsymbol{s}'|\xi)\, \alpha_\xi^4\, A(\boldsymbol{s}|\xi) \tag{45}$$

defines the RG transformation $\boldsymbol{s} \to \xi$ via the orthogonal matrix $A(\boldsymbol{s}|\xi)$. For example, $C^{(N)}$ can be renormalized as

$$\tilde{C}^{(N)}(\xi'|\xi) = \sum_{\boldsymbol{s}'\boldsymbol{s}} A(\boldsymbol{s}'|\xi')C^{(N)}(\boldsymbol{s}'|\boldsymbol{s})A(\boldsymbol{s}|\xi) = \delta_{\xi'\xi}\, \alpha_\xi\,, \tag{46}$$

where ξ' and ξ are m-state block variables. One thing one has to keep in mind is that ρ_c is block diagonal, as $C^{(N)}$ is, and therefore the block variable ξ implicitly includes the spin variable s_N, the central spin in Fig. 10. The variational partition function of the $2N-1$ by $2N-1$ square is then expressed as

$$\tilde{Z}^{(2N-1)} = \mathrm{Tr}\,\tilde{\rho}_c = \mathrm{Tr}\left(\tilde{C}_c^{(N)}\right)^4 = \sum_{\xi=1}^{m} \alpha_\xi^4\,. \tag{47}$$

As the infinite-system algorithm, the CTMRG consists of the successive mappings $\tilde{C}^{(N)} \to \tilde{C}^{(N+1)} \to \tilde{C}^{(N+2)}$ using the enlargement of the system (41) and the RG transformation (46). As a result, we obtain $\tilde{Z}^{(2N-1)} \to \tilde{Z}^{(2N+1)} \to \tilde{Z}^{(2N+3)} \to \cdots$ up to arbitrary system sizes, starting from a small N ($N = 2$ or 3).

Numerical data calculated by the CTMRG can be used for a finite-size scaling analysis of classical systems. Figure 11 shows the N-dependence of the local order parameter $\langle s\rangle$ of the Ising model at the center of a $(2N-1)$ by $(2N-1)$ square, when $T = T_c$. Fixed boundary conditions (all spins up at the boundary) are chosen. We plot representative data for both $m = 4$ and $m = 200$ up to $2N-1 = 19999$. At criticality, the local order parameter obeys the scaling formula

$$\langle s\rangle \propto (2N-1)^{-(d-2+\eta)/2}\,, \tag{48}$$

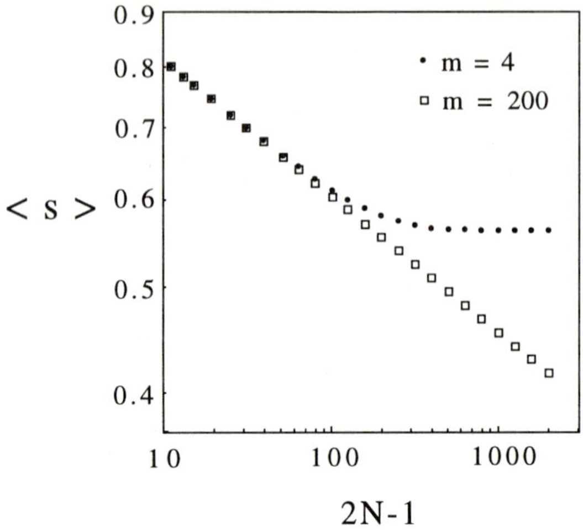

Fig. 11. Order parameter of the Ising model at the center of a $(2N-1)$ by $(2N-1)$ square when $T = T_c$.

where d is the spatial dimension ($d = 2$). Indeed, the calculated parameter $\langle s \rangle$ is almost proportional to $(2N-1)^{-1/8}$. The least-square fitting to the data in the range $199 \leq 2N-1 \leq 19999$ gives $\eta = 0.2501$, which is quite close to the exact one $\eta = 1/4$. In the same manner, we can determine another exponent ν from the nearest-neighbor spin correlation $E_{2N-1} = \langle s' s \rangle$ at the center of the $(2N-1)$ by $(2N-1)$ system, that obeys the scaling form

$$E_{2N-1} - E_\infty \propto (2N-1)^{1/\nu - d} . \tag{49}$$

From the calculated data for $199 \leq 2N-1 \leq 19999$, we obtain $\nu = 1.0006$. Again, the numerical result agrees with the exact exponent $\nu = 1$.

The CTMRG is also useful to detect a latent heat L. As an example, let us calculate L of the $q = 5$ Potts model [54], which shows a weak first-order transition. (It was impossible to determine L by Monte Carlo simulations because L is quite small.) Figure 12 shows the calculated local energy at the center of a $(2N-1)$ by $(2N-1)$ square system for both fixed (= ordered) and free (= disordered) boundary conditions, up to $2N-1 \leq 3999$ for $m = 40, 67$ and 200. Though there is a non-negligible m-dependence, it is clear that the model does not show a second-order transition. A double extrapolation with respect to N and m gives the latent heat $L = 0.027$ which agrees with the exact result $L \sim 0.0265$ [18,20]. This is the first quantitative numerical estimate of L for the $q = 5$ Potts model.

Finally we mention that Baxter's construction of the DSM can be generalized to any dimension [55–57]. For example, consider a cubic cluster of a 3D Ising model, and divide it into 8 subcubes. As a direct extension of the CTM in 2D, we can imagine 'corner tensors' $C^{(N)}(a|b|c)$ that correspond to the

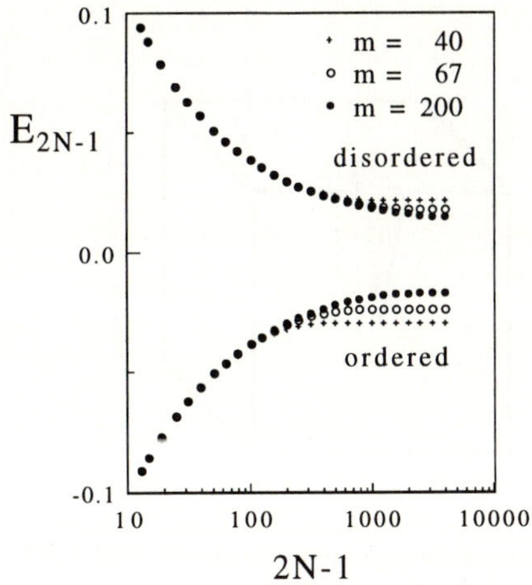

Fig. 12. Local energy E_{2N-1} of the $q = 5$ Potts model at the center of a $(2N - 1)$ by $(2N - 1)$ square, calculated for the ordered and the disordered phase. The zero of the energy has been shifted to the average of the two quantities [18,20].

Boltzmann weight for a subcube; the tensor indices $a = (a_{ij})$, $b = (b_{ij})$, and $c = (c_{ij})$ represent 2D spin arrays on the surfaces of the subcubes. Figure 13 shows the pictorial image of the DSM $\rho_{\text{cube}}(a|b)$, which is constructed as a contraction between 8 corner tensors. In 3D, the DSM corresponds to a cut in the cube. We obtain the RG transformation from the surfaces a, b, and c to m-state block spins by diagonalizing $\rho_{\text{cube}}(a|b)$. The enlargement of the corner cube can be performed via a 3D generalization of Fig. 9. Thus, as far as the formulation is concerned, we can extend CTMRG to 3D systems.

More generally, by breaking up a n-dimensional *hypercube* into 2^n-numbers of *hyper-corner-cubes*, we can define a DSM for a $(n-1)$-dimensional surface of the hyper-corner cube. Once we have obtained the DSM, we can define the block-spin transformation, and can apply RG transformations to the hyper-cubic system. Since the renormalized *hyper-corner tensor* has m^n numbers of elements, numerical calculations in higher dimension are much more difficult than in 2D. [58]

9 Discussion

We finally list several unsolved problems in classical DMRG.

- The matrix-product variational state for a finite-size system with periodic boundary conditions (= finite-size ring) should be translational invariant, however, the finite-system algorithm does not give such a translational invariant state [59].

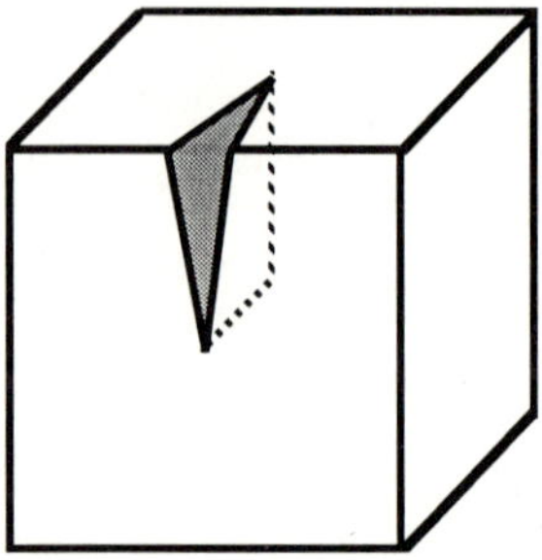

Fig. 13. The density matrix for a 3D cube, constructed from eight corner tensors. This is a generalization of Fig. 9.

- The way to apply classical DMRG to models with long-range interaction is not known; it is even difficult to define a transfer matrix for such models.
- It is not straightforward to treat random classical systems, because the transfer-matrix eigenvalue can be negative, and because the RG transformation depends on the position of block spins [60].
- To treat systems with continuous site variables or systems defined in continuous space (-time), we have to first discretize them [61]; we do not know the general principle for such a discretization that is suitable for DMRG.

We hope that these problems will be solved in the near future. [62]

The authors thank the organizers of the DMRG '98 workshop at the Max-Planck-Institut für Physik komplexer Systeme in Dresden. T. N. thanks I. Peschel, E. Carlon, G. Sierra, M. A. Martín-Delgado, X. Wang and T. Xiang for the discussions on classical DMRG. Numerical calculations were done using NEC SX-4 in the computer center of Osaka University. Y. H. is partly supported by the Sasakawa Scientific Research Grant from The Japan Science Society. K. O. and T. H. are partially supported by a Grant-in-Aid from the Ministry of Education, Science and Culture of Japan.

References

1. S.R. White, Phys. Rev. Lett. **69**, 2863 (1992); Phys. Rev. B **48**, 10345 (1993)
2. H.F. Trotter, Proc. Am. Math. Soc. **10**, 545 (1959)
3. M. Suzuki, Prog. Theor. Phys. **56**, 1454 (1976)
4. R.P. Feynmann and A. R. Hibbs, Quantum Mechanics and Path Integrals, McGraw-Hill (1965)

5. T. Nishino, J. Phys. Soc. Jpn. **64**, 3598 (1995)
6. E. Carlon and A. Drzewiński, Phys. Rev. Lett. **79**, (1997) 1591; Phys. Rev. E **57**, 2626 (1998)
7. E. Carlon and F. Igloi, Phys. Rev. B **57**, 7877 (1998); F. Igloi and E. Carlon, cond-mat/9805083
8. E. Carlon, A. Drzewiński and J. Rogiers, Phys. Rev. B **58**, 5070 (1998)
9. R.J. Bursill, T. Xiang and G.A. Gehring, J. Phys. Condensed Matter L583-L590 (1996)
10. X. Wang and T. Xiang, Phys. Rev. B **56**, 5061 (1997)
11. F. Naef, X. Wang, X. Zotos and W. van der Linden, cond-mat/9812117, to appear in Phys. Rev. B (1999)
12. N. Shibata, J. Phys. Soc. Jpn **66**, 2221 (1997)
13. B. Ammon, M. Troyer, T.M. Rice and N. Shibata, cond-mat/9812144
14. Y. Honda and T. Horiguchi, Phys. Rev. E **56**, 3920 (1997)
15. S. Östlund and S. Rommer, Phys. Rev. Lett **75**, 3537 (1995); S. Rommer and S. Östlund, Phys. Rev. B **55**, 2164 (1997); M. Andersson, M. Boman and S. Östlund, cond-mat/9810093
16. It is possible to fix the boundary spins to consider more general boundary conditions.
17. There is a bibliography of Ising in cond-mat/9605174.
18. R.J. Baxter, Exactly Solved Models in Statistical Mechanics, Academic Press, London, (1982)
19. R.B. Potts, Proc. Camb. Phil. Soc.**48**, 106
20. F.Y. Wu, Rev. Mod. Phys. **54**, 235 (1982) and references therein.
21. Baxter used another definition of the density submatrix in his variational method [18], where his density submatrix is block diagonal, see (42)-(44).
22. T. Nishino and K. Okunishi, in Strongly Correlated Magnetic and Superconducting Systems, Eds. G. Sierra and M.A. Martín-Delgado, Springer (1997)
23. Eigenvalues of the DSM can be negative when the system contains randomness.
24. I. Peschel, M. Kaulke and Ö. Legeza, Ann. Physik (Leipzig) **8**, 153 (1999); cond-mat/9810174
25. Strictly speaking, the density matrix eigenvalues do not decay exponentially; See K. Okunishi, Y. Hieida and Y. Akutsu, cond-mat/9810239
26. C. Lanczos: J. Res. Nat. Bur. Std. **45**, 255 (1950)
27. The Numerical recipes home page (`http://cfata2.harvard.edu/numerical-recipes/`) is useful to know about numerical linear algebra. Also it is worth reading J. Wilkinson, The Algebraic Eigenvalue Problem, Oxford (1965)
28. L. Onsager, Phys. Rev. **65**, 117 (1944)
29. H.A. Kramers and G.H. Wannier, Phys. Rev. **60**, 263 (1941)
30. R. Kikuchi, Phys. Rev. **81**, 988 (1951)
31. H.A. Bethe, Proc. Roy. Soc. A **150**, 552 (1935)
32. M.C. Gutzwiller, Phys. Rev. **137**, A1726 (1965)
33. J. Kanamori, J. Phys. Soc. Jpn. **30**, 275 (1963)
34. J. Hubbard, Proc. Roy. Soc. A **276**, 238 (1963); A **281**, 401 (1964)
35. R.J. Baxter, J. Math. Phys. **9**, 650 (1968); R.J. Baxter, J. Stat. Phys. **19**, 461 (1978)
36. N.P. Nightingale and H.W. Blöte: Phys. Rev. B **33**, 659 (1986)
37. I. Affleck, T. Kennedy, E.H. Lieb and H. Tasaki, Phys. Rev. Lett. **59**, 799 (1987)

38. M. Fannes, B. Nachtergale and R.F. Werner, Europhys. Lett. **10**, 633 (1989); M. Fannes, B. Nachtergale and R.F. Werner, Commun. Math. Phys. **144**, 443 (1992); M. Fannes, B. Nachtergale and R.F. Werner, Commun. Math. Phys. **174**, 477 (1995)
39. A. Klümper, A. Schadschneider and J. Zittarz, Z. Phys. B **87**, 281 (1992); H. Niggemann, A. Klümper and J. Zittartz, Z. Phys. B **104**, 103 (1997)
40. B. Derrida and M.R. Evans, J. Phys. A: Math. Gen. **26**, 1493 (1993)
41. N. Rajewsky, L. Santen, A. Schadschneider and M. Schreckenberg, cond-mat/9710316
42. A. Honecker and I. Peschel, J. Stat. Phys. **88**, 319 (1997)
43. Y. Hieida, J. Phys. Soc. Jpn. **67**, 369 (1998)
44. S.R. White, Phys. Rev. Lett. **77**, 3633 (1996)
45. T. Nishino and K. Okunishi, J. Phys. Soc. Jpn. **64**, 4084 (1995)
46. U. Schollwöck, Phys. Rev. B **58**, 8194 (1998)
47. A modification of the whole variational state is more time consuming. The situation is similar to the 'Order N' problem in the density functional formalism.
48. S.R. White, D.J. Scalapino, R.L. Sugar, E.Y. Loh, J.E. Gubernatis and R.T. Scalettar, Phys. Rev. B **40**, 506 (1989)
49. M.E. Fisher, in Proc. Int. School of Physics 'Enrico Fermi', Ed. M.S. Green, Academic Press, **51**, p 1 (1971)
50. M.N. Barber, in Phase Transitions and Critical Phenomena, Ed. C. Domb and J.L. Lebowitz, Academic Press, **8**, p. 146 (1983) and references therein.
51. K. Okunishi, Thesis, Osaka University 1996 (in Japanese); Thesis, Osaka University 1999 (in English). (Contact to `okunishi@godzilla.phys.sci.osaka-u.ac.jp`)
52. T. Nishino and K. Okunishi, J. Phys. Soc. Jpn. **65**, 891 (1996); T. Nishino and K. Okunishi, J. Phys. Soc. Jpn. **66**, 3040 (1997)
53. T. Nishino, K. Okunishi, and M. Kikuchi, Physics Letters A **213**, 69 (1996)
54. T. Nishino and K. Okunishi, J. Phys. Soc. Jpn. **67**, 1492 (1998)
55. T. Nishino and K. Okunishi, J. Phys. Soc. Jpn. **67**, 3066 (1998)
56. G. Sierra and M.A. Martín-Delgado, cond-mat/9811170.
57. Y. Hieida, K. Okunishi and Y. Akutsu, cond-mat/9901155
58. A way to decrease the computational effort in higher dimension is to perform RG transformations *before* creating the new RG transformations; it is possible, because the (infinite-system) DMRG is a self consistent method. [1,45]
59. The numerical precision in DMRG for a system with periodic boundary conditions is lower than that for a system with open boundary conditions. The reason can be understood by looking at the variational state written as a matrix product.
60. Position dependence *in the quantum Hamiltonian* can be treated by quantum DMRG; K. Hida, J. Phys. Soc. Jpn. **65**, 895 (1996)
61. S.G. Chung, J. Phys.: Cond. Matt. **9**, L619 (1997); Current Topics in Physics, Ed. Y.M. Cho, J.B. Hong, and C.N. Yang, World Scientific, **1**, p 295 (1998)
62. The authors will list up the latest information about DMRG in this URL: `http://quattro.phys.sci.kobe-u.ac.jp/dmrg.html`

6 Quantum Transfer-Matrix and Momentum-Space DMRG

Tao Xiang[1] and Xiaoqun Wang[2]

[1] Institute of Theoretical Physics, Academia Sinica
Beijing 100080, P. R. China
[2] Institut Romand de Recherche Numerique en Physique des Materiaux
PPH-EPFL, CH-1015 Lausanne, Switzerland

The quantum transfer-matrix and momentum-space density-matrix renormalization group (DMRG) are two important tools for studying physical properties of quantum lattice models. Their development has greatly extended the field of application of the DMRG. This chapter is devoted to the technical aspects of both methods, emphasizing those features which are not present in the standard DMRG discussed in Chap. 2(I). From a technical point of view, there is no direct connection between these two methods. Readers who are interested in just one of them need to read the relevant section only.

The DMRG is a systematical numerical method for studying highly correlated systems [1]. It uses a small number of optimized basis states, say m states, to accurately expand the ground state (or other targeted states). In conventional numerical renormalization-group methods, one keeps the m lowest energy eigenstates of a block Hamiltonian [2–4]. In the DMRG, however, one keeps the m most probable basis states of a subblock in the ground state of a superblock. The standard DMRG has been successfully applied to many quasi one-dimensional systems at zero temperature. A detailed discussion has been given by Noack and White in Chap. 2(I).

The quantum transfer-matrix DMRG (TMRG) is by far the most accurate method for studying thermodynamic properties of quasi one-dimensional (1D) quantum lattice models with short range interactions. This method was first explored by Bursill, Xiang and Gehring [5] based on the classical TMRG method discussed in Chap. 3(I) [6]. Wang and Xiang [7] then introduced a non-symmetric density matrix in the context of a non-symmetric transfer-matrix representation of the partition function. In Sect. 1, we will discuss this method in detail. It has now been successfully applied to several interesting quasi one-dimensional systems [8–18]. Some applications can also be found in Chaps. 4.1-4.4 (II). Recently, the TMRG was extended by Mutou *et al.* and Naef *et al.* independently for the investigation of dynamic properties at finite temperatures [19,20]. The corresponding discussions can be found in the next chapter and in Chap. 4.2 (II).

The momentum-space DMRG is an extension of the standard DMRG method [1]. The application of the method is still in its early stage. The key technical problem in the momentum-space DMRG was solved by Xiang in 1996 [21]. He introduced the concept of composite operators to reduce the number of independent operators whose matrix elements need to be stored and updated, from the order of N^3 to the order of N. This has made DMRG iteration in momentum space practically possible. The development of this method provides a solution to the application of the DMRG in two or higher dimensions. Section 2 will focus on the detailed technical aspects of this method. The concept of composite operators can be used in other DMRG applications, such as the quantum-chemistry calculations discussed in Chaps. 2.1(II) and 2.2(II), respectively.

1 Transfer-Matrix DMRG

In this section, we illustrate how the TMRG works using a periodic isotropic spin-$\frac{1}{2}$ Heisenberg chain described by

$$H = J \sum_i \mathbf{S}_i \cdot \mathbf{S}_{i+1} - BS_i^z, \tag{1}$$

where J is the exchange constant and B is an external magnetic field. We will first discuss the quantum transfer-matrix theory [23] and show that all thermodynamic quantities can be obtained from the largest and the second-largest eigenvalue and corresponding eigenvectors of the transfer matrix. We will then discuss the problems encountered in applying the DMRG to the quantum transfer matrix. We define a non-symmetric reduced density matrix and show that the optimized basis vectors retained in the truncation of the Hilbert space are the eigenvectors of this non-symmetric density matrix.

Quantum transfer-matrix theory

The quantum transfer-matrix method is based on a Trotter-Suzuki decomposition of the partition function, which maps a d-dimensional quantum system to a $(d+1)$-dimensional classical one [24,25]. This decomposition has been widely used in quantum Monte Carlo simulations. It was first introduced by Trotter in 1959 [24] and justified by Suzuki in 1976 [25].

The first step of the quantum transfer-matrix method is to decompose the Hamiltonian into two parts, $H = H_1 + H_2$, with each part a sum of commuting terms:

$$\begin{aligned} H_1 &= \sum_{i=odd} h_{i,i+1}, \\ H_2 &= \sum_{i=even} h_{i,i+1}, \end{aligned} \tag{2}$$

where

$$h_{i,i+1} = J\mathbf{S}_i \cdot \mathbf{S}_{i+1} - \frac{B}{2}\left(S_i^z + S_{i+1}^z\right), \tag{3}$$

for the Heisenberg model (1).

The TMRG uses the second-order approximation of the Trotter formula [26]

$$Z = \mathrm{Tr} e^{-\beta H} = \mathrm{Tr}\left(V_1 V_2\right)^M + o(\epsilon^2), \tag{4}$$

where $\epsilon = \beta/M$ and M is the Trotter number .

$$V_1 = e^{-\epsilon H_1} = \prod_{i=odd} e^{-\epsilon h_{i,i+1}} = \prod_{i=odd} v^{i,i+1}, \tag{5}$$

$$V_2 = e^{-\epsilon H_2} = \prod_{i=even} e^{-\epsilon h_{i,i+1}} = \prod_{i=even} v^{i,i+1}, \tag{6}$$

where $v^{i,i+1}$ are local transfer matrices defined by $v^{i,i+1} = e^{-\epsilon h_{i,i+1}}$. By inserting $2M$ times the identity

$$\sum |s^1 \cdots s^N\rangle\langle s^1 \cdots s^N| = 1$$

between the V_1 and V_2 operators in (4) and labelling successively the complete bases with $k \in [1, 2M]$ (so-called times slices), the partition function can then be expressed as

$$\begin{aligned} Z = \lim_{\epsilon\to 0} \sum_{\{s_k^i\}} &\langle s_1^1 \cdots s_1^N|V_1|s_2^1 \cdots s_2^N\rangle\langle s_2^1 \cdots s_2^N|V_2|s_3^1 \cdots s_3^N\rangle \cdots\cdots \\ &\langle s_{2M-1}^1 \cdots s_{2M-1}^N|V_1|s_{2M}^1 \cdots s_{2M}^N\rangle\langle s_{2M}^1 \cdots s_{2M}^N|V_2|s_1^1 \cdots s_1^N\rangle \\ = \lim_{\epsilon\to 0} \sum_{\{s_k^i\}} &\left(v_{1,2}^{1,2} v_{1,2}^{3,4} \cdots v_{1,2}^{N-1,N}\right)\left(v_{2,3}^{2,3} v_{2,3}^{4,5} \cdots v_{2,3}^{N,1}\right) \cdots\cdots \\ &\left(v_{2M-1,2M}^{1,2} \cdots v_{2M-1,2M}^{N-1,N}\right)\left(v_{2M,1}^{2,3} \cdots v_{2M,1}^{N,1}\right), \end{aligned} \tag{7}$$

where $v_{k,k+1}^{i,i+1} = \langle s_k^i s_k^{i+1}|v^{i,i+1}|s_{k+1}^i s_{k+1}^{i+1}\rangle$ represents the matrix element of $v^{i,i+1}$. The superscripts i and subscripts k for s and v (subsequently σ and τ below) represent the spin coordinates in the space and Trotter directions, respectively. Equation (7) can be interpreted as an *evolution* of the product of V_1 and V_2 *in the Trotter direction*, as shown graphically in Fig. 1(a).

In the quantum transfer-matrix theory, the partition function is reformulated in terms of transfer matrices, *evolving along the spatial direction*, as illustrated in Fig. 1(b). If we collect all $v_{k,k+1}^{i,i+1}$ with the same $(i, i+1)$, then the partition function becomes

$$Z = \sum_{\{s_i^k\}} \left(v_{1,2}^{1,2} v_{3,4}^{1,2} \cdots v_{2M-1,2M}^{1,2}\right)\left(v_{2,3}^{2,3} v_{4,5}^{2,3} \cdots v_{2M,1}^{2,3}\right) \cdots\cdots$$

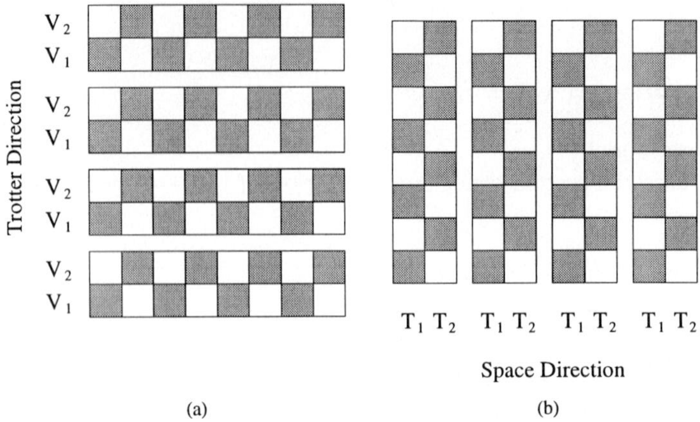

Fig. 1. Schematic representation of the partition function. Shaded squares denote local matrices $v^{i,i+1}$ in (**a**) and $\tau_{k,k+1}$ in (**b**).

$$
\begin{aligned}
&\left(v_{1,2}^{N-1,N} v_{3,4}^{N-1,N} \cdots v_{2M-1,2M}^{N-1,N}\right)\left(v_{2,3}^{N,1} v_{4,5}^{N,1} \cdots v_{2M,1}^{N,1}\right) \\
&= \mathrm{Tr}\left[\left(\tau_{1,2}\tau_{3,4}\cdots\tau_{2M-1,2M}\right)\left(\tau_{2,3}\tau_{4,5}\cdots\tau_{2M,1}\right)\right]^{N/2} \\
&= \mathrm{Tr}\mathcal{T}_M^{N/2}, \qquad (8)
\end{aligned}
$$

where $\mathcal{T}_M = T_1 T_2$ is called a quantum transfer matrix [23], T_1 and T_2 are products of local transfer matrices:

$$T_1 = \tau_{1,2}\tau_{3,4}\cdots\tau_{2M-1,2M}, \qquad (9)$$

$$T_2 = \tau_{2,3}\tau_{4,5}\cdots\tau_{2M,1}. \qquad (10)$$

The local transfer matrix $\tau_{k,k+1}$ is defined through the matrix $v^{i,i+1}$:

$$\tau(\sigma_k^i \sigma_{k+1}^i | \sigma_k^{i+1} \sigma_{k+1}^{i+1}) \equiv \langle s_k^i, s_k^{i+1} | v^{i,i+1} | s_{k+1}^i, s_{k+1}^{i+1} \rangle, \qquad (11)$$

where $\sigma_k^i \equiv (-1)^{i+k} s_k^i$. σ_k^i differs from s_k^i only by a phase factor and is introduced for later convenience.

The transfer matrix $\mathcal{T}_M$ defined above is not symmetric. In the case where both T_1 and T_2 are symmetric and semi-positive (i.e. all eigenvalues of T_1 and T_2 are larger than or equal to zero), the transfer matrix $\mathcal{T}_M$ can be symmetrized. To see this more clearly, we rewrite (8) as

$$Z = \lim_{M\to\infty} \mathrm{Tr}\left(T_1^{1/2} T_2 T_1^{1/2}\right)^{N/2}. \qquad (12)$$

Since $T_1^{1/2}$ is space symmetric when T_1 is symmetric and semi-positive, $\mathcal{T}_M^S = T_1^{1/2} T_2 T_1^{1/2}$ is a real symmetric transfer matrix. This symmetrized transfer matrix, however, may not be of any practical use in the DMRG computations since it is a product of three matrices and the computer memory space and CPU time needed for storing and updating the matrix elements of $\mathcal{T}_M^S$ is much larger than for a non-symmetric transfer matrix. For a spin-S Heisenberg model with either ferromagnetic or antiferromagnetic coupling, the transfer matrix can be symmetrized in the absence of an external field. But in general T_1 and T_2 are neither symmetric nor semi-positive and the transfer matrix cannot be symmetrized.

Free energy: The free energy per site can be obtained directly from the partition function. In the thermodynamic limit, it is given by

$$\begin{aligned} F &= -\lim_{N\to\infty}\frac{1}{\beta N}\ln Z = -\lim_{N\to\infty}\lim_{\epsilon\to 0}\frac{1}{\beta N}\ln \mathrm{Tr}\mathcal{T}_M^{N/2} \\ &= -\lim_{N\to\infty}\lim_{\epsilon\to 0}\frac{1}{\beta N}\ln\sum_l \langle\psi_l^L|\mathcal{T}_M^{N/2}|\psi_l^R\rangle \\ &= -\lim_{N\to\infty}\lim_{\epsilon\to 0}\frac{1}{\beta N}\ln\lambda_{max}^{N/2}\sum_l(\lambda_l/\lambda_{max})^{N/2}, \\ &= -\lim_{\epsilon\to 0}\frac{\ln\lambda_{max}}{2\beta} \end{aligned} \tag{13}$$

where λ_l is the eigenvalue of $\mathcal{T}_M$, and $|\psi_l^R\rangle$ and $\langle\psi_l^L|$ are the corresponding right and left eigenvectors, respectively. We use λ_{max} to denote the largest eigenvalue of $\mathcal{T}_M$. The normalization condition $\langle\psi_l^L|\psi_l^R\rangle = 1$ is assumed hereafter. In obtaining the above formula, the property that the order of the limits $N \to \infty$ and $\epsilon \to 0$ is interchangeable [27,28] is used. The free energy is determined solely by the *largest eigenvalue* of the quantum transfer matrix $\mathcal{T}_M$. In principle, one can obtain all thermodynamic quantities from various derivatives of F, i.e. only the largest eigenvalue of $\mathcal{T}_M$ is needed to study the full thermodynamics of a system. This is an important property in the transfer-matrix theory, which allows us to adopt the idea of the density-matrix renormalization to approximately find λ_{max}.

Conservation law: The local Hamiltonian $h_{i,i+1}$ conserves the total spin at sites i and $i+1$, i.e. $s_k^i + s_k^{i+1} = s_{k+1}^i + s_{k+1}^{i+1}$. In terms of the spins σ introduced before, the same relation holds if one moves in the space direction: $\sigma_k^i + \sigma_{k+1}^i = \sigma_k^{i+1} + \sigma_{k+1}^{i+1}$. From this local conservation law it can be shown that the total spin, $\sum_k \sigma_k^i$, is also conserved in the Trotter space: $\sum_k \sigma_k^i = \sum_k \sigma_k^{i+1}$. Thus $\mathcal{T}_M$ is block-diagonal according to the value of $\sum_k \sigma_i^k$. For the spin-1/2 Heisenberg model, it was shown rigorously that the largest eigenvector is nondegenerate and lies in the $\sum \sigma_i^k = 0$ subspace, irrespective of the sign of J and the value of B [29]. For larger spins, it was found numerically that the largest eigenvectors of $\mathcal{T}_M$ are also in the $\sum_k \sigma_k^i = 0$ subspace [13], but a rigorous proof for this is still not available.

The above discussion can be readily extended to an interacting electron model, for example the Hubbard model. In this case charge is conserved in addition to the spin conservation. If we use $n^i_{k\sigma}$ to denote the number of electrons with spin σ at site i and time slice k, then the spin and charge conservation laws for this system are given by the equation $n^i_{k\sigma} + n^{i+1}_{k\sigma} = n^i_{k+1\sigma} + n^{i+1}_{k+1\sigma}$ in the real space. In the Trotter space, the corresponding conservation law can be expressed as $\tilde{n}^k_{i\sigma} + \tilde{n}^{k+1}_{i\sigma} = \tilde{n}^k_{i+1\sigma} + \tilde{n}^{k+1}_{i+1\sigma}$, where $\tilde{n}^i_{k\sigma} = (-1)^{i+k}(n^i_{k\sigma} - 1/2)$. When $i+k$ is odd, this transformation between n^i_σ and $\tilde{n}^i_\sigma$ is equivalent to a particle-hole transformation: $c^k_{i\sigma} \leftrightarrow (c^k_{i\sigma})^\dagger$.

Form of local transfer matrix τ: Before we discuss the properties of thermodynamic quantities, let us illustrate the symmetry property of the local transfer matrix explicitly using the spin-$\frac{1}{2}$ Heisenberg model (1). To obtain $\tau_{1,2}$, we first calculate the matrix elements of $v^{1,2}$ by diagonalizing the local Hamiltonian $h_{1,2}$. If the basis states of the two spins are $(|\uparrow\uparrow\rangle, |\uparrow\downarrow\rangle, |\downarrow\uparrow\rangle, |\downarrow\downarrow\rangle)$, then it is simple to show that

$$v^{1,2} = \begin{pmatrix} e^{-(\frac{J}{4}-\frac{B}{2})\epsilon} & 0 & 0 & 0 \\ 0 & e^{\frac{J\epsilon}{4}}\cosh\frac{J\epsilon}{2} & -e^{\frac{J\epsilon}{4}}\sinh\frac{J\epsilon}{2} & 0 \\ 0 & -e^{\frac{J\epsilon}{4}}\sinh\frac{J\epsilon}{2} & e^{\frac{J\epsilon}{4}}\cosh\frac{J\epsilon}{2} & 0 \\ 0 & 0 & 0 & e^{-(\frac{J}{4}+\frac{B}{2})\epsilon} \end{pmatrix}, \tag{14}$$

$v^{1,2}$ is a block-diagonal and symmetric matrix. Substituting this equation into (11), we then obtain

$$\tau_{1,2} = \begin{pmatrix} -e^{\frac{J\epsilon}{4}}\sinh\frac{J\epsilon}{2} & 0 & 0 & 0 \\ 0 & e^{\frac{J\epsilon}{4}}\cosh\frac{J\epsilon}{2} & e^{-(\frac{J}{4}-\frac{B}{2})\epsilon} & 0 \\ 0 & e^{-(\frac{J}{4}+\frac{B}{2})\epsilon} & e^{\frac{J\epsilon}{4}}\cosh\frac{J\epsilon}{2} & 0 \\ 0 & 0 & 0 & -e^{\frac{J\epsilon}{4}}\sinh\frac{J\epsilon}{2} \end{pmatrix}. \tag{15}$$

One can immediately see that $\tau_{1,2}$ is still block-diagonal. But it is asymmetric when B is non-zero.

Thermal average of local quantities: As mentioned above, all thermodynamic quantities can be calculated from the derivatives of F. However, for the internal energy or any other quantity which can be written as a sum of local operators, we can evaluate the thermal average of these quantities directly from the largest eigenvalue and the corresponding eigenvectors $\{|\psi^R_{max}\rangle, \langle\psi^L_{max}|\}$ of $\mathcal{T}_M$. This can eliminate the error resulting from the numerical derivatives.

Let us consider the thermal average of a local operator, $A_{1,2}$ which is defined at sites 1 and 2 and *commutes with* the local Hamiltonian $h_{1,2}$. Following the steps leading to (7), it can be shown that

$$\langle A_{1,2}\rangle = \lim_{N\to\infty} \frac{1}{Z}\mathrm{Tr}\left(A_{1,2}e^{-\beta H}\right)$$

$$= \lim_{N\to\infty} \lim_{\epsilon\to 0} \frac{1}{Z} \mathrm{Tr} \left(\mathcal{T}_M(A_{1,2}) \mathcal{T}_M^{N/2-1} \right)$$
$$= \lim_{\epsilon\to 0} \frac{\langle \psi^L_{max} | \mathcal{T}_M(A_{1,2}) | \psi^R_{max} \rangle}{\lambda_{max}} \tag{16}$$

where

$$\mathcal{T}_M(A_{1,2}) = (\tilde{\tau}_{1,2}(A_{1,2}) \tau_{3,4} \cdots \tau_{2M-1,2M})\, \mathcal{T}_2, \tag{17}$$

and the elements of $\tilde{\tau}_{1,2}(A_{1,2})$ are given by

$$\langle \sigma^1_1 \sigma^1_2 | \tilde{\tau}_{1,2}(A_{1,2}) | \sigma^2_1 \sigma^2_2 \rangle \equiv \langle s^1_1, s^2_1 | A_{1,2} v^{1,2} | s^1_2, s^2_2 \rangle. \tag{18}$$

The internal energy is the thermal average of the Hamiltonian H. Using the translation symmetry, the internal energy per site is equal to the thermal average of $h_{1,2}$. Thus

$$U = \langle h_{1,2} \rangle = \lim_{\epsilon\to 0} \frac{\langle \psi^L_{max} | \mathcal{T}_M(h_{1,2}) | \psi^R_{max} \rangle}{\lambda_{max}} \tag{19}$$

Similarly the uniform magnetization is given by

$$M_z = \lim_{N\to\infty} \frac{1}{N} \langle \sum_i S^z_i \rangle = \lim_{\epsilon\to 0} \frac{\langle \psi^L_{max} | \mathcal{T}_M(\frac{S^z_1 + S^z_2}{2}) | \psi^R_{max} \rangle}{\lambda_{max}} \tag{20}$$

Specific heat and spin susceptibility: From the results for U and M_z, one can evaluate the specific heat

$$C_v = \frac{\partial U}{\partial T}, \tag{21}$$

and the magnetic susceptibility

$$\chi = \frac{\partial M_z}{\partial B}, \tag{22}$$

by numerical derivatives. The magnetic susceptibility is simply given by M_z/B for sufficiently small magnetic field since $M_z = 0$ at $B = 0$. Therefore, the numerical errors for χ at $B = 0$ are generally smaller than those at $B \neq 0$ or those for the specific heat.

Correlation length: The correlation length at finite temperatures can also be calculated from the eigenvalues of $\mathcal{T}_M$. To do this let us consider the correlation function of the fluctuation of a quantity around its mean value, $\delta A_i = A_i - \langle A_i \rangle$, between sites 1 and $r+1$. For even r, the thermal average of $\delta A_1 \delta A_{r+1}$ in the thermodynamic limit $N \to \infty$ is found to be

$$\langle \delta A_1 \delta A_{r+1} \rangle = \frac{\langle \psi^L_{max} | \mathcal{T}_M(\delta A_1) \mathcal{T}_M^{r/2-1} \mathcal{T}_M(\delta A_{r+1}) | \psi^R_{max} \rangle}{\lambda_{max}^{r/2+1}}$$
$$= \sum_l \frac{\langle \psi^L_{max} | \mathcal{T}_M(\delta A_1) | \psi^R_l \rangle \langle \psi^L_l | \mathcal{T}_M(\delta A_{r+1}) | \psi^R_{max} \rangle}{\lambda_{max} \lambda_l} \left(\frac{\lambda_l}{\lambda_{max}} \right)^{r/2}. \tag{23}$$

In the limit $r \to \infty$, the above formula becomes

$$\begin{aligned}&\lim_{r\to\infty} \langle \delta A_1 \delta A_{r+1}\rangle \\ &= \lim_{\epsilon\to 0} \frac{\langle \psi^L_{max}|\mathcal{T}_M(\delta A_1)|\psi^R_\alpha\rangle\langle\psi^L_\alpha|\mathcal{T}_M(\delta A_{r+1})|\psi^R_{max}\rangle}{\lambda_{max}\lambda_\alpha} e^{-(\xi^{-1}-i\kappa)r}, \end{aligned} \tag{24}$$

where $\lambda_\alpha = |\lambda_\alpha| e^{2i\kappa}$ is the largest eigenvalue of $\mathcal{T}_M$ satisfying the condition

$$\langle \psi^L_{max}|\mathcal{T}_M(\delta A_1)|\psi^R_\alpha\rangle\langle\psi^L_\alpha|\mathcal{T}_M(\delta A_{r+1})|\psi^R_{max}\rangle \neq 0. \tag{25}$$

If the vector $\mathcal{T}_M(\delta A_1)|\psi^R_{max}\rangle$ is in the same invariant subspace as $|\psi^R_{max}\rangle$ (for example, when $A_i = S^z_i$), then λ_α is usually the second largest eigenvalue of $\mathcal{T}_M$ in that subspace. If, on the other hand, the vector $\mathcal{T}_M(\delta A_1)|\psi^R_{max}\rangle$ is in a different subspace $|\psi^R_{max}\rangle$, (for example, when $A_i = S^x_i$), then λ_α is the largest eigenvalue of $\mathcal{T}_M$ in the subspace which contains $\mathcal{T}_M(\delta A_1)|\psi^R_{max}\rangle$.

In (24), ξ is the thermal correlation length determined by λ_{max} and λ_α

$$\xi^{-1} = \lim_{\epsilon\to 0} \frac{1}{2} \ln \left| \frac{\lambda_{max}}{\lambda_\alpha} \right| = -\lim_{r\to\infty} \frac{1}{r} \ln\langle \delta A_1 \delta A_{r+1}\rangle, \tag{26}$$

and κ is the characteristic wave vector of the most dominant fluctuations which is determined by the phase of λ_α

$$\kappa = \lim_{\epsilon\to 0} \frac{1}{2} \arg\left(\frac{\lambda_\alpha}{\lambda_{max}}\right) + n\pi \qquad (n = 0 \,\text{or}\, 1). \tag{27}$$

The value of κ is not uniquely determined because $\mathcal{T}_M$ crosses over two sites and cannot distinguish between κ and $\kappa + \pi$. This ambiguity in determining the value of κ can be removed by considering the behavior of κ at high temperature and the continuity of κ. At high temperature λ_α is generally real, κ is therefore either 0 or π. From physical considerations, one can readily determine whether κ is 0 or π.

DMRG implementation for $\mathcal{T}_M$

The transfer matrix $\mathcal{T}_M$, rotated by $\pi/2$ in the rest of this section, is shown again in Fig. 2. It corresponds to a one-dimensional quantum operator con-

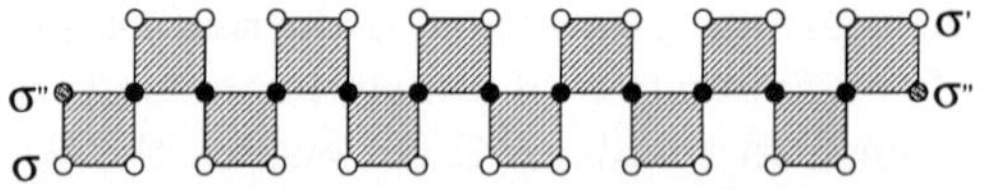

Fig. 2. One-dimensional structure of the transfer matrix $\mathcal{T}_M$ along the Trotter direction which is taken horizontal here. $\mathcal{T}_M$ is formed by connecting local transfer matrices τ at solid circles. At each solid circle, the common states of two τ are summed over. Either the upper open circles or the lower ones individually construct the Hilbert space to represent $\mathcal{T}_M$. Two shaded circles represent the same states σ'' to be summed over to form a transfer-matrix ring.

sisting of the local transfer matrices. As such it can be treated as the usual quantum spin chains. There is a difference, however, to the usual Hamiltonians which are sums of local terms. In $\mathcal{T}_M$, the local transfer matrices are connected at the sites shown as solid circles in Fig. 2 and one has several possibilities of separating $\mathcal{T}_M$ into two parts and enlarging them in the density-matrix renormalization procedure. In each iteration of the infinite-system algorithm, two time slices are added, i.e. M is increased by one. It turns out that the temperature is lowered as $T = 1/\epsilon M$ for given ϵ. Moreover the asymmetry of the transfer matrix $\mathcal{T}_M$ intrinsically also leads to a non-symmetric reduced density matrix. Below we discuss these aspects.

Superblock I: This construction, slightly different from the one proposed for systems which are spatial reflection symmetric [7], is generally suitable for (quasi) one-dimensional systems with local interactions. The two typical configurations of the superblock are shown in Fig. 3. The superblock consists of the two blocks in the dashed frames, which we call renormalized blocks, plus two additional time slices. The system contains a renormalized block and one slice on the right. The rest is thus its environment. We use n_s and n_e to label

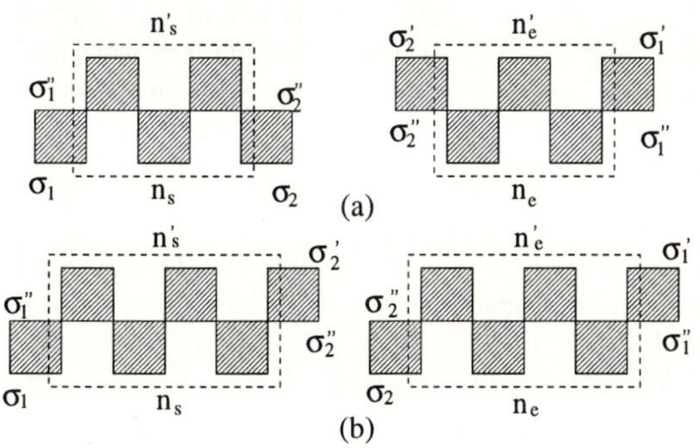

Fig. 3. Configurations of the superblock: (**a**) $M = 5$ (odd) and (**b**) $M = 6$ (even). The left and right transfer matrices are connected by summing over the internal states σ_1'' and σ_2'' at slices σ_1 and σ_2 to form a periodic local transfer matrix chain for $\mathcal{T}_M$.

the basis states of the renormalized blocks in the system and the environment, respectively. The states of two slices are represented by σ_1, σ_1'', σ_1' and σ_2 σ_2'', σ_2' correspondingly. Then the elements of the left transfer matrices are denoted as $\mathcal{S}_o(\sigma_1'', n_s', \sigma_2''; \sigma_1, n_s, \sigma_2)$ or $\mathcal{S}_e(\sigma_1', n_s', \sigma_2''; \sigma_1'', n_s, \sigma_2)$ and those of the right ones as $\mathcal{E}_o(\sigma_2', n_e', \sigma_1'; \sigma_2'', n_e, \sigma_1')$ or $\mathcal{E}_e(\sigma_2'', n_e', \sigma_1'; \sigma_2, n_e, \sigma_1'')$, where the subscripts o or e correspond to M being odd or even for $\mathcal{T}_M$. Now let us see how $\mathcal{S}_{e,o}$ and $\mathcal{E}_{e,o}$ can be constructed and renormalized in an iterative way.

•• Initially, $M = 2$ is *even*, the number of slices is $2M = 4$, and $\mathcal{T}_M$ contains four local transfer matrices τ. One then has $\mathcal{S}_e(\sigma_1', \sigma', \sigma_2''; \sigma_1'', \sigma, \sigma_2) = \sum_{\sigma''} \tau(\sigma_1', \sigma' | \sigma_1'', \sigma'')\, \tau(\sigma'', \sigma_2'' | \sigma, \sigma_2)$ and $\mathcal{E}_e = \mathcal{S}_e$.

•• When we construct $\mathcal{T}_M$ with an *odd* M, one slice is added to the right of the renormalized system block in $\mathcal{S}_e$ and one to the left of the renormalized environment block in $\mathcal{E}_e$. The elements of the enlarged matrices, respectively, are given by

$$\mathcal{S}_o(\sigma_1'', \tilde{n}_s', \sigma_2''; \sigma_1, \tilde{n}_s, \sigma_2) = \sum_{\sigma''} \mathcal{S}_e(\sigma_1', n_s', \sigma''; \sigma_1'', n_s, \sigma)\tau(\sigma'', \sigma_2''|\sigma, \sigma_2)$$
$$\mathcal{E}_o(\sigma_2', \tilde{n}_e', \sigma_1''; \sigma_2'', \tilde{n}_e, \sigma_1) = \sum_{\sigma''} \tau(\sigma_2', \sigma'|\sigma_2'', \sigma'')\mathcal{E}_e(\sigma'', n_s', \sigma_1''; \sigma, n_s, \sigma_1).$$

As a consequence, the basis states of the enlarged renormalized blocks become $|\tilde{n}_s\rangle = |n_s\rangle \otimes |\sigma\rangle$ and $|\tilde{n}_e\rangle = |\sigma\rangle \otimes |n_e\rangle$.

•• When $\mathcal{T}_M$ with *even* M is constructed, $\mathcal{S}_e$ and $\mathcal{E}_e$ can be considered to be identical. Then one can actually carry out the manipulations on $\mathcal{S}_e$ only. As illustrated in Fig. 4 (a), this involves the enlargenment via

$$\mathcal{S}_e(\sigma_1'', \tilde{n}_s', \sigma_2'; \sigma_1, \tilde{n}_s, \sigma_2'') = \sum_{\sigma''} \mathcal{S}_o(\sigma_1'', n_s', \sigma''; \sigma_1, n_s, \sigma)\tau(\sigma', \sigma_2'|\sigma'', \sigma_2'').$$

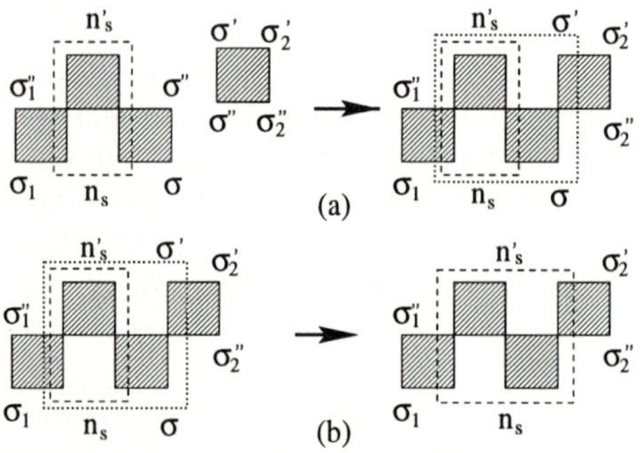

Fig. 4. (**a**) The renormalized block in the dashed frame is enlarged, by adding one slice $(\sigma, \sigma'', \sigma')$, into the one in the dotted frame and the basis $|n_s\rangle$ into $|\tilde{n}_s\rangle = |n_s\rangle \otimes |\sigma\rangle$. (**b**) The block in the dotted frame is further renormalized into the one in the dashed frame and $|\tilde{n}_s\rangle$ into $|n_s\rangle$

•• As soon as the number of states in $|\tilde{n}_\gamma\rangle$ $(\gamma = s, e)$ exceeds m, the enlarged transfer matrices $\mathcal{S}_{e,o}$ and $\mathcal{E}_o$ are renormalized by

$$\mathcal{A}_\alpha(n_\gamma', n_\gamma) = \sum_{\tilde{n}_\gamma', \tilde{n}_\gamma} O_\gamma^l(n_\gamma', \tilde{n}_\gamma')\mathcal{A}_\alpha(\tilde{n}_\gamma', \tilde{n}_\gamma)O_\gamma^r(\tilde{n}_\gamma, n_\gamma) \tag{28}$$

where $\mathcal{A}_\alpha(\tilde{n}_\gamma', \tilde{n}_\gamma)$ stand for the enlarged matrices: $\mathcal{S}_e(\sigma_1', \tilde{n}_s', \sigma_2''; \sigma_1'', \tilde{n}_s, \sigma_2)$, or $\mathcal{S}_o(\sigma_1'', \tilde{n}_s', \sigma_2''; \sigma_1, \tilde{n}_s, \sigma_2)$, or $\mathcal{S}_e(\sigma_2', \tilde{n}_e', \sigma_1'; \sigma_2'', \tilde{n}_e, \sigma_2)$ for $\alpha = e$ or o, while $\mathcal{A}_\alpha(n_\gamma', n_\gamma)$ with $n_\gamma, n_\gamma' \in [1, m]$ are the renormalized ones which are used to construct $\mathcal{T}_M$. An example for the renormalization of $\mathcal{S}_e$ is shown in Fig. 4(b). The transformation matrices $O_\gamma^{l,r}$ are constructed from the m largest left and right eigenvectors of the reduced density matrices for the system $(\gamma = s)$ or the environment $(\gamma = e)$. The reduced density matrices are represented in terms of both left and right eigenvectors of $\mathcal{T}_{M-1}$ and are discussed later on.

•• The transfer matrix $\mathcal{T}_M$ for the superblock is formally written as

$$\mathcal{T}_M(\sigma_1', n_e', \sigma_2', n_s'; \sigma_1, n_e, \sigma_2, n_s)$$
$$= \begin{cases} \sum_{\sigma_1'', \sigma_2''} \mathcal{S}_o(\sigma_1'', n_s', \sigma_2''; \sigma_1, n_s, \sigma_2) \times \mathcal{E}_o(\sigma_2', n_e', \sigma_1'; \sigma_2'', n_e, \sigma_1''), \\ \sum_{\sigma_1'', \sigma_2''} \mathcal{S}_e(\sigma_1'', n_s', \sigma_2'; \sigma_1, n_s, \sigma_2'') \times \mathcal{E}_e(\sigma_2'', n_e', \sigma_1'; \sigma_2, n_e, \sigma_1'') \end{cases} \tag{29}$$

for M odd or even as shown in Fig. (3).

In practical computations, one does not have to construct $\mathcal{T}_M$ explicitly. Rather, instead of direct matrix-vector manipulations between $\mathcal{T}_M$ and $|\psi_{max}\rangle$, one can successively operate with $\mathcal{S}_{e,o}$ and $\mathcal{E}_{e,o}$ on $|\psi_{max}\rangle$. In this case, a high computational performance can be achieved by an efficient optimization [31]. Moreover, one can also easily make a parallelization with respect to the indices of the states for two spins (σ_1'' and σ_2'') connecting left and right transfer matrices.

Superblock II: An alternative construction of superblocks was proposed in [13] to improve the numerical performance and is shown in Fig. 5(a).

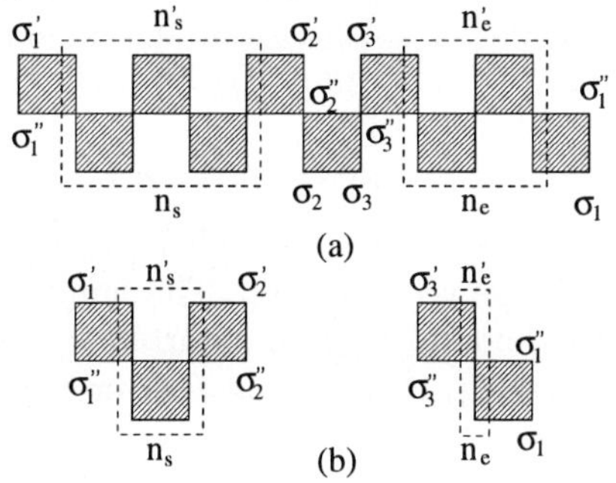

Fig. 5. An alternative construction of the superblock. (**a**) A superblock configuration for $M = 5$ is constructed in the basis $|\sigma_1\rangle \otimes |n_s\rangle \otimes |\sigma_2\rangle \otimes |\sigma_3\rangle \otimes |n_e\rangle$; (**b**) Initialization with $M = 3$.

•• Initially, the (renormlized) block in the system contains two slices and the one in the environment does only one slice as shown in Fig. 5(b). The renormalized system and environment blocks are connected by summing over the three additional spins σ_1'', σ_2'' and σ_3'' to form $\mathcal{T}_M$ as seen in Fig. 5(a).
•• In each iteration, M is increased by one and the number of slices in the two renormalized blocks always differs by one. Because of this, only the larger renormalized block is enlarged and then renormalized. This is a kind of alternative renormalization procedure between the renormalized system and environment blocks as illustrated in Fig. 6. Moreover, this transfer matrix

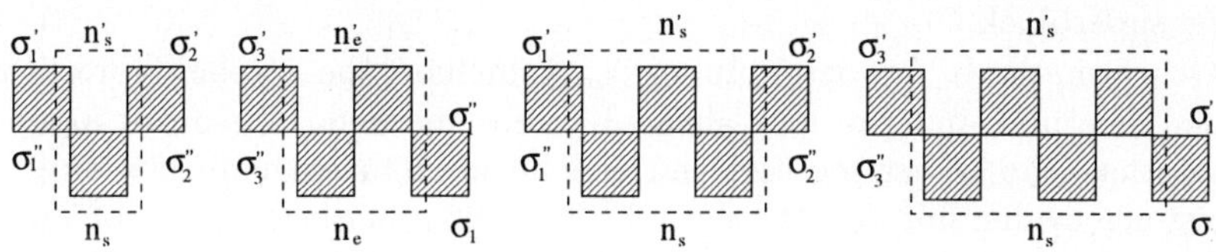

Fig. 6. An alternative renormalization procedure. From left to right: $\mathcal{S}_o(M = 3) \rightarrow \mathcal{E}_e(M = 4) \rightarrow \mathcal{S}_o(M = 5) \rightarrow \mathcal{E}_e(M = 6)$, ... The renormalized block on the left side of each arrow is enlarged and then renormalized to the one on the right side.

containing the larger renormalized block can also be directly taken as another transfer matrix for the construction of $\mathcal{T}_M$ in the next iteration, i.e. $\mathcal{S}_e(M) =$

$\mathcal{S}_o(M-1)$ and $\mathcal{E}_o(M) = \mathcal{E}_e(M-1)$. The corresponding bases are given by

$$|n_e(M)\rangle \leftarrow |\tilde{n}_e(M)\rangle = |n_s(M-1)\rangle \otimes |\sigma\rangle \tag{30}$$
$$|n_s(M)\rangle = |n_s(M-1)\rangle \tag{31}$$

for even M, and

$$|n_e(M)\rangle = |n_e(M-1)\rangle \tag{32}$$
$$|n_s(M)\rangle \leftarrow |\tilde{n}_s(M)\rangle = |n_e(M-1)\rangle \otimes |\sigma\rangle \tag{33}$$

for odd M. In the above expressions, the left arrows indicate renormalization.

The advantage of forming such a type of superblock is that the transfer matrix $\mathcal{T}_M$ in this case can always be factorized into a product of two sparse matrices which are block diagonal with respect to $|n_s\rangle$ and $|\sigma_3\rangle \otimes |n_e\rangle$, respectively. To treat these two sparse matrices instead of $\mathcal{T}_M$ itself allows one to save both computer memory space and CPU time.

Methods for calculating λ_{max}: There are several methods for computing λ_{max} and other extreme eigenvalues and eigenvectors of $\mathcal{T}_M$. The simplest one is the so-called power method. This method starts from an arbitrary initial vector $|\psi_0^R\rangle$ which is not orthogonal to $|\psi_{max}^R\rangle$. By applying $\mathcal{T}_M$ iteratively to the vector which is generated in the previous iteration, i.e. $|\psi_K^R\rangle = \mathcal{T}_M|\psi_{K-1}^R\rangle$, it will finally project out $|\psi_{max}^R\rangle = |\psi_K^R\rangle$ for sufficiently large K. The value of K needed depends on the ratio between the second-largest eigenvalue λ_2 and the largest eigenvalue $\eta = |\lambda_2/\lambda_{max}|$. The smaller η is, the smaller K is needed. If δ is the relative error required for λ_{max}, then K required is of order $\ln\delta/\ln\eta$. For the spin one-half Heisenberg model, we find that the value of K needed for producing an eigenvalue with a relative error less than 10^{-16} is generally less than 20, but it increases with decreasing temperature since η or the correlation length increases with decreasing temperatures. The left eigenvector $\langle\psi_{max}^L|$ can be similarly determined for both constructions of the superblock. When a Hamiltonian has a reflection symmetry, one can obtain the left eigenvector directly from the right one by a symmetric construction of the superblock [7].

Other methods for determining λ_{max} include the implicitly restarted Arnoldi method and the look-ahead Lanczos method. These are more efficient than the power method and can be used to evaluate not only the largest eigenvalue but also the second-largest or other eigenvalues of $\mathcal{T}_M$.

Both the implicitly restarted Arnoldi method and the look-ahead Lanczos method are well developed and tested. The source codes for both can be down-loaded freely from the internet [30].

Reduced density matrix: The density matrix can generally be defined as a semi-positive matrix whose trace is equal to the partition function up to a constant factor. For a Hamiltonian system, the thermodynamic density matrix is known to be

$$\rho_{th} = e^{-\beta H}. \tag{34}$$

The reduced density matrix for the system block is defined from ρ_{th} by integrating out all degrees of freedom of the environment block

$$\rho_{th,s} = \mathrm{Tr}_e e^{-\beta H}, \tag{35}$$

where s and e denote the system and environment blocks, respectively. It is straightforward to show that the traces of ρ_{th} and $\rho_{th,s}$ are equal to the partition function Z. At zero temperature, $\beta \to \infty$, $\rho_{th,s}$ is equal to $\mathrm{Tr}_e(|\psi_0\rangle\langle\psi_0|)$ up to a normalization constant, where $|\psi_0\rangle$ is the ground state eigenstate of the Hamiltonian.

In the Trotter space, the density matrix for the superblock can be defined as

$$\rho = \mathcal{T}_M^{N/2}. \tag{36}$$

This is an extension of the thermodynamic density matrix ρ_{th} in the Trotter space. Similarly, by integrating out all degrees of freedom of the environment, we define the reduced density matrix of the system block as

$$\rho_s = \mathrm{Tr}_e \mathcal{T}_M^{N/2}. \tag{37}$$

In the thermodynamic limit, ρ_s is equal to $\mathrm{Tr}_e(|\psi^R_{max}\rangle\langle\psi^L_{max}|)$ up to a normalization constant. This definition of the reduced density matrix is similar to that of $\rho_{th,s}$. But unlike $\rho_{th,s}$, ρ_s is non-symmetric. Physically, ρ_s should be a semi-positive matrix. If both $T_1$and T_2 are symmetric and T_1or T_2 is semi-positive, we can rigorously prove that all the eigenvalues of ρ_s are larger than or equal to zero. A general proof for the semi-positivity of ρ_s is still not available although we found numerically that the eigenvalues of ρ_s are semi-positive definite when the Trotter space is not truncated.

Suppose that ω_l is an eigenvalue of ρ_s and $|\phi^{R,L}_{s,l}\rangle$ are the corresponding right and left eigenvectors. Then one can write

$$\omega_l = \sum_j \langle\psi^L_{max}|\phi^R_{s,l}\rangle|\psi^R_{e,j}\rangle\langle\phi^L_{s,l}|\langle\psi^L_{e,j}|\psi^R_{max}\rangle, \tag{38}$$

where $\{|\psi^R_{e,j}\rangle, \langle\psi^L_{e,j}|\}$ is a complete set of basis vectors of the environment block. In obtaining (38), the normalization conditions, $\langle\phi^L_{s,l}|\phi^R_{s,l'}\rangle = \delta_{l,l'}$, $\langle\phi^L_{e,j}|\phi^R_{e,j'}\rangle = \delta_{j,j'}$, and $\langle\psi^L_{max}|\psi^R_{max}\rangle = 1$ are used. (38) indicates that ω_l is the projection probability of the state $|\phi^R_{s,l}\rangle\langle\phi^L_{s,l}|$ in $|\psi^R_{max}\rangle\langle\psi^R_{max}|$. Thus the optimal states which should be retained are the eigenvectors of the reduced density matrix ρ_s corresponding to the largest eigenvalues ω_l.

Variational principle: The zero-temperature DMRG method satisfies the variational principle because one is dealing with a hermitian operator H and works in a truncated Hilbert space. Thus the ground-state energy E_0 obtained using this method is always an upper bound for the true value, i.e. $E_0 \geq E_0^{\mathrm{True}}$. The systematic error for the ground-state energy decreases as the number of retained states increases. The truncation error is generally smaller than the true systematic error of the result.

At finite temperature, the TMRG method also uses a small set of basis vectors to approximate the full Hilbert space. We expect that the value of λ_{max} obtained by the TMRG is a lower bound to the true value, namely $\lambda_{max} \leq \lambda_{max}^{\mathrm{True}}$. Substituting this inequality into (13), we then have

$$F(T) \geq F^{\mathrm{True}}(T), \tag{39}$$

which can be taken as a generalized variational principle.

Numerical aspects

Reduced density matrix: In practical TMRG calculations, due to round-off errors and especially the truncation of the Trotter space, some eigenvalues of the reduced density matrices $\rho_{s,e}$ may become complex, which violates the semi-positivity of $\rho_{s,e}$. However, we found that the moduli of the complex eigenvalues are generally small and are supposed to be zero if no truncation was made. There exist some ways to cure this violation of the semi-positivity. Here we discuss an economic way which we have used extensively in the numerical calculations as follows.

First, since ρ_s is real, complex eigenvalues and corresponding eigenvectors always appear in pairs. For instance, let us consider

$$\begin{cases} \rho_s|\phi_{s,l}^R\rangle = \omega_l|\phi_{s,l}^R\rangle \\ \rho_s^T|\phi_{s,l}^L\rangle = \omega_l^*|\phi_{s,l}^L\rangle \end{cases} ; \quad \begin{cases} \rho_s|\phi_{s,l+1}^R\rangle = \omega_{l+1}|\phi_{s,l+1}^R\rangle \\ \rho_s^T|\phi_{s,l+1}^L\rangle = \omega_{l+1}^*|\phi_{s,l+1}^L\rangle \end{cases} , \tag{40}$$

where the superscripts T and $*$ represent the transpose and complex conjugate, respectively. When $\omega_{l+1} = \omega_l^*$, thus $|\phi_{s,l+1}^{R,L}\rangle = |\phi_{s,l}^{R,L}\rangle^*$.

Secondly, as we expected, the imaginary part is usually negligible. Then $|\phi_{s,l+1}^{L,R}\rangle$ and $|\phi_{s,l}^{L,R}\rangle$ can be regarded as two "degenerate" left/right eigenvectors corresponding to $\bar{\omega}_l = \mathrm{Re}\ \omega_l = \mathrm{Re}\ \omega_{l+1}$. In this sense, we can combine these two complex eigenvectors $|\phi_{s,l+1}^{L,R}\rangle$ and $|\phi_{s,l}^{L,R}\rangle$ into two real ones $|\bar{\phi}_{s,l+1}^{L,R}\rangle$ and $|\bar{\phi}_{s,l}^{L,R}\rangle$.

Finally, ρ_s is block diagonalized by a standard library routine such as DGEEV from the Lapack subroutines. We note that the complex eigenvector pair $|\phi_{s,l}^{R,L}\rangle$ output by the subroutine is generally normalized to *a complex number*, i.e.

$$\langle\phi_{s,l}^L|\phi_{s,l}^R\rangle = \langle\phi_{s,l+1}^L|\phi_{s,l+1}^R\rangle = \mathrm{e}^{i\alpha}. \tag{41}$$

We simply choose

$$\begin{cases} \bar{\phi}^L_{s,l} = \text{Re}\ \phi^L_{s,l} \\ \bar{\phi}^L_{s,l+1} = \text{Im}\ \phi^L_{s,l} \end{cases} ; \quad \begin{cases} \bar{\phi}^R_{s,l} = \text{Re}\ \mathrm{e}^{-i\alpha}\phi^R_{s,l} \\ \bar{\phi}^R_{s,l+1} = \text{Im}\ \mathrm{e}^{-i\alpha}\phi^R_{s,l} \end{cases} . \tag{42}$$

This procedure does not suffer from round-off errors destroying the orthogonality between the right and left basis of ρ_s and has been used for (frustrated) spin as well as electron systems. The above discussions for ρ_s are also valid for ρ_e which is defined in a similar way and has to be used for the first construction of $\mathcal{T}_M$ shown in Fig. 3.

Comparison to exact results: The spin one-half Heisenberg model in one dimension is integrable by Bethe ansatz. Many of its thermodynamic properties, for example the specific heat and the spin susceptibility, can be evaluated by solving the Bethe-ansatz equations. This model therefore provides a good opportunity for testing the TMRG method.

Figures 7 and 8 show the TMRG results for the specific heat C_v/T, the spin susceptibility χ, and the correlation length ξ of the spin one-half Heisenberg antiferromagnet at zero temperature. The exact results are also shown in these figures for comparison. The agreement between the TMRG results

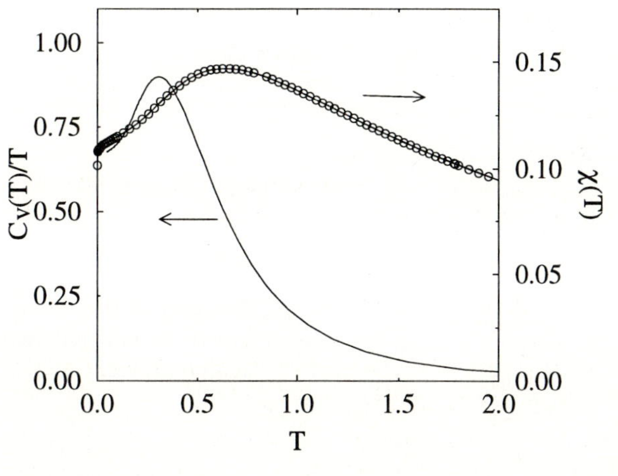

Fig. 7. Zero-field TMRG results for C_v/T and χ of the spin one-half Heisenberg model, calculated with $\epsilon = 0.1$ and $m = 81$. The circles are the exact results.

and the exact ones is apparently very good. For the results shown in these figures, $m = 81$ and $\epsilon = 0.1$ are used. The errors of the TMRG results are less than 10^{-4} even at very low temperatures. More accurate results can be obtained if more states are retained or if a systematical extrapolation with respect to both m and ϵ is taken.

A more detailed discussion in this respect can be found in [13].

Error analysis: In the TMRG calculation, we always deal with spatially infinite systems. This is different from zero-temperature and classical-system

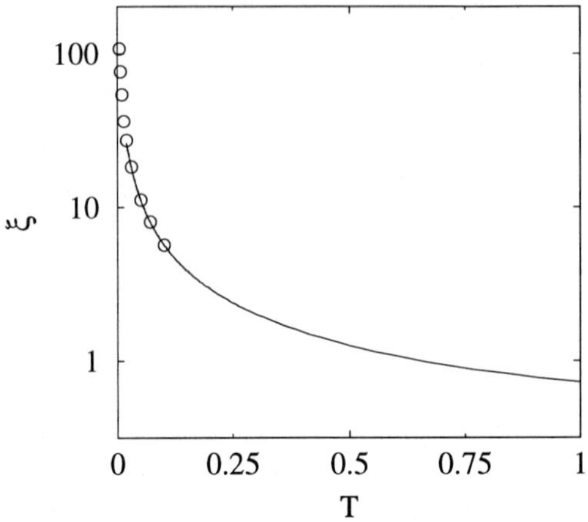

Fig. 8. TMRG result for the thermal correlation length of the spin one-half Heisenberg model at zero field, calculated with $\epsilon = 0.1$ and $m = 81$. The circles are thermal Bethe-ansatz results.

calculations. On the other hand, in addition to the error resulting from the basis truncation, there is a systematic error caused by the Trotter-Suzuki decomposition.

This error as shown in (4), is of the order ϵ^2. At high temperatures, or more precisely when the size M of superblock is large, the truncation error is very small, the error comes only from the Trotter-Suzuki approximation and varies as ϵ^2 for fixed m as shown in Fig. 9. At low temperature, the accumulated truncation error becomes important and the total error may not decrease with decreasing ϵ. This is also shown in Fig. 9.

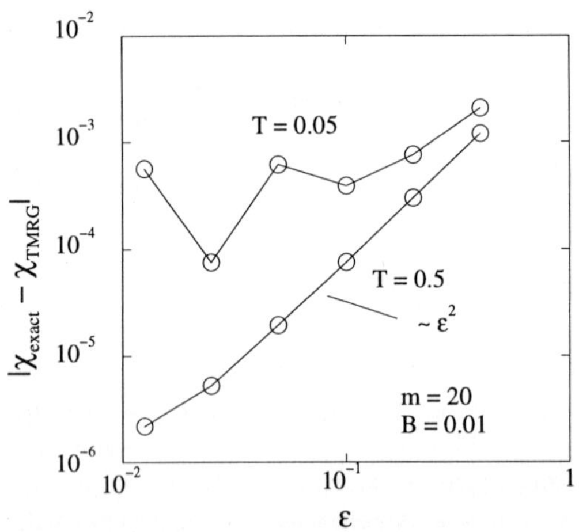

Fig. 9. Error in the susceptibility as a function of the parameter ϵ at two different temperatures.

In Fig. 10, the error is shown as a function of m. Given ϵ, we find that the systematical errors drop exponentially with increasing m initially and reach a value which is only determined by the Trotter-Suzuki approximation when m becomes large enough. This is similar to the zero-temperature DMRG method. Physically this is because at finite temperatures the correlation length is always finite and there is a gap between the largest eigenvalue and the next largest eigenvalues of the transfer matrix $\mathcal{T}_M$. If we assume that the ratio between the largest eigenvalue and the next largest eigenvalues of $\mathcal{T}_M$ is $\exp(\Delta_T)$, the error is expected to drop as $\exp(-\Delta_T m)$. Since Δ_T decreases with decreasing temperature, the systematical error drops faster at high temperature than at low temperature.

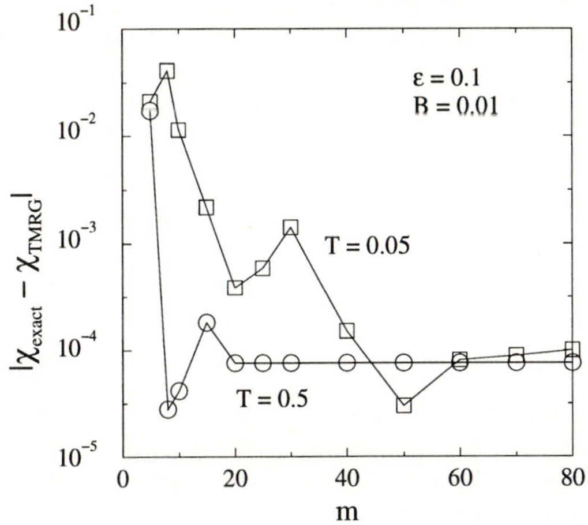

Fig. 10. Error in the susceptibility as a function of the number m of kept states at two different temperatures.

From the above discussion, we see that reducing the value of ϵ (or equivalently increasing the initial temperature at $2M = 4$) would reduce the error of the Trotter-Suzuki approximation, but the accumulated truncation error at low temperature will become large. Thus simply choosing a very small ϵ in the calculation would not improve the results at low temperatures.

The truncation errors for the second-largest eigenvalue of $\mathcal{T}_M$ are generally larger than those of λ_{max}. Thus the correlation lengths determined from (26) are generally expected to be less accurate than the free energy or other thermodynamic quantities.

2 Momentum-Space DMRG

The standard DMRG was originally proposed to treat quantum lattice models with short-range interactions in real space. It has been used successfully to investigate the low-energy properties of many quasi one-dimensional systems. However, its application in 2D is still not satisfactory. This has been the motivation for the extension of the DMRG method to momentum space. In principle, the standard DMRG can be applied directly in momentum space. However, as the interactions are non-diagonal in the momentum space and the DMRG iterations involve operator manipulations of the order N^3, a straightforward implementation is not possible for realistic calculations. In order to resolve this difficulty, so-called composite operators were introduced in [21].

Then the number of operators whose matrix elements are stored and updated is reduced from the order N^3 to the order N. This makes the DMRG iteration in momentum space practically possible. Since momentum is now a good quantum number, the matrices becomes more sparse and many more states can be kept in momentum space than in the real space.

In this section, we first use the Hubbard model to demonstrate how to apply the DMRG in momentum space. Later we will discuss how to extend the discussions to other interacting electron systems.

DMRG implementation

The Hubbard model is defined by the Hamiltonian

$$H = -t \sum_{<ij>\sigma} c_{i\sigma}^{\dagger} c_{j\sigma} + U \sum_{i} n_{i\uparrow} n_{i\downarrow}, \tag{43}$$

where $< ij >$ means summation over nearest neighbors. In momentum space, this model is

$$H = \sum_{k\sigma} \epsilon_k c_{k\sigma}^{\dagger} c_{k\sigma} + \frac{U}{N} \sum_{k_1,k_2,k_3} c_{k_1\uparrow}^{\dagger} c_{k_2\uparrow} c_{k_3\downarrow}^{\dagger} c_{k_1-k_2+k_3\downarrow}, \tag{44}$$

where ϵ_k is the energy dispersion of the electrons and periodic boundary conditions are assumed. With each point in k-space one can associate four degrees of freedom, namely the states $\{ \, |0\rangle, \, c_{k\uparrow}^{\dagger}|0\rangle, \, c_{k\downarrow}^{\dagger}|0\rangle, \, c_{k\uparrow}^{\dagger}c_{k\downarrow}^{\dagger}|0\rangle \, \}$, where $|0\rangle$ denotes the vacuum. A k-point with these four states can be taken as the basic unit. However, in calculations it is more convenient to treat the spin degree of freedom as an extra spatial coordinate and to consider a momentum-spin point $(k\sigma)$, which has only two degrees of freedom $\{|0\rangle, c_{k\sigma}^{\dagger}|0\rangle\}$, as a basic unit site. In this case, the spin rotation symmetry is not manifest, but the total number of degrees of freedom of a superblock is reduced and more states can be retained in the truncation of the Hilbert space. This can reduce truncation errors and save computer time.

As in real space, we define a superblock as an assembly of a system block A, an environment block B, and two unit sites $(k_1\sigma_1)$ and $(k_2\sigma_2)$. To carry out the DMRG iteration, we need to order the $(k\sigma)$ points. As all $(k\sigma)$ points are connected by interactions, the ordering of $(k\sigma)$ is not as important as in real space. However, the rate of convergence for the ground-state energy does depend on the ordering. A rule of thumb for the ordering of the $(k\sigma)$ points is that those $(k\sigma)$ pairs which have the strongest interaction should be arranged as close as possible in the ordered $(k\sigma)$ chain. In the Hubbard model, however, the interaction strength does not depend on k and one may choose any ordering.

The Hubbard interaction is local in real space but non-local in momentum space — it contains terms which link two or more k points in momentum space. The summation in the second term of (44) contains N^3 terms. To evaluate the matrix elements of the Hamiltonian, we need to evaluate the matrix elements of these N^3 terms. If one treated them separately, the calculation would be very difficult, if not impossible. However, after a careful analysis, one finds that the three-fold summations in the second term of H can actually be reduced to a summation of some terms, whose number is the order of N. To do this, one has to define the following composite operators in block A

$$\begin{aligned} a_0(p\sigma) &= c_{p\sigma}\delta_{(p\sigma)\in A}, \\ a_1(p\sigma) &= \sum_q a_0^\dagger(q\sigma)a_0(p+q\sigma), \\ a_2(p) &= \sum_q a_0^\dagger(q\uparrow)a_0(p+q\downarrow), \\ a_3(p\sigma) &= \sum_{q_1q_2} a_0^\dagger(q_1\bar{\sigma})a_0(q_2\bar{\sigma})a_0(p+q_1-q_2\sigma)\delta_{(p\sigma)\notin A}, \\ a_4(p) &= \sum_q a_0(q\downarrow)a_0(p-q\uparrow), \end{aligned} \tag{45}$$

where $\bar{\sigma} = -\sigma$. In block B, the corresponding operators can be obtained by changing (a, A) into (b, B). In each block, the total number of composite operators is $6N$.

To simplify the discussion below, we define $A' = A\oplus(k_1\sigma_1)$ and $B' = B\oplus(k_2\sigma_2)$. The Hamiltonian can formally be written as $H = H_{A'}+H_{B'}+H_{A'B'}$, where $H_{A'}$ (or $H_{B'}$) contains all interactions among electrons belonging to A' (or B'), and $H_{A'B'}$ represents the interaction between A' and B'. $H_{A'}$ can be iteratively obtained from H_A through the equation

$$H_{A'} = H_A + n_{k\sigma}\epsilon_k + \frac{U}{N}\left[n_{k\sigma}a_1(0,\bar{\sigma}) + c_{k\sigma}^\dagger a_3(k\sigma) + a_3^\dagger(k\sigma)c_{k\sigma}\right]. \tag{46}$$

In a DMRG iteration, we can take $H_{A'}$ (similarly $H_{B'}$) as a single operator to be stored and updated. The composite operators in A' can also be obtained from those in A via

$$\begin{aligned} a_1'(p\sigma') &= a_1(p\sigma') + \delta_{\sigma,\sigma'}\left[a_0^\dagger(k-p\sigma)c_{k\sigma} + c_{k\sigma}^\dagger a_0(k+p\sigma) + n_{k\sigma}\delta_{p,0}\right], \\ a_2'(p) &= a_2(p) + \delta_{\sigma,\uparrow}c_{k\uparrow}^\dagger a_0(k+p\downarrow) + \delta_{\sigma,\downarrow}a_0^\dagger(k-p,\uparrow)c_{k\downarrow}, \\ a_3'(p\sigma') &= a_3(p\sigma') + \delta_{\sigma,\sigma'}a_1(p-k\bar{\sigma})c_{k\sigma} \\ &\quad +\delta_{\sigma',\bar{\sigma}}\left[n_{k\sigma}a_0(p\bar{\sigma}) - \sigma c_{k\sigma}^\dagger a_4(p+k) - c_{k\sigma}\tilde{a}_2(p-k)\right], \\ a_4'(p) &= a_4(p) + \sigma a_0(p-k,\bar{\sigma})c_{k\sigma}, \end{aligned}$$

where $\tilde{a}_2(p) = a_2(p)$ if $\sigma =\uparrow$ and $a_2^\dagger(p)$ if $\sigma =\downarrow$. With the above definitions we can show that

$$\begin{aligned} H_{A'B'} = \frac{U}{N}\sum_p\{\sum_{\sigma'}[\frac{1}{4}a_1'(p\sigma')b_1'(-p\bar{\sigma}') + b_0'^\dagger(p\sigma')a_3'(p\sigma') \\ +(a' \longleftrightarrow b')] + a_4'^\dagger(p)b_4'(p) - b_2'(p)a_2'^\dagger(p)\} + h.c.. \end{aligned} \tag{47}$$

Thus, in the DMRG iterations, only the matrix elements of H_A, H_B and the composite operators in both subblocks need to be evaluated and stored.

The truncation of the Hilbert space in the DMRG procedure may cause the loss of the symmetries of the problem. For example, in real space DMRG, the momentum conservation is lost. In momentum space, however, this symmetry is preserved. Combined with the conservation of the number of up spins $N_\uparrow$ and the number of down spins $N_\downarrow$, the momentum conservation can be used to block-diagonalize the Hamiltonian. This does not only save computer time, but also allows one to keep many more eigenstates in the truncation as the number of non-zero matrix elements is now significantly reduced. Clearly the basis states in each block can be classified by three quantum numbers $(N_\uparrow, N_\downarrow, P)$, where P is the total momentum. For the composite operators defined in (45) it can be shown that their matrix elements are non-zero only when the difference between the momenta of the initial and the final state is p.

In the DMRG method, the basis states for both the system A and environment B are incomplete. If the interaction between A and B, i.e. H_{AB}, is non-zero, the matrix elements of H_{AB} will be less accurately approximated than the terms defined just in each subblock after the truncation of Hilbert space. In a one-dimensional-space calculation, one can choose A and B so that no interactions exist between A and B, i.e. $H_{AB} = 0$, if the interaction is short ranged. In real space in 2D or in momentum space in any dimension, however, the interaction between A and B is always non-zero, no matter how A and B are constructed. Thus in general results obtained by the DMRG method in these cases will not be as accurate as, for example, the ground-state energy of the 1D spin-one Heisenberg model [32].

Initialization

In momentum space, lattices of different size have different k points. Thus the size of the lattice has to be fixed at the beginning. This means that only the finite-system algorithm of the DMRG can be used in momentum space. To use the finite-size approach, however, one has to first build up a series of initial system and environment blocks. Since one cannot use the standard infinite-system algorithm, one has to look for a different procedure. A simple way is to use the conventional renormalization-group method [3,4]. The initial blocks C_n for system ($C = A$) as well as environment ($C = B$) are built up as follows:

1. Start from a small block C_1, which can be handled without truncation of basis states.
2. Add a new $(k\sigma)$ point to C_1 to form a new system block C_2.
3. Diagonalize the Hamiltonian in the Hilbert space spanned by C_2.
4. Truncate the Hilbert space to retain the m lowest energy eigenstates.
5. Replace C_2 by C_1 and repeat the steps 2-5 until all the initial system blocks required are established.

All $(N_\uparrow, N_\downarrow, P)$ subspaces that a block can have, are determined solely by the $(k\sigma)$ points the block contains. In the above initialization procedure, not all $(N_\uparrow, N_\downarrow, P)$ subspaces in a block need be considered. If a subspace with the quantum numbers $(N_{1,\uparrow}, N_{1,\downarrow}, P_1)$ in a system (environment) block cannot be matched to a subspace with quantum numbers $(N_{2,\uparrow}, N_{2,\downarrow}, P_2)$ in the corresponding environment (system) block such that $(N_{1,\uparrow} + N_{2,\uparrow}, N_{1,\downarrow} + N_{2,\downarrow}, P_1 + P_2)$ is equal to the quantum numbers of the targeted ground state, then this $(N_{1,\uparrow}, N_{1,\downarrow}, P_1)$ subspace in the system (environment) block will have no contribution to the ground state and can be discarded.

After the above initialization step, one can then use the finite-system algorithm of the DMRG method to find out the eigenvalue and eigenvector of the ground state for a given filling factor and momentum.

Results

In the following, we discuss the performance of the momentum-space DMRG comparing some preliminary results, for the Hubbard model, with those results from real-space DMRG results in one dimension and from various calculation in two dimensions.

Table 1 compares the ground-state energies obtained by the DMRG in momentum space as well as in real space at half-filling for 16 sites. For a

Table 1. Ground-state energies for a 16-sites Hubbard chain at half-filling, calculated by DMRG in momentum space and in real space

	Momentum space		Real space	
m	U=1	U=4	U=1	U=4
400	-1.02925	-0.51316	-1.02958	-0.575896
600	-1.02944	-0.53574	-1.02969	-0.575900
800	-1.02952	-0.53724	-1.02972	-0.575901
1000	-1.02958	-0.54562		
1200	-1.02959	-0.55218		

finite chain, no exact diagonalization results are available. When 800 states were kept in the real-space calculation, absolute errors for the ground states were found to be of the order of 10^{-6} for $U = 4$ as examined by varying the number of kept states. Relative to this results, an error for the ground-state energy obtained by the momentum-space DMRG with $m = 1000$ is 5% when $U = 4$. When U is not too small, we found that the ground-state energy is lower in real space than in momentum space if the same number of states is kept. This means that for the Hubbard model the DMRG method works better in real space than in momentum space in one dimension. This is not surprising because the Hubbard interaction is a local interaction and at half filling the electrons tend to be localized in space.

Table 2 compares the momentum-space results with those from the exact diagonalization and various approaches such as quantum Monte-Carlo calculations (QMC), exact cluster diagonalization (CD) and stochastic diagonalization (SD) for square lattices. The largest lattice calculated by using momentum-space DMRG had 12×12 sites. Comparison with the exact results for the 4×4 lattice shows that the DMRG results are very accurate when U is small. The relative error for the DMRG result with $U = 2$ is 3×10^{-4}. The

Table 2. Ground-state energies for Hubbard square lattices obtained by various methods: DMRG (m=1000), Exact (4×4), CD (6×6), QMC and SD with N electrons.

$L_x \times L_y$	U	N	Exact [33]/CD [34]	DMRG	QMC [35]	SD [35]
4×4	2	16	-18.01757	-18.012		
4×4	4	14	-15.74459	-15.673		
4×4	4	16	-13.62185	-13.571	-13.6	-13.59
4×4	8	16	-8.46887	-8.263	-8.48	
6×6	4	26	41.49	41.108	41.98	40.77
8×8	4	10		-34.325	-34.3	-34.31
8×8	4	18		-54.394	-54.6	-54.37
8×8	4	26		-66.098	-66.8	-66.05
12×12	4	18		-64.107		

momentum-space DMRG works better in the weak-coupling limit because the electron basis states used here are plane waves. When U=0, the ground state is a filled Fermi sea of non-interacting electrons, and the momentum-space method gives the exact ground-state energy even if only one state is kept. For large U, the DMRG results are not as good as in the weak-coupling limit. However, they are still comparable with those obtained by the quantum Monte Carlo methods. Both the DMRG and stochastic diagonalization methods satisfy the variational principle while the projected quantum Monte Carlo does not satisfy it due to the importance of sampling. For an 8×8 lattice, the ground-state energies obtained by the DMRG method are systematically lower (hence better) than the stochastic diagonalization results.

As shown above, the relative error for the ground-state energy obtained by the momentum-space DMRG with $m = 1000$ is 5%, when $U = 4$ and the chain length is 16. The value of this error is higher than that in two dimensions with the same value of U and lattice size. Therefore one can expect that the momentum-space DMRG works better for the Hubbard model in two dimensions than one. Physically this is because the relative contribution of the kinetic energy term to the ground-state energy, which is treated rigorously in the momentum-space DMRG method, compared with the potential energy, is larger in two dimensions than in one.

Extension to other models

The above discussion for the momentum-space DMRG has focused on the Hubbard model. However, it can be easily extended to several other physically interesting models, for example the periodic Anderson model and the interacting fermion model with nearest-neighbor Coulomb potentials. For the periodic lattice model, the composite operators defined in (45) can be used without modification. In other cases, one has to generalize the definitions of the composite operators in (45). For example, a_3 in (45) should in general be defined as

$$a_3(p\sigma) = \sum_{q_1 q_2 \sigma'} V(q_1 - q_2) a_0^\dagger(q_1\sigma') a_0(q_2\sigma') a_0(p + q_1 - q_2\sigma) \delta_{(p\sigma)\notin A} \quad (48)$$

if the interaction is $\sum V(q-q')c^\dagger_{q\sigma}c_{q'\sigma}c^\dagger_{q''+q'\sigma'}c_{q''+q\sigma'}$. For an arbitrary interacting fermion model with an unscreened Coulomb potential, the number of composite operators required is generally of order N^2, which will limit the application of the momentum-space method to small lattices. However, if $V(q-q')$ can be factorized as a sum of products of a function of q and a function of q', i.e. $V(q-q') = \sum_l^n f_l(q) g_l(q')$, it can be shown that the total number of composite operators required can be reduced again to the order of the system size N. The Hubbard interaction is obviously a factorizable potential, corresponding to $f_1 = g_1 = 1$ and $n = 1$. A nearest-neighbor Coulomb potential is also factorizable since $V(q-q') \sim \cos(q-q') = \cos q \cos q' + \sin q \sin q'$. For this potential, the total number of composite operators is $5N$ in one dimension and $7N$ in two dimensions if the fermions are spinless.

3 Prospects

In this chapter, we discussed two different extensions of the density-matrix renormalization idea. The transfer-matrix DMRG, which has already been applied successfully to several interesting (quasi) one-dimensional systems, shows clearly its potential for the reliable determination of thermodynamic properties for more and more challenging systems, such as a periodic Anderson model, two-chain t-J model and Hubbard model, etc.

The momentum-space DMRG approach, although still at an earlier stage of its application, already serves as a guide how to treat more complicated systems using the composite operators. Its potential will become clearer as one succeeds in calculating other physical quantities.

References

1. S.R. White, Phys. Rev. Lett. **69**, 2863 (1992)
2. K.G. Wilson, Rev. Mod. Phys. **47**, 773 (1975)

3. T. Xiang and G.A. Gehring, J. Magn. Magn. Mater. **104-107**, 861 (1992), Phys. Rev. B **48**, 303 (1993)
4. S.R. White, Phys. Rev. B **45**, 5752 (1992)
5. R.J. Bursill, T. Xiang and G.A. Gehring, J. Phys. Conden. Matter **8**, L583 (1996)
6. T. Nishino, J. Phys. Soc. Jpn. **64**, 3598 (1995)
7. X. Wang and T. Xiang, Phys. Rev. B **56**, 5061 (1997)
8. X. Wang, Thermodynamic properties of a $S = 1$ anisotropic chain, unpublished (1997)
9. N. Shibata, J. Phys. Soc. Jpn. **66**, 2221 (1997)
10. K. Maisinger and U. Schollwöck, Phys. Rev. Lett. **81**, 445 (1998)
11. N. Shibata, B. Ammon, M. Troyer, M. Sigrist, K. Ueda J. Phys. Soc. Jpn. **67**, 1086 (1998)
12. D. Coombes, T. Xiang, G.A. Gehring, J. Phys. C, **10**, L159 (1998)
13. T. Xiang, Phys. Rev. B **58**, 9142 (1998)
14. S. Eggert and S. Rommer, Phys. Rev. Lett. **81**, 1690 (1998)
15. K. Maisinger, U. Schollwöck, S. Brehmer, H.J. Mikeska and S. Yamamoto, Phys. Rev. B **58**, R5908 (1998)
16. S. Yamamoto, T. Fukui, K. Maisinger and U. Schollwöck, (1998), to appear in J. Phys. C
17. X. Wang and T. Xiang, Thermodynamic properties of the Hubbard model unpublished (1998)
18. A. Klümper, R. Raupach and F. Schönfeld, cond-mat/9809224
19. T. Mutou, N. Shibata and K. Ueda, Phys. Rev. Lett. **81**, 4939 (1998)
20. F. Naef, X. Wang, X. Zotos and W. von der Linden, cond-mat/9812117 and to appear in Phys. Rev. B (1999)
21. T. Xiang, Phys. Rev. B **53**, R10445 (1996)
22. S. R. White and R. L. Martin, cond-mat/9808118
23. e.g. M. Suzuki, Phys. Rev. B **31**, 2957 (1985); H. Betsuyaku, Prog. Theor. Phys. **73**, 319 (1985); I. Morgenstern and D. Würtz, Phys. Rev. B **32**, 532 (1985)
24. H. F. Trotter, Proc. Am . Math. Soc. **10**, 545 (1959)
25. M. Suzuki, Commun. Math. Phys. **51**, 183 (1976)
26. R. M. Fye, Phys. Rev. B **33**, 6271 (1986)
27. M. Suzuki, Phys. Rev. B **31**, 2957 (1985)
28. M. Suzuki and M. Inoue, Prog. Theor. Phys. **87**, 787 (1997)
29. T. Koma, Prog. Theor. Phys. **81**, 783 (1989)
30. see `http://www.caam.rice.edu/software/ARPACK` for the restarted Arnoldi subrountines and `http://netlib.uow.edu.au/linalg/lalqmr` for the look-ahead Lanczos method.
31. X. Wang and S. Mallwitz, Phys. Rev. B **48**, R492 (1996)
32. S.R. White and D.A. Huse, Phys. Rev. B **48**, 2844 (1993)
33. A. Parola, S. Sorella, S. Baroni, R. Car, M. Parrinello, and E. Tosatti, Physica C **162-164**, 771 (1989); G. Fano, F. Ortolani, and A. Parola, Phys. Rev. B **42**, 6877 (1990)
34. J. Riera and E. Dagotto, Phys. Rev. B **48**, 9515 (1993)
35. H.D. Raedt and M. Frick, Phys. Rep. **231**, 107 (1993); S. Sorella, S. Baroni, R. Car, and M. Parrinello, Europhys. Lett. **8**, 663 (1989)

7 Calculation of Dynamical Properties

Xiaoqun Wang[1], Karen Hallberg[2], and Felix Naef[1]

[1] Institut Romand de Recherche Numérique en Physique des Matériaux
PPH-EPFL, CH-1015 Lausanne, Switzerland
[2] Centro Atomico Bariloche
8400 Bariloche, Argentina

In the study of many-body systems, dynamical properties play an important rôle. They are measured e.g. in nuclear magnetic resonance, neutron scattering, optical absorption, photo-emission and other experiments. Their calculation, however, is a challenging task even in cases where a system is exactly solvable via Bethe Ansatz. It is therefore natural to apply the DMRG technique which has proved very efficient in obtaining static quantities also to this kind problem.

White's original idea [1], namely to renormalize operators and wave function using the eigenvectors and corresponding eigenvalues (called eigenpair in the following) of the density matrix has been extended to the renormalization of transfer matrices for classical statistical systems [2,3] as well as (quasi) one-dimensional quantum systems at finite temperature [4,5]. These techniques have been elucidated in great details in the previous chapters. This chapter is devoted to reliable calculations of the dynamic properties at zero and finite temperature within the DMRG framework and the transfer-matrix DMRG (TMRG) framework, respectively.

In the zero-temperature case, both ground state and higher excitations are relevant to dynamical properties, whereas all operators are usually represented in a reduced Hilbert space which is determined variationally by the optimal representation of a so-called target state (or several target states, see below). The target state is usually the lowest energy state for a given symmetry. There are two fundamental questions for the DMRG calculation of dynamic properties:

- To what extent does the density of states for the superblock, which is constructed by only targeting a few states, approach the true one for the full Hilbert space in a wide energy range?
- In order to describe dynamical properties reliably, what is a criterion to select more target states?

A full investigation of these two points is still in progress. On the other hand, since the main difference between the DMRG and exact diagonalization is the reduction of the Hilbert space, it is straightforward to evaluate any spectral function by using the Lanczos-vector method and the spectral method [6,7]. In Sect. 1, we will briefly discuss these aspects.

In the finite-temperature case, all physical quantities are evaluated by taking a (grand) canonical ensemble average in the thermodynamic limit. For (quasi) one-dimensional lattice models, this average is carried out using a quantum transfer matrix, as discussed in the last chapter. Consequently, one only needs to find the maximal eigenpair of the transfer matrix rather than all eigenpairs of the corresponding Hamiltonian. In this sense, the maximal eigenpair of the transfer matrix accounts for the relevant states of the Hamiltonian for given temperature and couplings. As will be discussed in detail below, local imaginary-time correlation functions can be evaluated as accurately as any other thermodynamic quantities. For non-local imaginary-time correlation functions, reasonably accurate results can be obtained by a proposed approximation. In principle, one can extract frequency-dependent quantities from imaginary-time correlation functions using the maximum entropy method. Nevertheless, we stress that the quality of frequency-dependent results is not as good as that of the source data for the imaginary-time correlation functions because of the intrinsically ill-conditioned inverse problem. In Sect. 2. we will discuss the TMRG calculation of the imaginary-time correlations and numerical aspects of analytical continuation [8].

1 Zero-Temperature Dynamics

Dynamical correlation functions

Let us first discuss how one can evaluate dynamic quantities for finite systems. For clarity, we give an outline with some relevant equations which are used in a realistic calculation for finite systems and will also be used in the DMRG schemes. The detailed derivations can be found in [9–11]. For a given Hamiltonian H and its ground-state wave function $|\psi_0\rangle$, the correlation function of a dynamical variable A at zero temperature is defined by

$$C(t-t') = \langle\psi_0|A^\dagger(t)A(t')|\psi_0\rangle, \tag{1}$$

where $A(t) = \mathrm{e}^{iHt}A\mathrm{e}^{-iHt}$ is the Heisenberg representation of A. Experiments usually measure directly the Fourier transform, i.e. the spectral function $C(\omega)$ given by

$$C(\omega) = \int_{-\infty}^{+\infty} dt\, \mathrm{e}^{\mathrm{i}\omega t}\langle\psi_0|A^\dagger(t)A(0)|\psi_0\rangle. \tag{2}$$

$C(\omega)$ can be evaluated in either the Lehmann spectral representation or the continued-fraction representation.

Lehmann spectral representation: If one fully diagonalizes the Hamiltonian H

$$H|\psi_n\rangle = E_n|\psi_n\rangle, \tag{3}$$

and uses all the eigenvectors as a complete basis in (2), one obtains the spectral function $C(\omega)$

$$C(\omega) = \sum_n |\langle\psi_n|A|\psi_0\rangle|^2\delta(\omega - E_n + E_0), \tag{4}$$

which therefore depends on both the weight $|\langle\psi_n|A|\psi_0\rangle|^2$ and the positions of the δ-function.

This method is limited by the dimensionality of the Hilbert subspace. With presently available computers, a full diagonalization of a sparse matrix with size $N \sim 5000 - 10000$ is feasible.

Continued-fraction representation: The spectral function $C(\omega)$ is associated with the corresponding Green's function $G(z)$ defined by

$$G(z) = \langle\psi_0|A^\dagger(z - H)^{-1}A|\psi_0\rangle. \tag{5}$$

via

$$C(\omega) = -\frac{1}{\pi}\lim_{\eta\to 0^+} \mathrm{Im}\,\mathrm{G}(\omega + \mathrm{i}\eta + \mathrm{E}_0). \tag{6}$$

The Green's function $G(z)$ can be expressed as a continued fraction. For this one uses $A|\psi_0\rangle = |f_0\rangle$ as an initial vector in a Lanczos procedure and thereby generates a set of orthogonal basis vectors via

$$|f_{n+1}\rangle = H|f_n\rangle - a_n|f_n\rangle - b_n^2|f_{n-1}\rangle \tag{7}$$

with $n \geq 0$, $b_0 = 0$ and $\langle f_n|f_{n'}\rangle = 0$ for $n \neq n'$. In this f_n basis H is tridiagonal and the coefficients a_n and b_n are obtained as:

$$a_n = \langle f_n|H|f_n\rangle/\langle f_n|f_n\rangle \tag{8}$$
$$b_n = \langle f_n|f_n\rangle/\langle f_{n-1}|f_{n-1}\rangle \tag{9}$$

Representing $(z - H)^{-1}$ in this basis leads to

$$G(z) = \cfrac{\langle\psi_0|A^\dagger A|\psi_0\rangle}{z - a_0 - \cfrac{b_1^2}{z - a_1 - \cfrac{b_2^2}{z - \cdots}}}. \tag{10}$$

In practice, the continued fraction often converges to a limit already for finite n (a few $|f_n\rangle$) since b_n goes to zero or a_n becomes very large. This fast convergence is due to the Lanczos tridiagonalization. It turns out that one can obtain accurate results for $C(\omega)$ with less CPU time than in the Lehmann spectral representation. In principle one can deal with larger systems, typically $N \sim 10^6 - 10^7$. However, one should remark that (i) spurious peaks can be introduced in the spectral function $C(\omega)$ when the orthogonality of the Lanczos basis $\{f_n\}$ is lost and (ii) the convergence crucially depends upon the level spacing of a given system.

DMRG schemes

As discussed in Chap. 2 (I), the standard DMRG technique provides a reduced Hilbert space, in which the Hamiltonian, wave functions and operators are represented. Since the DMRG technique usually treats much larger systems than the exact diagonalization, the level spacing is much smaller. The number of states in the reduced Hilbert space for the superblock is much smaller than that in the original one. More important, density-matrix renormalization keeps track of successive eigenvalues of the density matrix rather than successive eigenvalues of the Hamiltonian. In principle, it is straightforward to evaluate spectral functions using (4) or (6). As far as the excitation properties and feasibility are concerned, some additional considerations on the target procedures beyond the standard DMRG techniques are required.

Lanczos-vector method: This procedure is based on the continued fraction representation described above. In order to keep information on the excitations of the model, it was proposed in [6] that besides the ground state ψ_0, also several vectors of the tridiagonal Lanczos sequence of vectors $\{f_n\}$ given by (7) should be included as target states in DMRG calculations, since these vectors already describe the low-lying excitations with good accuracy. In principle, one could also consider higher excitations as target states. Then the reduced density matrix is given by

$$\rho_{ii'} = \sum_{l=1}^{M} p_l \sum_j \phi^l_{ij} \phi^l_{i'j} \tag{11}$$

where M is the total number of target states, i and j denote the states of the system and the environment, respectively, and p_l is a weight for the target state ϕ^l ($\sum_l p_l = 1$ in order to keep the unitary trace in ρ). In (11) the vectors ϕ^l are defined by

$$|\phi^l\rangle = \begin{cases} |\Psi_0\rangle & \text{if } l = 1, \\ \dfrac{|f_{l-2}\rangle}{\langle f_{l-2}|f_{l-2}\rangle} & \text{otherwise.} \end{cases} \tag{12}$$

It is expected that states in the reduced Hilbert space relevant to the excitation states connected to the ground state via the operator of interest A, are included in this set. It was shown in [13] that $|f_0\rangle$ is an excellent trial state, in particular for the lowest triplet excitations of the 2D antiferromagnet.

It was pointed out by White that if the number m of states kept per block is fixed, the more target states one considers, the less precisely each one of them will be described. In practice, an optimal number of target states M used to construct the reduced density matrix and the number m of kept states in the reduced density matrix have to be found for each case. Due to this restriction, the algorithm can be applied only up to certain lengths, depending on the states involved. For longer chains, the higher energy excitations will become inaccurate. Proper sum rules have to be calculated to determine the errors in each case. In the Lanczos-vector DMRG calculation, one needs a little extra computational effort compared to the standard DMRG calculations for static quantities. Moreover, this method can also be implemented directly in the momentum-space DMRG approach developed by Xiang [14].

As a test, the method has been applied to the antiferromagnetic spin one-half Heisenberg chain with periodic boundary conditions. One is usually interested in the spin correlation function

$$S^{zz}(q,\omega) = \int_{-\infty}^{+\infty} dt\, \mathrm{e}^{\mathrm{i}\omega t} \langle \psi_0 | S^z_{-q}(t) S^z_q | \psi_0 \rangle = C(q,\omega). \tag{13}$$

In Fig. 1 the spectral function $S(q,\omega)$ is shown at $q = \pi$ for $N = 24$ where exact results are available for comparison. In the evaluation of $C(q,\omega)$ using (6), the infinitesimal parameter η is chosen to be 0.01 which leads to a broadening of the peaks. This result illustrates how increasing m considerably improves the precision of results for the higher excitations.

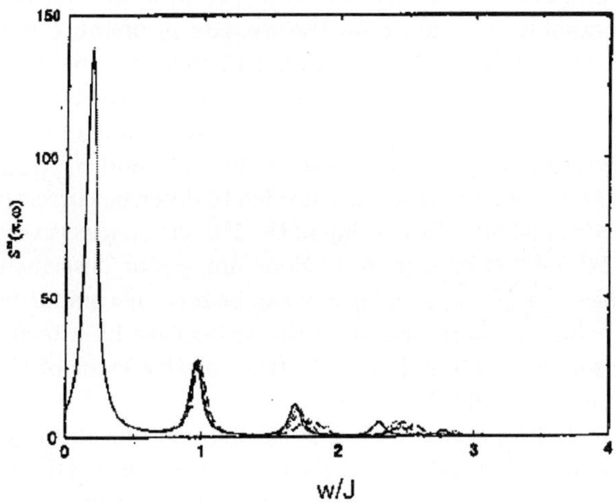

Fig. 1. Spectral function for a Heisenberg chain with $N = 24$ and $q = \pi$. Full line: exact result [12]. The rest is calculated using DMRG with $m = 100$ (long-dashed line), $m = 150$ (dashed line) and $m = 200$ (dotted line).

In Fig. 2, the spectral functions are shown for two sizes of the system and $q = \pi$ and $q = \pi/2$. For this case three target states, *i.e.* $|\psi_0\rangle$, $|f_0\rangle = S^z_\pi|\psi_0\rangle$

and $|f_1\rangle$ were used. Due to the truncation of the Hilbert space the spectrum shows some irregularities at high energies. We note that if one considers only the first (~ 10) coefficients a_n and b_n, the spectrum at low energies remains essentially unchanged. Minor differences arise at $\omega/J \simeq 2$. This is another indication that only the first $|f_n\rangle$ are relevant for the low-energy dynamical properties of finite systems.

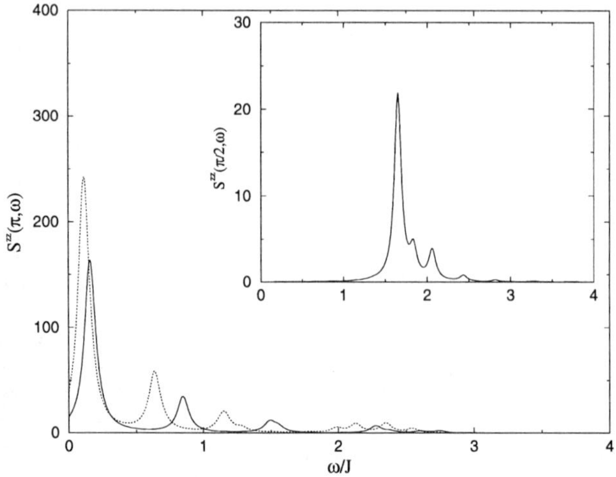

Fig. 2. Spectral densities for $q = \pi$, $N = 28$ (continuous line) and $N = 40$ (dotted line). Inset: Spectral density for $q = \pi/2$ for $N = 28$ ($\eta = 0.05$).

In the inset of Fig. 2 the spectrum for $q = \pi/2$ and $N = 28$ is shown. For this case five target states were used, *i.e.* $|\psi_0\rangle$, $|f_0\rangle = S^z_{\pi/2}|\psi_0\rangle$, $|f_n\rangle$ for $n = 1, 3$ and $m = 200$. Here, and for all the cases considered, it was verified that the results are very weakly dependent on the weights p_l of the target states. For lengths where $q = \pi/2$ is not an allowed momentum, we took the nearest value. The case $q = \pi/2$ is the most unfavourable one because it involves high-energy excitations. Although we have included some of these states as target states, the reduced Hilbert spaces related to $|\psi_0\rangle$ and $S^z_{\pi/2}|\psi_0\rangle$ have very small overlap and many more states are needed to describe correctly both target states. For $q = \pi$, instead, the overlap of the Hilbert spaces is very high and the target states and low-energy excitations are better described. The difference between the cases $q = \pi/2$ and $q = \pi$ can be seen by comparing the ground-state energy, which is more precise in the latter case by a factor of 3 in the relative error for $m = 200$ and $N = 28$ (the relative error of the ground state for the case $q = \pi$ is 10^{-6}).

Even though we are including states with a given momentum as target states, due to the particular real-space construction of the reduced Hilbert space, this translational symmetry is not present and the momentum is not fixed. To check how the reduction of the Hilbert space influences the momentum q of the target state $|f_0\rangle = S^z_q|\psi_0\rangle$, we calculated the overlap

$$Z = \langle\psi_0|S^z_{-q'}S^z_q|\psi_0\rangle \tag{14}$$

for different values of q'. For well-defined momenta of the states, this is proportional to $\delta_{q,q'}$ if $q \neq 0$. For $q = 0$, it vanishes since $S^z_{tot} = 0$. In Fig. 3, this overlap is shown for $q = \pi/2$ and different lengths with $S^z_{\pi/2}|\psi_0\rangle$ as one of the target states. As the system becomes larger (higher reduction of the Hilbert space), the q value becomes less defined, giving a wider distribution. The figure also shows a marked oscillation. This is due to the fact that the system is built from two identical blocks and that $|\langle\psi_0|S^z_\pi|\psi_0\rangle|$ is small but non-zero ($\simeq 10^{-3}$ for the largest system). These oscilations are expected to disappear when the finite-system algorithm is used.

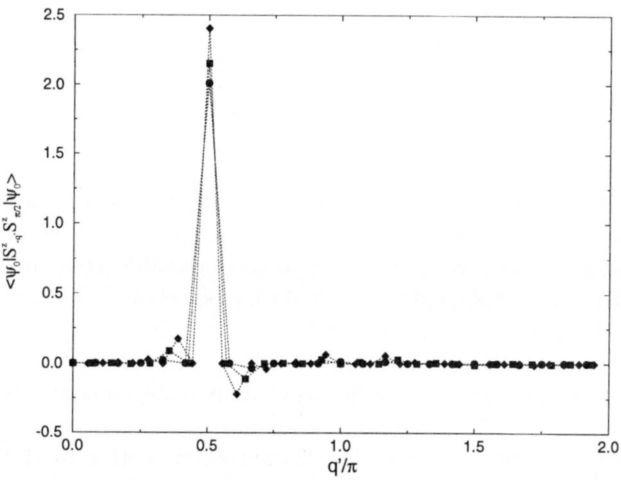

Fig. 3. Momentum weights (14) of a target state with $q = \pi/2$ and $N = 24$ (circles), $N = 28$ (squares) and $N = 36$ (diamonds).

The momentum distribution for $q = \pi$ is shown in Fig. 4 in a semilogarithmic plot. In this figure we have shifted the values by .003 so as to have well-defined logarithms. We have also neglected the points where Z is zero, *i. e.* between any two successive values in the figure there is a q' that leads to a zero expectation value. one can see that the momentum is better defined, even for much larger systems, but, as expected, more weight on other q' values arises for larger N. We also calculated (14) for $N = 28$ and $q = \pi/2$ but using $S^z_\pi|\psi_0\rangle$ as a target state. Of course, we find a very poorly defined momentum centered at $q' = \pi/2$. This is expected since the reduced Hilbert space targeted consists of states with $q = \pi$ in addition to the ground state.

As a check of the approximation, one can look at the sum rule

$$\frac{1}{4\pi^2}\int_0^\infty d\omega \int_{q=0}^{2\pi} S^{zz}(q,\omega) \equiv \langle\psi_0|(S^z_{r=0})^2|\psi_0\rangle = \frac{1}{4}. \tag{15}$$

For $N = 28$, 5 target states and $m = 200$, the relative error is about 0.86%.

According to the above results, one can expect that spectral functions can be reliably described in the low-energy regime by targeting the ground state as well as a few sequential Lanczos vectors and keeping a sufficient number of

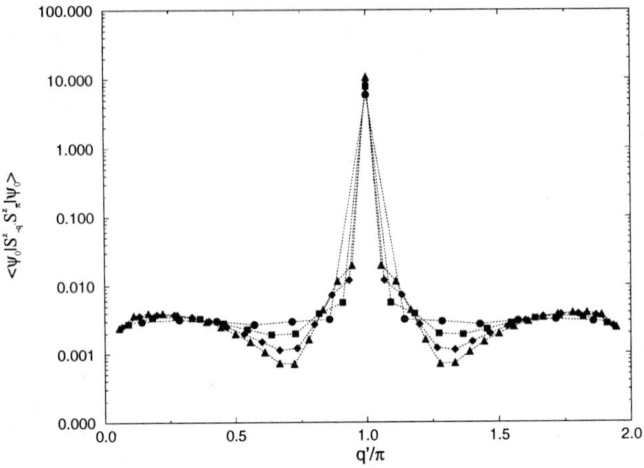

Fig. 4. Momentum weights (14) of a target state with $q = \pi$ for $N = 28$ (circles), $N = 44$ (squares), $N = 60$ (diamonds) and $N = 72$ (triangles). The dotted lines are a guide to the eye.

states in the density-matrix renormalization procedure. On the other hand, results for higher energy are to be improved.

In order to improve the target procedure, let us analyze possible problems inherently involved in this procedure and the evaluation of spectral functions. As emphasized by White, if the matrix representing the product of two or more operators in the same block is approximated by using the matrices individually representing these operators in the truncated space, some extra errors can be introduced.

To see this more clearly, one can write the Hamiltonian matrix in the form

$$H = H_e + H_s + H_{e-s}$$

where e indicates the environment, s the system and $e - s$ refers to the interaction between the two. When A is a non-local operator such as a spin operator for a given momentum or a current operator, one can also write it similarly

$$A = A_s + A_e + A_{e-s}. \tag{16}$$

Here the matrices $H_{s,e}$ and $A_{s,e}$ are represented in the truncated space. In the renormalization procedure, one has for instance

$$U^T A_s U U^T A_s U \neq U^T A_s^2 U \tag{17}$$

$$U^T A_s U U^T H_s U \cdots U^T H_s U U^T A U \neq U^T A_s H_s^n A_s U. \tag{18}$$

These two inequalities may hinder the accurate evaluation of the numerator and the denominator in (10). Pang, Akhlaghpour and Jarrell have addressed this problem and then evaluated the spectral function in an alternative way by calculating the different moments [15]. However, as the Lanczos-vector method is nothing but a Lanczos diagonalization in an excited subspace, one can expect the errors to be similar to those obtained using Lanczos to calculate the excitation spectra. One should also take into account that if only the first few $|f_n\rangle$ states are targeted, then high-energy excitations will be inaccurately described.

Most recently, Kühner and White [16] have investigated the Lanczos-vector method as applied to spin one-half and spin-one chains in more detail. They found indeed that this method works well for the low-energy regime. The results can even be improved when the finite-system algorithm is employed and more vectors are targeted. Nevertheless, higher excitation properties are still not improved significantly within this framework.

Spectral method: Now we discuss the spectral method where $\{E_n\}$, $\{|\psi_n\rangle\}$ and the matrix elements $\langle\psi_0|A|\psi_n\rangle$ are all explicitly calculated. In addition to the aspects relevant to the Lanczos-vector method discussed above, we have two more remarks:

- Since $C(\omega)$ is a superposition of the δ-functions at different excitation energies $E_n - E_0$, one can calculate spectral functions in a particular range of ω, where it is of particular interest for certain experiments such as optical absorption measurements.
- When one of the eigenstates of the superblock is targeted, one expects that a high accuracy of the targeted state intrinsically leads to a density of states for the system, the environment and the superblock, which fits the exact one quite accurately in a nearby energy range. Then the internal structure of the true spectra can be obtained by systematically targeting more eigenstates.

Taking into account these aspects and the feasibility of DMRG calculations, the following algorithm is proposed [7]

1. Target m_l lowest energy states and $m_h \geq 0$ higher excitations of the superblock using the finite-system algorithm up to certain length at which spectral functions are to be evaluated;
2. Renormalize $B \bullet \bullet B$ into BB using the density matrix constructed by m_l lowest energy states and m_h higher excitations;
3. Find eigenvalues and eigenvectors of the Hamiltonian and matrix elements $\langle\psi_n|A|\psi_0\rangle$ for the renormalized BB;
4. Evaluate dynamic correlation functions using (4).

In step 1, it is expected that m_l is larger than m_h, since one is usually interested in the dynamic properties for low and intermediate energies. The values of m_l and m_h have to be determined for each case. There is a technical step to be implemented for the higher excitations. The excited states $|\psi_n\rangle$ with E_n not smaller than a given energy E_t can be found by solving the following eigenproblem:

$$F = (H - E_t)^2$$
$$F|\phi_i\rangle = f_i|\phi_i\rangle; \quad f_i \geq 0$$

Numerically, $\{|\phi_n\rangle, f_n\}$ can be found by the use of the Davidson-Liu method [19]. Then one has:

$$E_n = E_t + \sqrt{f_i}$$
$$|\psi_n\rangle = \begin{cases} |\phi_i\rangle & \text{if } (H - E_d - \sqrt{f_i})|\phi_i\rangle = 0 \\ \dfrac{(H - E_d + \sqrt{f_i})|\phi_i\rangle}{\langle\phi_i|(H - E_d + \sqrt{f_i})^2|\phi_i\rangle} & \text{if } (H - E_d \pm \sqrt{f_i})|\phi_i\rangle \neq 0. \end{cases}$$

The renormalization of $B \bullet \bullet B$ into BB in step 2 has been proposed by Moukouri and Caron for thermodynamic calculations [18]. This makes the partial or full diagonalization of BB feasible.

The higher excitations targeted in this approach can be important for the description of dynamic properties in an intermediate energy regime. A full investigation of the spectral DMRG method is in progress [7].

2 Finite-Temperature Dynamics

As described in Chap. 6(I), thermodynamic quantities can be calculated via quantum transfer matrices using the TMRG approach. The method can be used even at very low temperature (see Chaps. 4.1-4.4(II) and [20–30]). In this Section, we discuss how to extend the TMRG to the evaluation of the imaginary-time correlation functions and how to use the TMRG data in the analytic continuation. As an application and a full test of the method, a systematic investigation of local spectral functions can be found in [8] for a prototype model: the anisotropic Heisenberg antiferromagnetic chain. An independent extension of the TMRG method to obtain local spectral functions has also been presented by Mutou, Shibata, and Ueda, who considered a one-dimensional Kondo lattice [28].

Transfer-matrix representation

Let us first outline the transfer-matrix representation. The Hamiltonian is divided into two parts:

$$H = H_o + H_e; \tag{19a}$$
$$H_o = h_1 + h_3 + h_5 + \cdots, \tag{19b}$$
$$H_e = h_2 + h_4 + h_6 + \cdots; \tag{19c}$$

In each part $H_{e,o}$, each term h_i commutes with the others. The partition function can be represented in terms of the quantum transfer matrix $\mathcal{T}_M$ [31,32]:

$$Z = \lim_{\epsilon \to 0} \mathrm{Tr}\left[e^{-\epsilon H_o} e^{-\epsilon H_e}\right]^M = \lim_{\epsilon \to 0} \mathrm{Tr}\mathcal{T}_M^{N/2} \tag{20}$$

where $\epsilon = \beta/M$, $\beta = 1/T$ and M is the Trotter number. $\mathcal{T}_M$ is asymmetric and its elements are determined by connecting $2M$ local transfer matrices at spins $\sigma_1^2, \sigma_2^2, \sigma_3^2, \cdots, \sigma_{2M}^2$ as follows

$$\langle \sigma_1^1 \cdots \sigma_{2M}^1 | \mathcal{T}_M | \sigma_3^1 \cdots \sigma_{2M}^3 \rangle$$
$$= \sum_{\{\sigma_k^2\}} \prod_{k=1}^{M} \tau(\sigma_{2k-1}^1 \sigma_{2k}^1 | \sigma_{2k-1}^2 \sigma_{2k}^2) \tau(\sigma_{2k}^2 \sigma_{2k+1}^2 | \sigma_{2k}^3 \sigma_{2k+1}^3), \tag{21}$$

where $\tau(\sigma_{2M}^2 \sigma_{2M+1}^2 | \sigma_{2M}^3 \sigma_{2M+1}^3) = \tau(\sigma_{2M}^2 \sigma_1^2 | \sigma_{2M}^3 \sigma_1^3)$ is used due to periodic boundary conditions in the Trotter direction. The local transfer matrix is given by

$$\tau(\sigma_k^i \sigma_k^i | \sigma_k^{i+1} \sigma_{k+1}^{i+1}) = \langle s_{k+1}^{i+1}, s_{k+1}^i | \exp(-\epsilon h_i) | s_k^i, s_k^{i+1} \rangle \tag{22}$$

where $\sigma_k^i = (-1)^{i+k} s_k^i$ and $|s_k^i\rangle$ is an eigenstate of S_i^z: $S_i^z |s_k^i\rangle = s_k^i |s_k^i\rangle$. The states $|\sigma_k^i\rangle \otimes |\sigma_{k+1}^i\rangle$ form a basis in the Trotter space, in which τ is represented. The superscripts $i \in [1, N]$ and subscripts $k \in [1, 2M]$ in $\mathcal{T}_M$ and τ represent the coordinates of spins in the space and the Trotter direction, respectively.

In the limit $N \to \infty$, one can evaluate thermodynamic quantities using the maximum eigenvalue λ_{max} and corresponding left $\langle \psi_{max}^L |$ and right $|\psi_{max}^R\rangle$ eigenvectors of the transfer matrix $\mathcal{T}_M$. For more details on the TMRG theory we refer to Sect. 1 in the last chapter.

Imaginary-time correlation function

We consider as an example the spin correlation function

$$G_{ij}(\tau) = \langle\, S_i^z(\tau)\; S_j^z(0) \rangle$$
$$= \frac{\mathrm{Tr}(\mathrm{e}^{-\beta H}\; S_i^z(\tau)\; S_j^z)}{\mathrm{Tr}(\mathrm{e}^{-\beta H})}, \tag{23}$$

where $S_i^z(\tau) = \mathrm{e}^{\tau H}\; S_i^z \mathrm{e}^{-\tau H}$ is the Heisenberg representation of S_i^z. Then one has

$$G_{ij}(\beta - \tau) = G_{ij}(\tau) = G_{ij}(-\tau) \tag{24}$$

due to the cyclic invariance of the trace. According to this property, one needs only to calculate $G_{ij}(\tau)$ for $0 \le \tau \le \beta/2$.

For convenience, we denote the numerator of G_{ij} by $\mathcal{N}_{ij}$. It is

$$\mathcal{N}_{ij}(\tau) = \mathrm{Tr}(S_i^z \mathrm{e}^{-\tau H} S_j^z \mathrm{e}^{-(\beta-\tau)H})$$
$$= \mathrm{Tr}[S_i^z (\mathrm{e}^{-\epsilon H})^k S_j^z (\mathrm{e}^{-\epsilon H})^{M-k}], \tag{25}$$

where $\tau = \epsilon k$ with integer $k \in [0, M-1]$. Using a symmetric decomposition

$$\mathrm{e}^{-\epsilon H} = \mathrm{e}^{-\frac{\epsilon}{2} H_o} \mathrm{e}^{-\epsilon H_e} \mathrm{e}^{-\frac{\epsilon}{2} H_o} + \mathcal{O}(\epsilon^2), \tag{26}$$

one can schematically represent $\mathcal{N}_{ij}(\tau)$ in the same way as the partition function Z. This is shown in Fig. 5. The only difference is that two of the local transfer matrices on the chequerboard for Z are replaced by the modified ones defined as

$$\tau_{i'}(\sigma_k^i, \sigma_{k+1}^i | \sigma_k^{i+1}, \sigma_{k+1}^{i+1}) = \langle s_k^i, s_k^{i+1} | \mathrm{e}^{-\frac{\epsilon}{2}h_i} S_{i'}^z \mathrm{e}^{-\frac{\epsilon}{2}h_i} | s_{k+1}^i, s_{k+1}^{i+1} \rangle \tag{27}$$

for $j > i+1$ or $k > 0$, or only one of them by the following modified one

$$\tau_{i'j'}(\sigma_k^i, \sigma_{k+1}^i | \sigma_k^{i+1}, \sigma_{k+1}^{i+1}) = \langle s_k^i, s_k^{i+1} | \mathrm{e}^{-\frac{\epsilon}{2}h_i} S_{i'}^z S_{j'}^z \mathrm{e}^{-\frac{\epsilon}{2}h_i} | s_{k+1}^i, s_{k+1}^{i+1} \rangle \tag{28}$$

for the case $j = i, i+1$ and $k = 0$. The modified local transfer matrices involve the spin operators $S_i^z(0)$ and $S_j^z(\tau)$ of $G_{ij}(\tau)$. Here i' is equal to i or $i+1$ and the prime is introduced due to the symmetric decomposition (26).

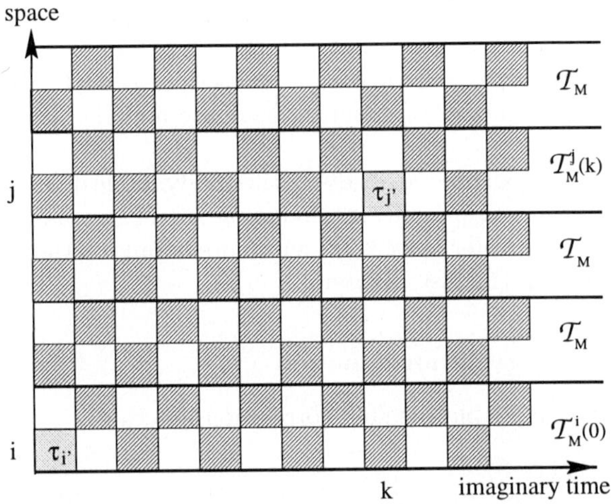

Fig. 5. The chequerboard representation of $\mathcal{N}_{ij}(\tau)$ for $j > i+1$. Each shaded plaquette indicates a local transfer matrix while the two labelled plaquettes correspond to $\tau_{i'}$ and $\tau_{j'}$. Between two thick horizontal lines, the local transfer matrices form a chain denoted by $\mathcal{T}_M$, or $\mathcal{T}_M^i(0)$, or $\mathcal{T}_M^j(k)$. The plaquette containing $S_i^z(0)$ is shifted to the origin by using the cyclic invariance of the traces.

In general, we can express $\mathcal{N}_{ij}(\tau)$ as follows [8]:

$$\mathcal{N}_{ij}(\tau) = \mathrm{Tr}[\mathcal{T}_M^{ij}(k)\ (\mathcal{T}_M)^{(\frac{N}{2}-[\frac{j}{2}]+[\frac{i}{2}]-1)}] + \mathcal{O}(\epsilon^2), \tag{29}$$

where $[\frac{i}{2}]$ is an integer equal to $i/2$ when i is even and $(i+1)/2$ otherwise; One can consider $\mathcal{T}_M^{ij}(k)$, whose form depends on the values of $j-i$ and k, as a generalized transfer matrix. We note that $\tau_{j'}$ is always considered for $k \geq M/2$ except $k = 0$ because of the periodic boundary condition in the Trotter direction for $G(\tau)$.

For the case $j > i+1$ shown Fig. 5, one has

$$\mathcal{T}_M^{ij}(k) = \mathcal{T}_M^{i}(0)\;(\mathcal{T}_M)^{[\frac{j}{2}]-[\frac{i}{2}]-1}\;\mathcal{T}_M^{j}(k), \tag{30}$$

where the modified matrices $\mathcal{T}_M^i(0)$ and $\mathcal{T}_M^j(k)$ contain a plaquette including the spin operators S_i^z and $S_j^z(\tau)$ at imaginary time $\tau = \epsilon k$, respectively. In this case, the whole object $\mathcal{T}_M^{ij}(k)$ is thus formed by these two modified transfer matrices and all the unmodified ones in between.

For $j = i$ and $i+1$, both S_i^z and S_j^z appear in the same modified transfer matrix $\mathcal{T}_M^{ij}(k)$ as shown in Fig. 6. In particular, at $k = 0$, they are located in the same local transfer matrix $\tau_{i'j'}$.

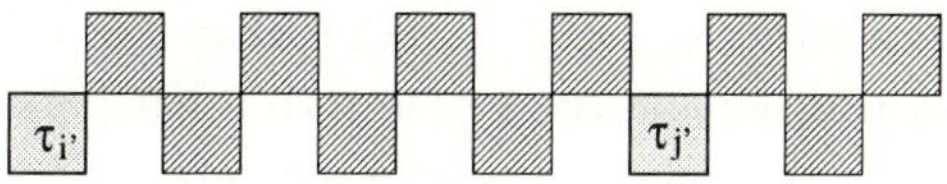

Fig. 6. A generalized transfer matrix $\mathcal{T}_M^{ij}(k)$ for $j = i, i+1$ and $k \neq 0$

Finally, the imaginary-time correlation function reads

$$G_{ij}(\tau) = \frac{\mathrm{Tr}[\mathcal{T}_M^{ij}(k)\;(\mathcal{T}_M)^{(\frac{N}{2}-[\frac{j}{2}]+[\frac{i}{2}]-1)}]}{\mathrm{Tr}(\mathcal{T}_M)^{\frac{N}{2}}} + \mathcal{O}(\epsilon^2). \tag{31}$$

After taking the limit $N \to \infty$, one has

$$G_{ij}(\tau) = \frac{\langle \psi^L | \mathcal{T}_M^{ij}(k) | \psi^R \rangle}{\lambda_{max}^{[\frac{j}{2}]-[\frac{i}{2}]+1}} + \mathcal{O}(\epsilon^2). \tag{32}$$

When $j \gg i$ the imaginary-time correlation function simply becomes independent of k:

$$G_{ij}(\tau) = \frac{\langle \psi_{max}^L | \mathcal{T}_M^i(0) | \psi_\alpha^R \rangle \langle \psi_\alpha^L | \mathcal{T}_M^j(k) | \psi_{max}^R \rangle}{\lambda_\alpha \lambda_{max}} \mathrm{e}^{(\xi^{-1}-ik)|j-i|} + \mathcal{O}(\epsilon^2), \tag{33}$$

with $\xi = \frac{1}{2}\ln\left|\frac{\lambda_\alpha}{\lambda_{max}}\right|$ being the correlation length at finite temperature as discussed in the last chapter and $\lambda_\alpha = |\lambda_\alpha|\mathrm{e}^{2ik}$ is the largest eigenvalue of $\mathcal{T}_M$ satisfying the condition that the above numerator is non-zero and $\psi_\alpha^{R,L} \neq \psi_{max}^{R,L}$. Therefore one only needs to calculate $G_{ij}(\tau)$ for $j - i \not\gg \xi$ according to (32). So far the above relations are exact except for the decomposition made in (26).

TMRG implementation

In Sect. 2 of the last chapter, for each β equal to ϵM and given ϵ, the transfer matrix $\mathcal{T}_M$ is considered as a superblock. As shown in Fig. 3 in the last chapter, it consists of two renormalized blocks and two time slices. The basis states for the renormalized system and environment are denoted by n_s and n_e, respectively, and those for two slices σ_1 and σ_2. $\mathcal{T}_M$ is separated into the left and right transfer matrices $\mathcal{S}_{e,o}$ and $\mathcal{E}_{e,o}$ in the density-matrix renormalization procedure. At each iteration, the renormalized block augmented by one slice is to be further renormalized by the truncated basis of the density matrix. Both $\mathcal{S}_{e,o}$ and $\mathcal{E}_{e,o}$ contain σ_1 and σ_2 which are summed over for the construction of $\mathcal{T}_M$.

In order to evaluate $G_{ij}(\tau)$ in (32), one needs to renormalize $\mathcal{T}_M^{ij}(k)$ and evaluate its expectation value in terms of the maximum eigenpair of $\mathcal{T}_M$. To be specific, let us consider the case $k > 0$ from which the discussion can be easily extended to $k = 0$. Because the density-matrix renormalization is performed directly for either $\mathcal{E}_{e,o}$ or $\mathcal{S}_{e,o}$ rather than $\mathcal{T}_M$, one has to separate also the generalized transfer matrix $\mathcal{T}_M^{ij}(k)$ into a left part $\mathcal{S}_{e,o}^i$ and a right part $\mathcal{E}_{e,o}^j(k)$. In this separation, the two spin operators S_i^z and S_j^z have been assigned to $\mathcal{S}_{e,o}^i$ and $\mathcal{E}_{e,o}^j(k)$, respectively. This is because one needs $\mathcal{T}_M^{ij}(k)$ only for $k \leq M/2$ according to $G_{ij}(\tau) = G_{ij}(\beta - \tau)$. Moreover, as seen from (30), the dimensionality of $\mathcal{E}_{e,o}^j(k)$ and $\mathcal{S}_{e,o}^i$ for $j > i + 1$ is larger than that of $\mathcal{E}_{e,o}$ and $\mathcal{S}_{e,o}$ by a factor of $(2S+1)^{2([\frac{j}{2}]-[\frac{i}{2}])}$, respectively. Therefore an accurate evaluation of $G_{ij}(\tau)$ is eventually limited for large values of $j - i$ because of the limited disk space for storing $\mathcal{E}_{e,o}^j(k)$ and $\mathcal{S}_{e,o}^i$. Below we discuss two different cases.

(1) When $j - i$ is small, one can construct $\mathcal{T}_M^{ij}$ in terms of $\mathcal{S}_{e,o}^i$ and $\mathcal{E}_{e,o}^j$ after taking into account some more internal states at time slices.

The simplest case $j = i$ or $i + 1$ is shown in Fig. 7 for $\mathcal{T}_M^{ij}(k)$. The left transfer matrix $\mathcal{S}_e^i$ and the right one $\mathcal{E}_e^j$ are connected as $\mathcal{T}_M^{ij}(k)$ as shown in Fig. 6. Following the recursion relations for $\mathcal{S}_{e,o}$ and $\mathcal{E}_{e,o}$ discussed in the last chapter, one can iteratively enlarge and renormalize $\mathcal{S}_{e,o}^i$ and $\mathcal{E}_{e,o}^j(k)$.

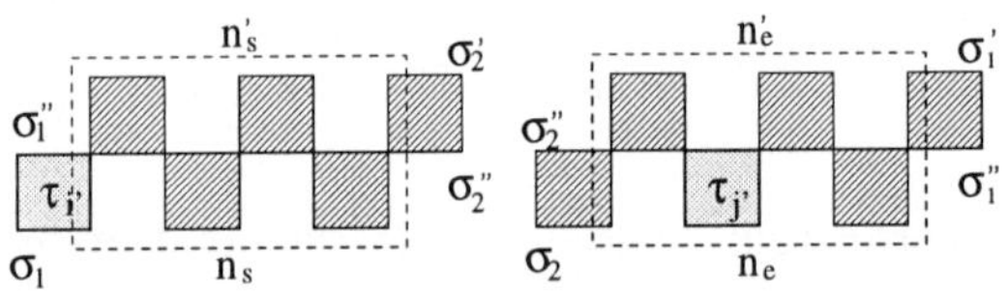

Fig. 7. $\mathcal{S}_e^i$ and $\mathcal{E}_e^j(k)$ for a modified transfer matrix $\mathcal{T}_M^{ij}(k)$ with $k = 2$, $M = 6$ and $j = i$ or $i + 1$.

For $j - i \geq 2$ but still small, $\mathcal{E}_{o,e}^j(k)$ and $\mathcal{S}_{o,e}^i$ contain more internal variables. An example is shown in Fig. 8. In this case, $\sigma_1^2, \sigma_1^3, \sigma_1^4$ and $\sigma_2^2, \sigma_2^3, \sigma_2^4$ which are internal variables are summed over when $\mathcal{E}_o^j(k)$ and $\mathcal{S}_o^i$ are used to construct the periodic transfer matrix $\mathcal{T}_M^{ij}(k)$.

(2) When $j - i$ is large enough, there is an approximate way to evaluate the expectation value of $\mathcal{T}_{ij}(k)$. One represents $\mathcal{T}_M^i(0)$, $\mathcal{T}_M^j(k)$, and $\mathcal{T}_M$ in (30) all in the truncated Trotter space. Then $\mathcal{T}_M^i(0)$ is constructed in terms of $\mathcal{E}_{e,o}$ and $\mathcal{S}_{e,o}^i$ as shown in Fig. 9(a), and $\mathcal{T}_M^j(k)$ in terms of $\mathcal{E}_{e,o}^j(k)$ and $\mathcal{S}_{e,o}$

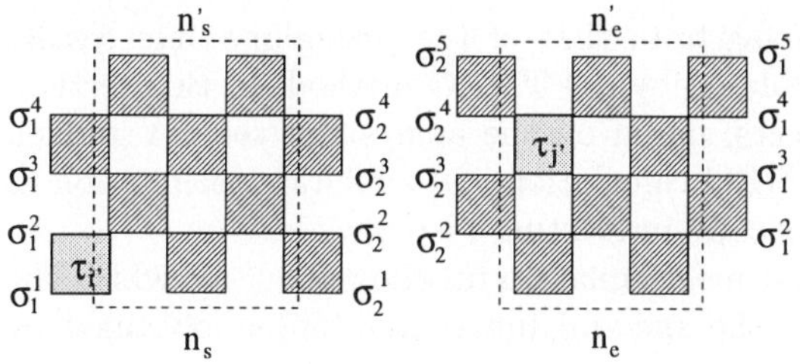

Fig. 8. A generalized transfer-matrix chain $\mathcal{T}_M^{ij}(k)$ with $k = 2$, $M = 5$ and $j = i+2$ or $i+3$. $\mathcal{E}_o^j(k)$ and $\mathcal{S}_o^i$ are shown on the right and left side, respectively.

as shown in Fig. 9(b). $\mathcal{S}_{e,o}^i$ and $\mathcal{E}_{e,o}^j(k)$ are the left and right transfer matrices defined for the case of $j = i$ and $j = i+1$. Numerically, it is straightforward to evaluate $\langle \psi_{max}^L | \mathcal{T}_M^{ij}(k) | \psi_{max}^R \rangle$. An optimal way is to operate first with the left and the right transfer matrices on $|\psi_{max}^R\rangle$ and then to form the inner product with $\langle \psi_{max}^L|$. In principle, the larger the values of $j - i$, the larger the errors. By keeping more states, one can reduce them systematically.

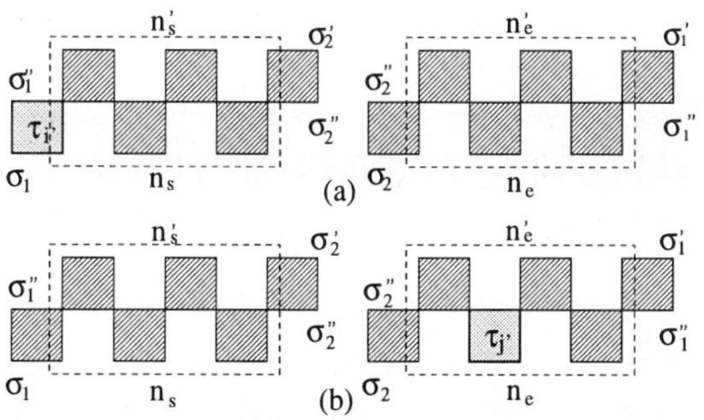

Fig. 9. Two transfer matrices $\mathcal{T}_M^i(0)$ (**a**) and $\mathcal{T}_M^j(k)$ (**b**), which are represented in the truncated space and used to construct $\mathcal{T}_M^{ij}(k)$ for large values of $j - i$. (**a**) On the left is $\mathcal{E}_e$ on the right $\mathcal{S}_e^i$ and (**b**) on the left is $\mathcal{E}_e^j(k)$ and on the right $\mathcal{S}_e$.

In practice, one has to store the transfer matrices $\mathcal{E}_{e,o}^j(k)$ for different k on the hard disk of the computer. To reduce computational effort, one can choose the imaginary-time interval to be 2ϵ or more depending on the values of the temperature.

Numerical aspects

As an application we have studied systematically the local function $G_{ii}(\tau)$ and the associated spectral function for an anisotropic Heisenberg chain described by

$$H = J \sum_i (S_i^+ S_{i+1}^- + h.c) + \Delta S_i^z S_{i+1}^z . \qquad (34)$$

This includes the XY model ($\Delta = 0$), the isotropic Heisenberg ($\Delta = 1$), and the gapped Heisenberg-Ising model($\Delta > 1$). The dynamic spectra were obtained by analytical continuation of the Fourier transform of the local imaginary-time correlation function [8]. We emphasize that the quality of the spectral function is often far from that of the imaginary-time results which are very accurately calculated by the TMRG method. In this section, we discuss some relevant aspects, based on the analysis of the XY model, concerning the reliability of TMRG calculation and analytic continuation as generally applied to other physically interesting systems.

For a given local imaginary-time correlation function $G(\tau)$, for which the subscript is omitted hereafter, the spectral function $A^{\alpha}(\omega)$ is obtained by inverting the relation

$$G(\tau) = \frac{1}{2\pi}\int_0^{+\infty} K(\tau,\omega)A(\omega)\,d\omega \tag{35}$$

where the integral kernel is given by

$$K(\tau,\omega) = \mathrm{e}^{-\tau\omega} + \mathrm{e}^{-(\beta-\tau)\omega}. \tag{36}$$

For negative frequencies $A(\omega)$ is given by

$$A(-\omega) = \mathrm{e}^{-\beta\omega}A(\omega). \tag{37}$$

Analytical continuation: After discretizing the variables τ and ω, one finds the spectral function $A(\omega_j)$ by solving the following linear equations

$$G(\tau_i) = \sum_{j=1}^{N_\omega} K_{ij}A(\omega_j) \tag{38}$$

with $i \in [1, N_\tau]$ and $j \in [1, N_\omega]$. According to the singular-value decomposition of a rectangular matrix, one can write $K_{ij} = U_{il}\lambda_l V_{lj}$ where U is a $N_\tau \times N_\omega$ column-orthogonal, V a $N_\omega \times N_\omega$ row-orthogonal matrix and $\lambda_l \geq 0$ with $l \in [1, N_\omega]$ are the singular values [33]. Then the spectral function reads

$$A(\omega_j) = \sum_{l=1}^{N_\tau} \frac{1}{\lambda_l} V_{ij} \sum_{i=1}^{N_\omega} G(\tau_i)U_{il}. \tag{39}$$

Although this solves the problem, one sees that small errors in G may be amplified drastically via small values of λ_l. In practice, one has to introduce a cutoff λ_c for λ_l in (39), which cuts a major part of the vector space for $A(\omega_j)$ [8]. For $\beta = 2, 8, 20$, one finds only $5, 6, 7$ largest λ_l which are larger than 10^{-8}. Therefore, this method does not work and one has to proceed differently. An alternative is the maximum-entropy (ME) method.

The maximum-entropy method has been the main tool for the analytic continuation (38). Since the spectral function is positive definite, one can interpret it as a probability function. According to information theory, one introduces the following information entropy

$$S = \sum_j [A(\omega_j) - m(\omega_i) - A(\omega_j)\ln(\frac{A(\omega_j)}{m(\omega_j)})]K_{0j} \tag{40}$$

where K_{0j} consistently accounts for the negative frequency contribution and m_i is a reference spectral function. We have used a flat $m(\omega_j)$ for $\omega_j > 0$ in our analysis.

One finds $A^\alpha(\omega_i)$ which maximizes the function

$$\phi(\{A(\omega_i)\}) = \alpha S(\{A(\omega_i)\}) - \frac{1}{2}\chi^2(\{A(\omega_i)\}), \tag{41}$$

where

$$\chi^2(\{A(\omega_i)\}) = \sum_i \left(\frac{G(\tau_i) - \sum_j K_{ij} A(\omega_j)}{\sigma_i}\right)^2 \tag{42}$$

Details for the maximum-entropy method can be found in [34,35].

In the TMRG calculation, errors σ_i for $G(\tau_i)$ usually are not given. In particular, they are not statistical as in quantum Monte Carlo simulations but systematic due to the truncation as well as the Trotter-Suzuki decomposition. It turns out that the solution is strongly dependent on the input σ_i values. From the systematic study of the anisotropic Heisenberg spin one-half model, we have found the following recipe to be best: start with a small value of σ_i (typically 10^{-8}) and change it until the solution becomes insensitive to it over a wide range (at least one order of magnitude, typically $\sigma_i \sim 10^{-6}$ for $\beta = 16$). Finally, the reconstructed imaginary-time correlation function obtained by using (38) agrees with the TMRG data up to the 6th digit or better. Care should be taken that too small values of σ_i overemphasize χ and tend to produce spurious peaks in $A^\alpha(w_j)$, especially at low temperatures.

Numerical analysis: The XY model, which can be mapped onto a spinless non-interacting fermion model via the Jordan-Wigner transformation, is useful for tests. Its longitudinal correlations can be expressed in closed form at any temperature T [37].

One obtains the following longitudinal local imaginary-time Green's function

$$G^z_{ex}(\tau) = \left(\frac{1}{\pi}\int_0^\pi \frac{\mathrm{e}^{\tau J \cos q}}{1 + \mathrm{e}^{\beta J \cos q}} dq\right)^2 \tag{43}$$

In Fig. 10, we compare this exact result with the TMRG data for $\beta = 2, 8, 20$ on a logarithmic scale. (The minima appear because the approximate results oscillate around the exact ones). For all the calculations, we have kept $m = 100$ optimal states from the density matrix so that the truncation error $(1 - \sum_m \rho_m)$ is smaller than 10^{-7} for the largest Trotter number M=800 ($\beta = 20$ when $\epsilon = 0.025$). In order to reduce the Trotter error, we have done a linear extrapolation with respect to ϵ^2 (this requires congruent values of ϵ). This procedure is justified as long as the systematic errors induced by the truncation are negligible compared to the Trotter errors, which is the case typically at high temperatures (small truncation error) in Fig. 10(d),

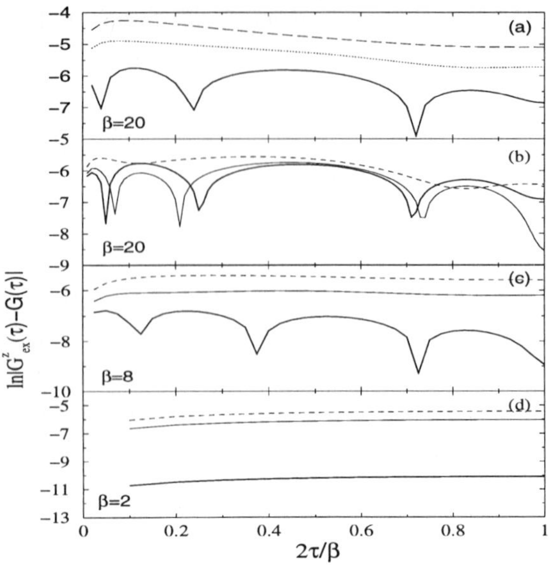

Fig. 10. Difference between $G^z_{\mathrm{ex}}(\tau)$ from (43) and TMRG results with m=100 states. The thick solid line denotes the results at each τ extrapolated from the two values of ϵ. In (**a**), the dot-dashed line corresponds to $\epsilon = 0.2$ and the dotted line to $\epsilon = 0.1$. In (**b**)-(**d**), the dashed lines are for $\epsilon = 0.025$, and the solid lines for $\epsilon = 0.05$.

or when ϵ is large (large Trotter error) in Fig. 10(a). Figure 10(b) is an example where the ϵ^2-extrapolation fails because the systematic errors become comparable to the Trotter error itself ($M = 800$ Trotter steps are needed to reach $\beta = 20$ when $\epsilon = 0.025$). In such a case, nothing can be gained from the extrapolation and it is justified to use the data for the smallest ϵ for the analytical continuation. Notice that in Fig. 10(a) the result obtained from the $\epsilon = 0.1, 0.2$ data is as precise as the $\epsilon = 0.025$ calculation. This should be systematically exploited for models with more degrees of freedom per site, where small ϵ calculations probably cannot be afforded and values $\epsilon = 0.1$ or larger may be used to reach the low-temperature regime.

Before doing the continuation to real frequencies, we note that because the XY model is a free-fermion model, the longitudinal correlation function may not be generic of a true interacting system and exhibit some peculiar behavior. For instance, it has a sharp step-like cutoff at $\omega = 2$ and a logarithmic divergence at $\omega = 0$ for all temperatures. As we are not interested in reproducing step discontinuities, which are unexpected in interacting systems at non-zero temperature, we artificially cutoff the integral at $\omega = 2$ in (35) to investigate how the line-shape in the interval $\omega \in [0, 2]$ can be reconstructed.

In Fig. 11, we compare the ME results obtained from the TMRG data as well as the exact $G^z_{ex}(\tau)$, see (43), and the exact $A^z_{ex}(\omega)$. We found that ME cannot resolve the low-frequency divergence at $\beta = 2$ since there is not enough spectral weight under the logarithmic divergence. The ME results from the TMRG data and those from $G_{ex}(\tau)$ cannot be distinguished for $\beta = 2$ and $\beta = 8$.

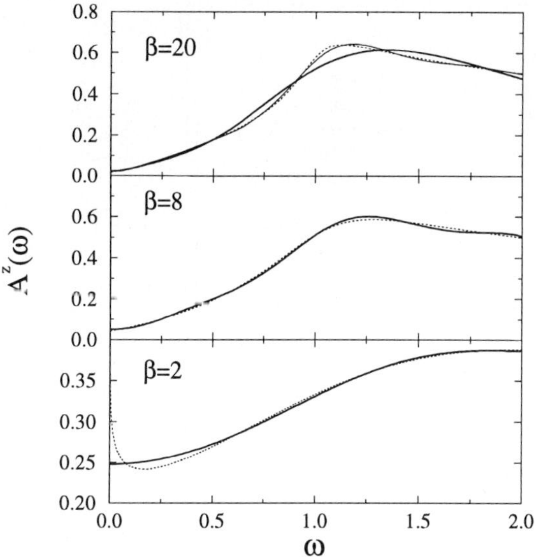

Fig. 11. The spectral function $A^z(\omega)$ for different temperatures $T = 1/\beta$. The thin solid and the solid lines are obtained by using the ME method for analytical continuation from $G^z_{ex}(\tau)$ and from TMRG $G^z(\tau)$. The dotted line is the exact solution $A^z_{ex}(\omega)$.

Now let us discuss ME results for the transverse correlations. After a Jordan-Wigner transformation, the transverse correlation function becomes a complicated many-fermion quantity which is expressible as a Pfaffian [36] for all temperatures. Simple expressions are only known at $T = 0, \infty$ [38]. The exact $T = 0$ autocorrelations have an infinite series of singularities at all integer frequencies $\omega = m$, $m = 0, 1, 2, \ldots$. $A^{\perp}(\omega)$ diverges as $\omega^{-1/2}$ for $\omega \to 0^+$ and as $\ln|\omega - 1|$ for $\omega \to 1$. The higher singularities are cusps, for $\omega \to 2^+$, $A^{\perp}(\omega) \sim (\omega - 2)^{1/2}$. At infinite T, the function becomes a Gaussian, $A^{\perp}(\omega) = \sqrt{\pi} e^{-w^2}$.

In Fig. 12 we show the transverse spectral function $A^{\perp}(\omega)$. In the intermediate to low temperature range $\beta = 6 - 16$, $A^{\perp}(\omega)$ is obtained from the TMRG data by using the ME. One can see nicely the emergence of two peaks at $\omega = 1$ and near $\omega = 0$ as the temperature is lowered. Also, one may see some sign of the cusp at $\omega = 2$. However, it is beyond the ability of the method to characterize and even identify the $T = 0$ divergences. When reducing the temperature further, the peak near $\omega = 0$ continues to grow, whereas that at $\omega = 1$ seems to increase a little. On the other hand, the peak at $\omega = 1$ starts to shift slightly to lower frequencies. We found this feature to be typical for all systems we have investigated. It may signify that the accuracy of the data becomes too poor. A more detailed discussion can be found in [8].

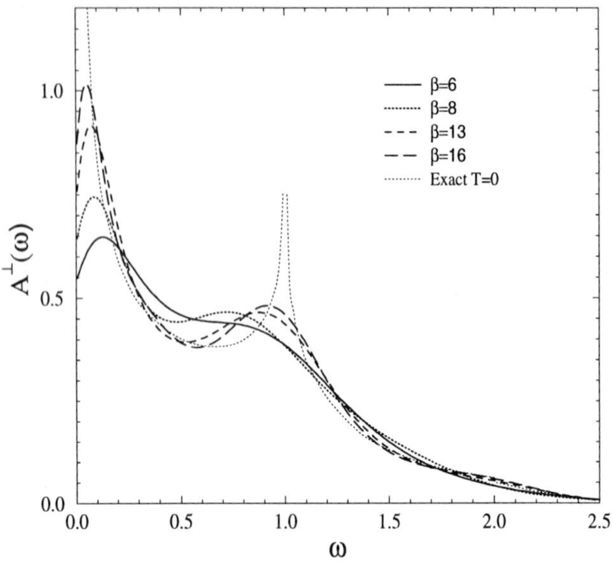

Fig. 12. Spectral function $A^{\perp}(\omega)$ obtained from the TMRG results using the ME method. $m = 160$ states were kept and $\epsilon = 0.025$.

3 Remarks

In this chapter, we have discussed how one can calculate dynamic properties at zero temperature and finite temperature within the DMRG framework and the TMRG framework, respectively.

In the zero-temperature case, two different targeting procedures have been discussed. For the Lanczos-vector method we have shown that by considering only a 0.1% of the total Hilbert space (for $N = 28$ around 40000 total states are kept), a reasonable description of the low-energy excitations is obtained. We also showed that it is possible to obtain states with well defined momenta if the appropriate target states are used. In principle, one can additionally consider the higher excitations as target states in the continued-fraction method. However, a more systematical analysis is required for realistic calculations of spin as well as electron systems.

Most recently, Kühner and White [16] also proposed to calculate dynamic properties using a correction vector as one of the target states. This vector, which was originally proposed by Ramasesha et al. [17] (Chap. 2.2(II)), is essentially a state next to one of the excited states. The range of the frequency under consideration is centered at the corresponding excitation energy. To some extent, this method can reproduce the higher energy structure of the dynamical structure factors and the investigation indicates the importance of targeting the higher exciations.

In the finite-temperature case, the TMRG method provides a framework for the evaluation of imaginary-time correlation functions. Especially the evaluation of local ones is accurate. Using the maximum-entropy method for analytical continuation, one can find that some qualitative features such as location of the peaks and gaps can be determined reliably. However, more quantitative features as precise line-shapes or the nature of divergences are beyond the method. This is due to the severe intrinsic limitation to the analytic continuation. Reliable real-time methods need to be developed in order to overcome this.

X. Wang and N. Felix are very grateful to I. Peschel, A. Sandvik, W. von der Linden, and X. Zotos for fruitful discussion and kind help.

References

1. S.R. White, Phys. Rev. Lett. **69**, 2863 (1992)
2. T. Nishino, J. Phys. Soc. Japan. **64** 3958 (1995)
3. T. Nishino and K. Okunishi, J. Phys. Soc. Japan. **67** 3066 (1998)
4. R.J. Bursill, T. Xiang, G.A. Gehring: J. Phys. Cond. Matter **8** L583 (1996)
5. X. Wang and T. Xiang, Phys. Rev. B **56** 5061 (1997)
6. Karen Hallberg, Phys. Rev. B **52**, R9827 (1995)
7. X. Wang, The spectral DMRG method for dynamic properties, unpublished
8. F. Naef, X. Wang, X. Zotos and W. von der Linden, cond-mat/9812117, to appear in Phys. Rev. B, July 1999
9. E.R. Gagliano and C.A. Balseiro, Phys. Rev. Lett. **59**, 2999 (1987)
10. V.S. Visvanath and G. Müeller in The Recursion Method, Springer (1994)
11. P. Fulde, Electron Correlations in Molecules and Solids in Springer Series in Solid-State Sciences, **100**, p 453 (1995)
12. S. Haas, J. Riera and E. Dagotto, Phys. Rev. B **48**, 3281 (1993)
13. P. Horsch and W. von der Linden, Z. Phys. B **72**, 181 (1988)
14. T. Xiang, Phys. Rev. B **53**, 10445 (1996)
15. H.B. Pang, H. Akhlaghpour and M. Jarrell, Phys. Rev. B **53**, 5086 (1995)
16. T.D. Kühner, S. White, cond-mat/9812372
17. Z.G. Soos and S. Ramasesha, J. Chem. Phys. **90**, 1067 (1989); S. Ramasesha, S.K. Pati, H.R. Krishnamurthy, Z. Shuai and J.L. Bredas, Synth. Metals **85**, 1018 (1997)
18. S. Moukouri and L.G. Caron, Phys. Rev. Lett. **77**, 4640 (1996)
19. E.R. Davidson, J. Comput. Phys. **17**, 87 (1975); B. Liu, in Numerical Algorithms in Chemistry: Algebraic methods, eds. C. Moler and I Shavitt, LBL-8158 Lawrence Berkeley Laboratory (1978)
20. N. Shibata, J. Phys. Soc. Japan. **66**, 2221 (1997)
21. K. Maisinger and U. Schollöck, Phys. Rev. Lett. **81**, 445 (1998)
22. N. Shibata, B. Ammon, M. Troyer, M. Sigrist and K. Ueda, J. Phys. Soc. Japan **67**, 1086 (1998)

23. D. Coombes, T. Xiang, and G.A. Gehring, J. Phys. C **10**, L159 (1998)
24. T. Xiang, Phys. Rev. B **58**, 9142 (1998)
25. S. Eggert and S. Rommer, Phys. Rev. Lett. **81**, 1690 (1998)
26. K. Maisinger, U. Schollwöck, S. Brehmer, H.J. Mikeska and S. Yamamoto, Phys. Rev. B **58**, R5908 (1998)
27. S. Yamamoto, T. Fukui, K. Maisinger and U. Schollwöck, (1998), to appear in J. Phys.: Cond. Mat.
28. T. Mutou, N. Shibata and K. Ueda, Phys. Rev. Lett. **81**, 4939 (1998)
29. X. Wang and T. Xiang, Thermodynamic properties of the Hubbard model, (1998), unpublished
30. A. Klümper, R. Raupach and F. Schönfeld, cond-mat/9809224
31. H.F. Trotter, Proc. Am. Math. Soc. **10**, 545 (1959); M. Suzuki, Prog. Theor. Phys. **56**, 1454 (1976)
32. H. Betsuyaku, Prog. Theor. Phys. **73**, 320 (1985)
33. W.H. Press, S.A. Teukolsky, W.T. Vetterling, and B.P. Flannery, Numerical Recipes, Cambridge University press.
34. M. Jarrell and J.E. Gubernatis, Phys. Rep. **269**, 133 (1996)
35. W. von der Linden, Appl. Phys. A **60**, 155-165 (1995)
36. J. Stolze, A. Nöppert and G. Müller, Phys. Rev. B **52**, 4319 (1995)
37. T. Niemeijer, Physica **36**, 377 (1967); S. Katsura, T. Horigushi and M.Suzuki, Physica **46**, 67 (1970)
38. G. Müller and R. Shrock, Phys. Rev. B **29**, 288 (1994)
39. A. R. Its, A. G. Izergin, V.E. Korepin and N.A. Slavnov, Phys. Rev. Lett. **70**, 1704 (1993)

Part II

Physical Applications

1 Chains and Ladders

1.1 Properties of the Hubbard Chain

Reinhard M. Noack[1], Stéphane Daul[2], and Sebastian Kneer[3]

[1] Institut de Physique Théorique, Université de Fribourg
CH-1700 Fribourg, Switzerland
[2] Department of Physics, University of California
Santa Barbara, CA 93106-9530, USA
[3] Institut für Theoretische Physik, Universität Würzburg
Am Hubland, D–97074 Würzburg, Germany

The Hubbard chain with nearest-neighbor hopping is the prototypical model for one-dimensional strongly interacting electrons. For repulsive interaction, it is a Luttinger liquid, characterized by gapless, bosonic low-lying charge and spin excitations, for all interaction strengths at almost all fillings. The only exception occurs at half-filling, where Fermi surface nesting leads to a gap in the charge excitation at all interaction strengths. Its behavior is quite well understood through a Bethe Ansatz exact solution and well-developed weak-coupling continuum techniques involving the renormalization group and bosonization. Bosonization determines the asymptotic form of the correlation functions and the Bethe Ansatz quantitatively determines the Luttinger liquid parameters, completely specifying the low-temperature, long distance behavior [1,2].

The possible behavior of the Hubbard chain becomes much richer and more interesting when a simple addition is made to the band structure of the model: an additional hopping to next-nearest neighbors. As we shall explore in this chapter, the phase diagram then includes more exotic behavior such as itinerant metallic ferromagnetism, spin-gapped pairing, and disordered, antiferromagnetic insulating behavior [3–5]. However, with the additional hopping, the model is no longer integrable [6] (and thus not Bethe-Ansatz solvable), the weak-coupling treatments no longer yield the correct phases at strong coupling, and the Lieb-Mattis theorem is no longer applicable.

Therefore, an accurate numerical tool is crucial in exploring the low-energy behavior of this system. The Density Matrix Renormalization Group (DMRG) technique [7] provides such a tool. Used in combination with other numerical techniques such as exact diagonalization, and analytical techniques such as the weak-coupling renormalization group, it can be used to obtain a fairly complete and accurate picture of the ground-state phase diagram. Here we will use the finite-system version of the ground-state algorithm for fermion systems as described in detail in Chap. 2(I), in which the convergence for the nearest-neighbor case was already investigated in detail. In the following we keep up to $m = 800$ states in the system block, and treat systems containing up to 128 sites. In all cases, the sum of the discarded density matrix eigenvalues is less than 10^{-5}.

We label the nearest-neighbor and next-nearest-neighbor hopping integrals t_1 and t_2, respectively, and study the Hamiltonian

$$H = -t_1 \sum_{i,\sigma} \left(c^\dagger_{i+1\sigma} c_{i\sigma} + h.c. \right) - t_2 \sum_{i,\sigma} \left(c^\dagger_{i+2\sigma} c_{i\sigma} + h.c. \right) + U \sum_i n_{i\uparrow} n_{i\downarrow} \,, \tag{1}$$

where, in the usual notation, $c^\dagger_{i\sigma}$ creates an electron at site i with spin σ, $n_{i\sigma} = c^\dagger_{i\sigma} c_{i\sigma}$, and U is the Coulomb repulsion. We will set $t_1 = 1$ and treat systems with N electrons on L sites so that the average density $n = N/L$. It is useful to picture the configuration as in Fig. 1, so that the geometry is that of two chains with a "zigzag" coupling.

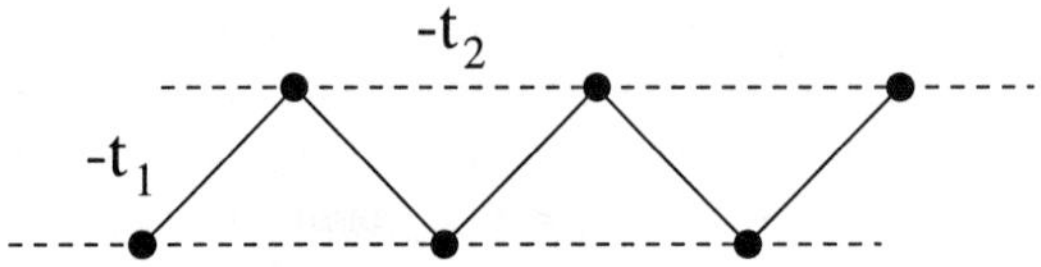

Fig. 1. The t_1–t_2 chain.

In the following, we will discuss how to determine the Luttinger liquid parameters in the $t_2 = 0$ case and compare with the exact solution in Sect. 1. We will investigate the phase diagram of the t_1–t_2 Hubbard chain for small interaction strength, U, in Sect. 2, and for large U in Sect. 3. Finally, in Sect. 4, we will discuss the phase diagram of the half-filled system.

1 Nearest-Neighbor Hopping

The nearest-neighbor case, $t_2 = 0$, is interesting in that a direct comparison can be made between numerical DMRG results and exact results. Since a direct comparison of the ground-state energies has already been made in Chap. 2 (I), here we will discuss the determination of the asymptotic behavior of the correlation functions and compare the Luttinger liquid parameters. In a Luttinger liquid, the low-lying excitations consist of bosonic spin and charge modes which are massless and noninteracting and can thus be described by two independent velocities, the spin velocity u_σ and the charge velocity u_ρ. The long-distance behavior of the correlation functions can be determined by calculations in the bosonic representation [2]. The spin-spin, charge-charge, and singlet superconducting correlation functions decay as

$$\langle \boldsymbol{S}_{i+r} \cdot \boldsymbol{S}_i \rangle = -\frac{1}{(\pi r)^2} + \frac{A}{r^{1+K_\rho}} \cos(2k_F r) \ln^{1/2}(r) + \ldots \tag{2}$$

$$\langle \delta n_{i+r} \delta n_i \rangle = -\frac{K_\rho}{(\pi r)^2} + \frac{B}{r^{1+K_\rho}} \cos(2k_F r) \ln^{-3/2}(r) + \dots \tag{3}$$

$$\langle \Delta(i+r) \Delta^\dagger(i) \rangle = \frac{C}{r^{1+\frac{1}{K_\rho}}} , \tag{4}$$

respectively, where A, B, and C are undetermined constants, $\delta n_i = n_i - \langle n_i \rangle$ with $n_i = n_{i\uparrow} + n_{i\downarrow}$, and $\Delta^\dagger(i) = c^\dagger_{i\uparrow} c^\dagger_{i\downarrow}$ creates a singlet pair at site i. Here $k_F = n\pi$ is the Fermi wave vector. Note that the leading minus sign in the terms proportional to $1/r^2$ is absent in [2]. This sign can be checked by examining the $U = 0$ limit [8]. The behavior is that of a system at the critical point in that the dominant decay of all of these correlation function is a power law whose strength is determined by a single nonuniversal parameter K_ρ.

It would thus be interesting to confirm that the form given by (2)–(4) is valid and applicable to finite-size lattice systems by explicitly calculating the correlation functions and extracting the exponents. There are, however, a number of technical problems that must be overcome in order to determine K_ρ accurately. First, the open boundary conditions lead to oscillations in the correlation functions which depend on distance from the boundary, i.e. (2)–(4) oscillate with i. We have found that a good way to minimize these effects is to average a correlation function such as $\langle \boldsymbol{S}_{i+r} \cdot \boldsymbol{S}_i \rangle$ over a number of i pairs for each r, where the number is larger than the typical size of the oscillations. Note, however, that additional information is contained in the Friedel oscillations due to the open boundaries and can be used to directly determine the boundary critical exponents [9].

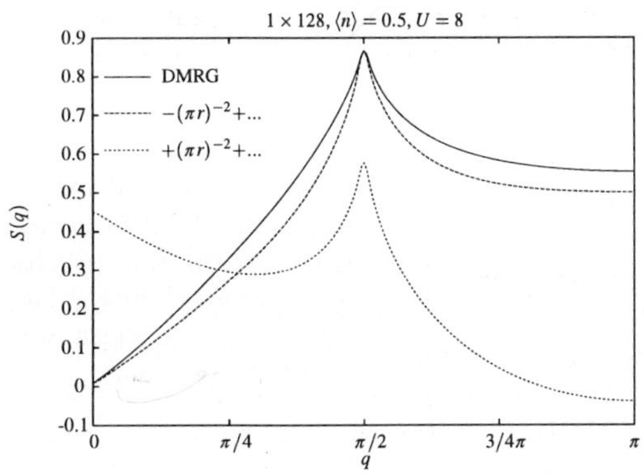

Fig. 2. Fourier transform, $S(q)$ of the spin-spin correlation function $\langle S^z(r) S^z(0) \rangle$ for the quarter-filled system. The dashed and dotted lines are Fourier transforms of the asymptotic form of (2) with the indicated sign of the $1/r^2$ term.

The Fourier transform of the averaged spin-spin correlation function is plotted in Fig. 2. For a finite system with open boundary conditions, we approximate the continuous Fourier transform by performing a discrete Fourier transform on a large lattice and assuming that the transformed function vanishes for distances larger than the actual system size [10]. As can be seen, the agreement with the asymptotic form, (2), (once the leading minus sign is included) is reasonably good. There is a well-defined peak at wave vector

$q = 2k_F$ as one would expect. It is nevertheless difficult to extract an accurate exponent by performing a fit to the asymptotic form either in momentum space or in real space, since the number of fit parameters is relatively large, and since higher order terms also contribute.

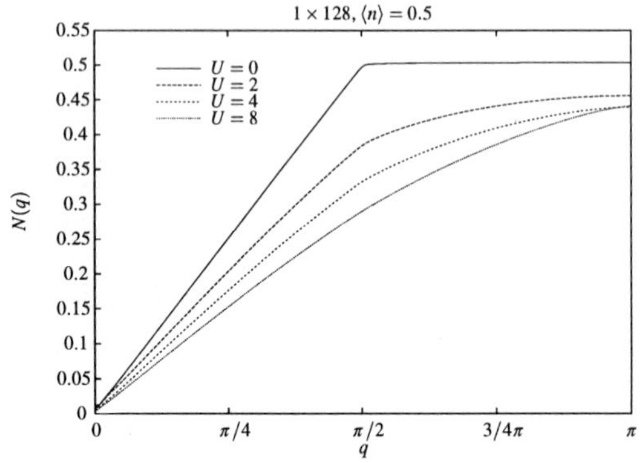

Fig. 3. Fourier transform, $N(q)$, of the density-density correlation function, $\langle \delta n_{i+r} \delta n_i \rangle$, for the quarter-filled system for various U.

The situation is even worse for the density-density correlation function, Fig. 3, since the $2k_F$ peak becomes less distinct with increasing U. However, K_ρ can also be extracted directly from the coefficient of the $1/r^2$ term in (3) using [11]

$$\lim_{q \to 0} \frac{\partial N(q)}{\partial q} = \frac{K_\rho}{\pi} . \tag{5}$$

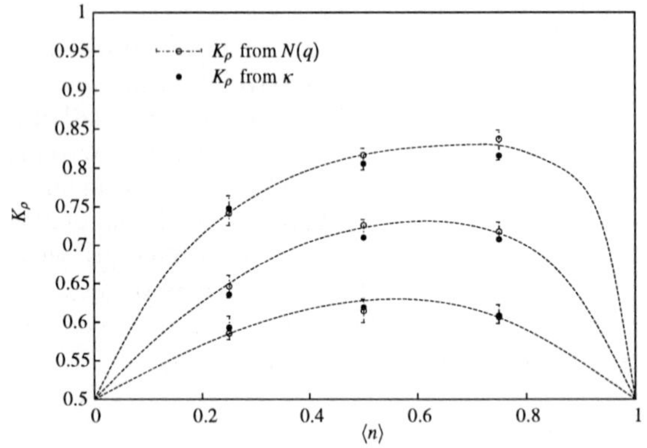

Fig. 4. The Luttinger parameter K_ρ calculated using the DMRG, compared with the results of [2] (dashed line). The error bars are estimates of the fitting error.

The parameter K_ρ calculated in this way is compared with K_ρ calculated using the Bethe Ansatz [2] in Fig. 4. We find that this method yields the most accurate K_ρ values. In the Bethe Ansatz, K_ρ is generally calculated from the compressibility

$$\kappa^{-1} = \frac{1}{L} \frac{\partial E_0(n)}{\partial n^2} , \tag{6}$$

where $E_0(n)$ is the ground-state energy at filling $n = \langle n \rangle$, and the relation

$$\frac{\pi}{2}\frac{u_\rho}{K_\rho} = \frac{1}{n^2 \kappa}\ , \qquad (7)$$

valid for a Luttinger liquid [2]. Using the DMRG, K_ρ can also be calculated using this method once u_ρ is known (see below), and as shown in Fig. 4, this method also yields accurate results.

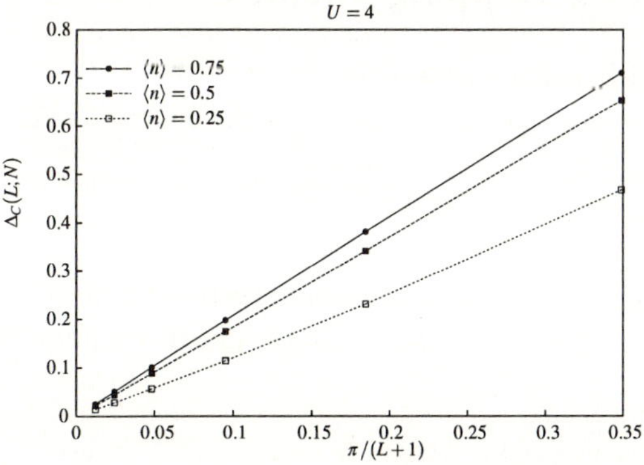

Fig. 5. The charge gap Δ_c plotted as a function of inverse system size $\pi/(L+1)$.

One can also calculate the remaining Luttinger liquid parameters, the charge velocity u_ρ and the spin velocity u_σ, using the DMRG. The idea is to utilize the finite-size dependence of the lowest spin and charge excitations. For open boundary conditions, the lowest possible wave vector of a spin or

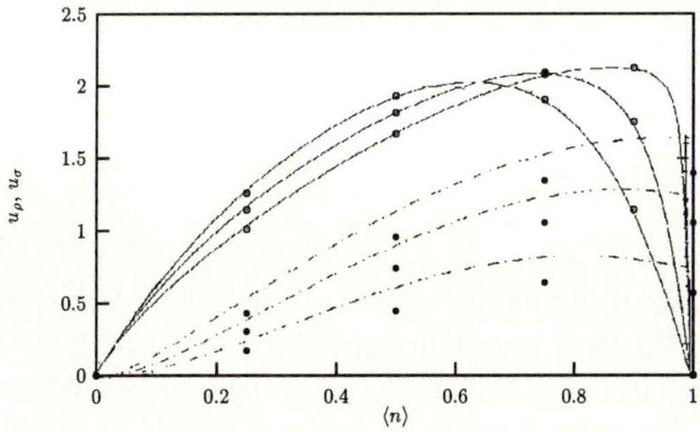

Fig. 6. Spin velocity u_σ (solid symbols) and charge velocity u_ρ (open symbols) calculated using the DMRG plotted as a function of band filling $\langle n \rangle$ for various U. The dashed (dotted) lines are spin (charge) velocities taken from [2].

charge excitation is given by $\Delta q = \pi/(L+1)$. The charge velocity is then given by $u_\rho = \lim_{L\to\infty} \Delta_c/\Delta q$ where $\Delta_c = E_1(L;N,S=0) - E_0(L;N,S=0)$ is the gap to the lowest charge excited state [12] on a system of size L, and $E_m(L;N,S)$ is the energy of the m-lowest state with N electrons and spin S. The linear dependence of Δ_c on $\Delta q = \pi/(L+1)$ is seen in Fig. 5, with different slopes corresponding to different u_ρ for three different band fillings. The spin velocity u_σ can be calculated in a similar way from the spin gap

$\Delta_s = E_0(L;N,S=1) - E_0(L;N,S=0)$. A comparison of the DMRG results and the Bethe Ansatz results from [2] is shown in Fig. 6. The agreement for the charge velocity is quite good, but there is a significant deviation for the spin velocity. The deviation for the spin velocity is due to uncontrolled higher order finite-size corrections in the scaling of the spin gap, which can be understood using the Bethe Ansatz [12].

2 Next-Nearest-Neighbor Hopping, Small U

In order to understand the small-U behavior for general t_2, it is first necessary to discuss the $U=0$ band structure. Diagonalization of the kinetic energy terms leads to the dispersion

$$\varepsilon(k) = -2t_1 \cos k - 2t_2 \cos 2k. \tag{8}$$

The behavior of the band structure for the noninteracting system [3] in terms of the band filling, n, and $-t_2$ is depicted in Fig. 7. Note that it is sufficient

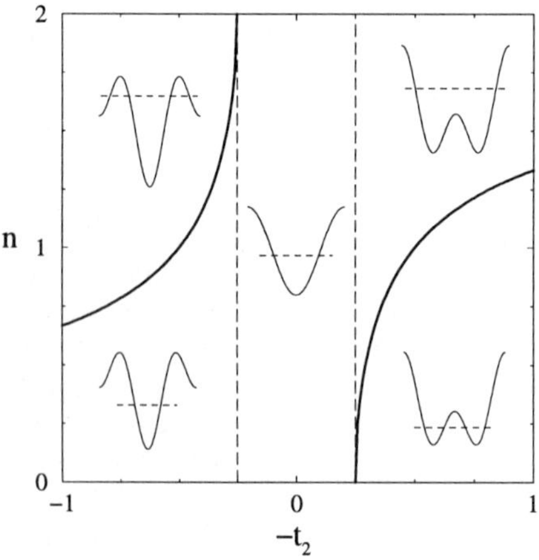

Fig. 7. The qualitative behavior of $\varepsilon(k)$ (solid line in the insets) and E_f (dashed line in the insets) for the $U=0$ system in different regions of the n–($-t_2$) plane.

to consider the right half ($t_2 < 0$) of the plane, since a particle-hole transformation maps $t_2 \to -t_2$ and $n < 1$ to $n > 1$. For $|t_2| < 0.25$, the qualitative behavior of $\varepsilon(k)$ does not change from that of the $t_2 = 0$ case, monotonically increasing from a minimum at $k = 0$ to maxima at $k = \pm\pi$. For $t_2 < -0.25$, however, $\varepsilon(k)$ has two minima. This portion of the n–t_2 plane can then divided into two parts: one in which the Fermi level E_f is above the local maximum, $\varepsilon(k=0)$ and there are two Fermi points, and one in which $E_f < \varepsilon(k=0)$ and there are four Fermi points. The qualitative behavior of $\varepsilon(k)$ and E_f is sketched in the appropriate portions of Fig. 7.

For finite but small U, one can start with the $U=0$ band structure, linearize around the appropriate Fermi points, and then obtain the phase diagram at small U analytically using the weak-coupling renormalization group and bosonization [13]. This approach can be applied quite generally to multiple-chain interacting electron systems and has been developed primarily for "ladder" geometries, i.e. coupled chains with a square interchain hopping. Since the t_1–t_2 model has either two or four Fermi points, the weak-coupling phase diagram can be mapped onto that of the two-chain ladder [4]. Here we adapt the two-chain phase diagram developed by Balents and Fisher [14] to the t_1–t_2 model, adopting their notation in which CαSβ denotes a phase with α massless bosonic charge modes and β massless bosonic spin modes. When there are two Fermi points (i.e. $|t_2/t_1|$ is sufficiently small), the weak-coupling picture is that of a single chain with a renormalized bare Fermi velocity. This leads to Luttinger liquid behavior, C1S1, at all fillings except $n = 1$, where Umklapp processes lead to a charge gap and therefore a C0S1 phase, just as in the $t_2 = 0$ case. When there are four Fermi points, the generic behavior is a spin-gapped C1S0 phase, a Luther-Emery liquid in which superconducting and charge-density-wave (CDW) correlations coexist. Other phases can occur when the Fermi velocities are very different and where special values of the Fermi wave vectors lead to Umklapp processes.

The resulting weak-coupling phase diagram in the $t_2 < 0$, $n \leq 1$ quadrant is shown in Fig. 8. (The behavior is less interesting for $n > 1$ since there are four Fermi points only at large $|t_2|$.) The weak-coupling phases expected

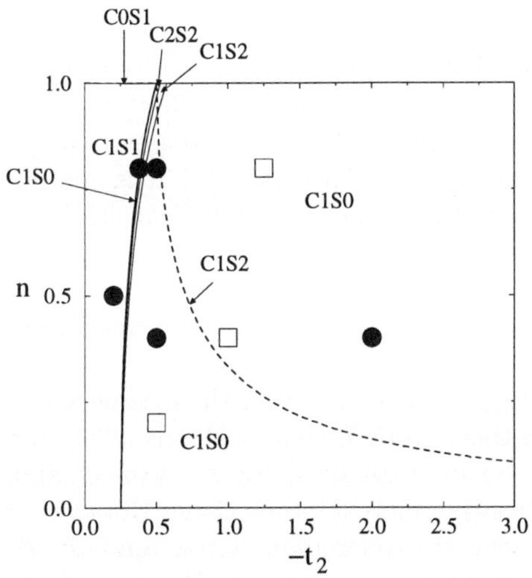

Fig. 8. Low-U phase diagram restricted to negative t_2 and $n < 1$. The filled circles are for parameters for which no spin gap has been found with DMRG at $U = 2$, while the open squares are for parameters where a spin gap has been found.

from an adaptation of the phase diagram of [14] are marked, with the solid lines marking phase boundaries near the transition from two Fermi points to four Fermi points, (see also Fig. 7). Along the dashed line, the larger Fermi vector, k_{F_2}, takes on the value $k_{F_2} = \pi$, resulting in Umklapp processes which can lead to a non-spin-gapped C1S2 phase along some of this line.

We examine the validity of the weak-coupling phase diagram at small but finite U with the DMRG. We take $U = 2$ and calculate the spin gap Δ_s and the charge gap $\Delta'_c = [E_0(L; N+2, S=0) + E_0(L; N-2, 0) - 2E_0(L; N, 0)]/2$

for finite-size systems of length L and then extrapolate to $L \to \infty$. Note that Δ'_c is different from Δ_c defined in Sect. 1, and is obtained from the chemical potential for adding pairs of particles. The results are signified by different symbols in the phase diagram. We find a vanishing charge gap for all $n < 1$. For small $|t_2|$, we find a vanishing spin gap, in agreement with the weak-coupling prediction, but with a phase boundary shifted to slightly larger $|t_2|$ than the weak-coupling prediction. Such a shift of the phase boundary is seen in the two-chain Hubbard ladder as U is increased [15]. At intermediate t_2, we find a spin gap, in agreement with the C1S0 phase predicted by the weak-coupling analysis. At larger $|t_2|$, we find that the spin gap vanishes to within our resolution. This is a region in which the system consists of two chains with a weak zigzag interchain coupling. In the weak-coupling analysis, a spin gap is still present in this parameter regime [4], but it should be small, perhaps too small to resolve in our numerical calculations.

3 Strong Coupling

One of the original motivations for formulating the Hubbard model was to study metallic ferromagnetism in the transition metals [16,17]. The mean-field picture is encouraging because the phase diagram has a relatively large region in which the ferromagnetic phase is stable. However, as was already known by Hubbard [16], fluctuations tend to destabilize the ferromagnetic phase. There is a proof by Nagaoka which shows that for $U = \infty$ and one hole in the half-filled band, the ground state is fully polarized for bipartite lattices in $d \geq 2$ and for fcc and hcp lattices with negative hopping integrals [18]. It is unclear, however, to what extent the Nagaoka phase extends to finite densities and physical values of interaction. For example, for the two-dimensional square lattice, variational methods [19] and numerical calculations [20,21] restrict the fully polarized phase to densities close to half-filling and to quite large U.

In one dimension, an even more rigorous constraint on the existence of a ferromagnetic phase is provided by a theorem by Lieb and Mattis [22] which proves that the ground state must be unmagnetized for nearest-neighbor hopping and a general class of particle-symmetric interactions. However, if next-nearest-neighbor hopping is present, the theorem no longer applies, and a ferromagnetic phase is possible. A local picture of the tendency to ferromagnetism in the t_1–t_2 model is provided by small clusters. On a three-site cluster with two electrons and open boundary conditions, the ground state is a triplet when $t_2 < 0$ [23]. A similar tendency to a polarized ground state is present in the four-site cluster [3].

It is first useful to examine the limit of strong interaction, $U = \infty$. In the t_1–t_2 Hubbard chain, there are three separate limiting cases for which the ground state has the maximum spin (i.e. is fully polarized) at $U = \infty$. The Nagaoka proof for one hole away from half-filling is applicable and yields a fully polarized ground state when $t_2 < 0$ [24]. For $t_2 = 0^-$, a perturbative treatment of t_2 using the exact $t_2 = 0$ wave function yields a ferromagnetic spin interaction and thus a fully polarized ground state [25]. Finally, when the dispersion has two minima (i.e. $t_2 < -0.25$), Müller-Hartmann has shown that the ground state is fully polarized in the limit of low density [26]. These three limits are depicted in the n–t_2 phase diagram in Fig. 9(a), along with the $t_2/t_1 \to -\infty$ limit, in which the chains become uncoupled and the ground state is paramagnetic.

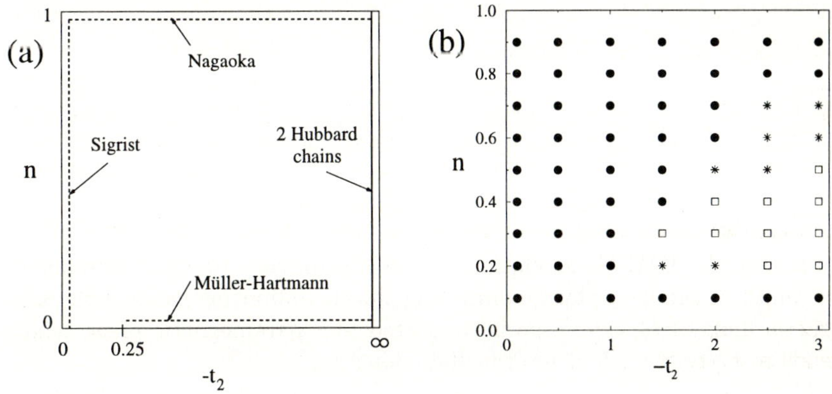

Fig. 9. The $U = \infty$ phase diagram showing **(a)** the exact limits discussed in the text and **(b)** the DMRG results, with the filled circles indicating a fully polarized ground state, the open squares a nonmagnetic ground state, and the stars a partially polarized ground state.

The DMRG can directly calculate the energy, E_0, and the expectation value of the total spin of the ground state, $\langle S^2 \rangle$, on a finite-size lattice. The total spin can also be determined by examining the S_z degeneracy of the low-lying states. Comparison of E_0 with the energy of the fully polarized state, E_{ferro}, which is exactly known since it is an eigenstate of the Hamiltonian, can then determine whether the system is in a ferromagnetic phase. The total spin of the ground state can be used to confirm the polarization. This is illustrated in Fig. 10. It is clear that the ground state energy goes to that of the fully polarized state at the same U value at which the total spin jumps to a value close to the fully polarized value, $S(S+1) = 30$ for $N = 10$.

In order to make contact with the exact $U = \infty$ limits discussed above, we have calculated the $U = \infty$ phase diagram using the DMRG, Fig. 9(b). The solid points mark parameter values for which the DMRG energy, $E_{\text{DMRG}} > E_{\text{ferro}}$, and for which $S(S+1)$ is approximately the maximum value and are therefore classified as fully polarized. When $E_{\text{DMRG}} < E_{\text{ferro}}$ and $S \approx 0$, we classify the phase as paramagnetic (open squares). On the border between these phases, we sometimes find $E_{\text{DMRG}} < E_{\text{ferro}}$, but $S > 0$, marked by stars. These partially polarized points could either be due to a second order transition between the fully polarized and the paramagnetic phases, or could

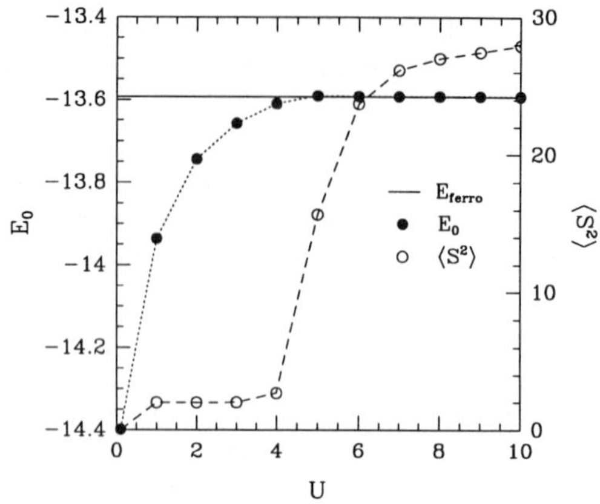

Fig. 10. E_0 and $\langle S^2 \rangle = S(S+1)$ for a system of size $L = 30$ with $N = 10$ electrons and $t_2 = -0.5$. The energy E_{ferro} is exactly known for the fully polarized state.

be an artifact of numerical inaccuracies in the DMRG which can occur when states of different total spin S are nearly degenerate.

As can be seen in Fig. 9(b), the exact limits shown in Fig. 9(a) are reproduced in the DMRG calculation, and the ferromagnetic phase extends to cover much of the n–t_2 plane, connecting the three limiting cases. The paramagnetic limit at $t_2/t_1 \to -\infty$ is connected to a paramegnetic phase which extends to $t_2/t_1 \approx -1.5$ at intermediate densities.

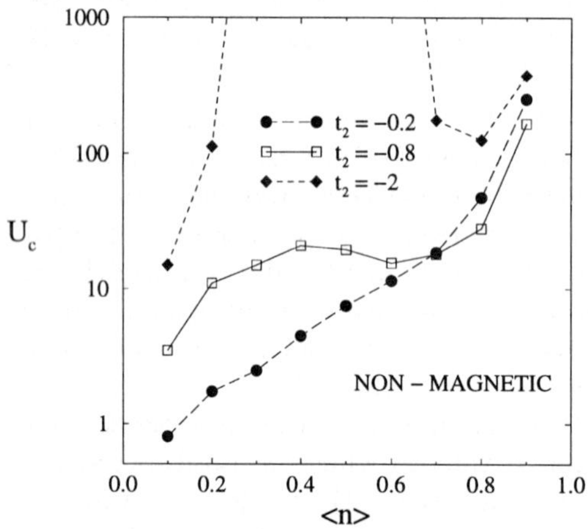

Fig. 11. Semilog plot of U_c as a function of density for $L = 50$ and three t_2 values.

At finite U, there are three different behaviors of the critical interaction strength, U_c, as a function of density, as depicted in Fig. 11. For $-0.25 < t_2 < 0$, U_c increases monotonically with density, as seen for $t_2 = -0.2$. In this region, $\varepsilon(k)$ has one minima (see Fig. 7), and the noninteracting density of states smoothly decreases with ε. For $-1.5 < t_2 < -0.25$, the system is ferromagnetic for all n, but there is a local minimum in U_c which corresponds to a filling in which the E_F for the fully polarized state lies at the local maxima, $\varepsilon(k = 0)$, corresponding to a singularity in the noninteracting density of states. Finally, for $t_2 < -1.5$, there is paramagnetic region at intermediate n, and a reentrance into the ferromagnetic phase at larger n with a minimum in U_c corresponding to the singularity in the noninteracting density of states at $\varepsilon(k=0)$.

The one-dimensional Hubbard model with next-nearest-neighbor hopping therefore has a large region of ferromagnetism in the strong-coupling phase diagram, with critical interaction strength which can extend down to intermediate values. The sign of the next-nearest-neighbor hopping t_2 is crucial, with ferromagnetism occuring only when $t_2 < 0$ for hole-doping of the half-filled system, which is equivalent to particle doping and $t_2 > 0$. The qualitative behavior of the noninteracting density of states in the ferromagnetic regime is in agreement with a general picture in which an asymmetric form with stronger peaks in the occupied portion is argued to be favorable for ferromagnetism [27].

4 Half-Filling

Distinctive features at both weak coupling and strong coupling make the phase diagram for the half-filled model interesting to study separately. At weak coupling there is a insulator-metal transition that occurs with increasing t_2, as is seen along the $n=1$ axis of Fig. 8. For small t_2, there are two Fermi points and $2k_F = \pi$, leading to a C0S1 (half-filled Luttinger liquid) phase. When $t_2 < -0.5$, however, there are no relevant Umklapp processes and the metallic C1S0 phase is obtained [4]. For strong U, one can map to the frustrated Heisenberg model (J_1–J_2 model) via a strong-coupling expansion. The J_1–J_2 model has a nontrivial phase diagram as a function of J_2/J_1 [28]. At small J_2, the system is in a 1D Heisenberg phase with gapless spin excitations and critical correlations. At intermediate J_2, there is a spin-gapped "dimer" phase which contains the exactly solvable Majumdar-Gosh point at $J_2/J_1 = 0.5$, whose exact ground can be obtained as a product of dimerized singlets. For $J_2/J_1 > 0.5$ there is an incommesurate, spin-gapped "spiral" phase.

We have also carried out a Hartree-Fock calculation [5] in which we have assumed constant density and magnetic order $\mathbf{m} = m(\cos qr_i, \sin qr_i, 0)$, i.e. spiral order at wavevector q. The strong-coupling (J_1–J_2 model) and the weak-coupling behavior are qualitatively reproduced in the Hartree-Fock phase diagram, as seen in Fig. 12(a). There is an ordered $q = \pi$ phase at small t_2 and all U, a metallic phase at small U and $t_2 < -1/2$, and an incommensurate spiral phase at larger U and $|t_2|$, with $q \to \pi/2$ as $t_2 \to -\infty$. However, the magnetic phases in Hartree-Fock theory are ordered, whereas fluctuations destroy long-range antiferromagnetic order in the real one-dimensional system.

The DMRG phase diagram, Fig. 12 (b), confirms the strong and weak-coupling pictures and is qualitatively similar to the Hartree-Fock phase dia-

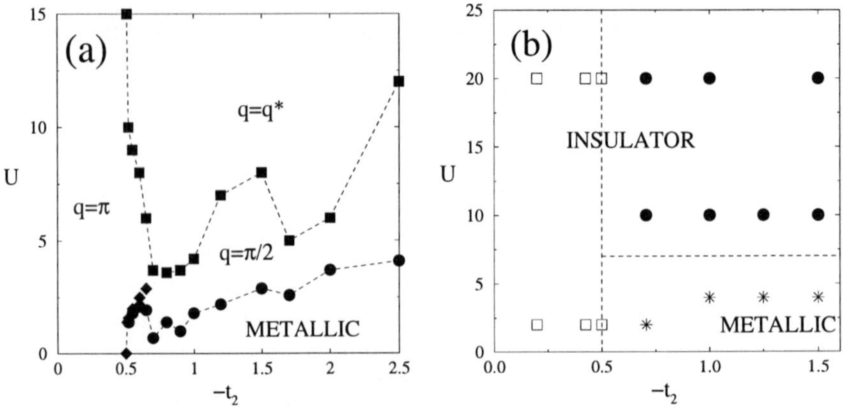

Fig. 12. (**a**) The unrestricted Hartree-Fock phase diagram with the insulating phases labelled by the wave vector q when $m > 0$, and (**b**) the DMRG phase diagram with the open squares representing a C0S1 phase, the filled circles a C0S0 phase, and the stars a C1S0 phase. The lines are a guide to the eye.

gram. At low $|t_2|$, we obtain a C0S1 phase with $q = \pi$ for all U; at larger $|t_2|$ and small U, we find the metallic C1S0 phase expected from weak-coupling, and at large U, there is a spin gap whose magnitude agrees well with that calculated for the J_1–J_2 model and whose peak wave vector $q \to \pi/2$ as $t_2 \to -\infty$.

References

1. V.J. Emery in Highly Conducting One-Dimensional Solids, edited by J. Devreese, R. Evrand and V. Van Doren, Plenum Press, p 247 (1979); J. Solyom, Adv. Phys. **28**, 201 (1979); F.D.M Haldane, J. Phys. C Solid State Phys. **14**, 2585 (1981)
2. H.J. Schulz, Int. J. Mod. Phys. B **5**, 57 (1991)
3. S. Daul and R.M. Noack, Phys. Rev. B **58**, 2635 (1998)
4. M. Fabrizio, Phys. Rev. B **54**, 10054 (1996)
5. S. Daul and R.M. Noack, unpublished
6. S. Daul, Ph.D. dissertation, Université de Fribourg, 1998, unpublished
7. S.R. White, Phys. Rev. Lett. **69**, 2863 (1992); Phys. Rev. B **48**, 10345 (1993)
8. S. Kneer, Diploma Thesis, Universität Würzburg (1997)
9. G. Bedürftig, B. Brendel, H. Frahm and R. M. Noack, Phys. Rev. B **58**, 10225 (1998)
10. For a discussion of the approximation of a continuous Fourier transform via a discrete Fourier transform see, for example, W.H. Press, S.A. Teukolsky, W.T. Vetterling, and B.P. Flannery, Numerical Recipes in C, second edition, Cambridge University press, chaps. 12–13 (1992)
11. M. Dzierzawa, in The Hubbard model, Ed. by D. Baeriswyl et al., New York (1995)
12. For the nearest-neighbor chain, the finite-size corrections are well-understood from the Bethe Ansatz. In particular, the lowest $S = 0$ is often a "string" excited

state, so that the second $S = 0$ excited state is the lowest charge excitation. The gap $\Delta_s = E_0(S = 1) - E_0(S = 0)$ has higher order corrections which only converge at very large system sizes, but an alternative gap $\Delta_s' = E_1(S_z = 1) - E_0(S_z = 1)$, with $E_1(S_z = 1)$ the first spin-particle-hole-excitation, converges more quickly to the $L = \infty$ result (G. Bedürftig, private communication).
13. C. Varma and A. Zawadowski, Phys. Rev. B **32**, 7399 (1985); H.J. Schulz, Phys. Rev. B **53**, 2959 (1996)
14. L. Balents and M.P.A. Fisher, Phys. Rev. B **53**, 12 133 (1996)
15. R.M. Noack, S.R. White and D.J. Scalapino, Physica C **270**, 281 (1996)
16. J. Hubbard, Proc. Roy. Soc. A **276**, 238 (1963)
17. M. C. Gutzwiller, Phys. Rev. Lett. **10**, 159 (1963); J. Kanamori, Prog. Theor. Phys. **30**, 275 (1963)
18. Y. Nagaoka, Phys. Rev. **147**, 392 (1966)
19. W. von der Linden and D.M. Edwards, J. Phys.: Condens. Matter **3**, 4917 (1991); T. Hanisch, G.S. Uhrig and E. Müller-Hartmann, Phys. Rev. B **56**, 13 960 (1997)
20. R. Hirsch, Dissertation, Universität Köln, (1994): Zum Magnetismus stark korrelierter Fermionsysteme, Verlag Shaker, Aachen (1994)
21. S. Liang and H. Pang, Europhys. Lett. **32**, 173 (1995)
22. E. Lieb and D. Mattis, Phys. Rev. **125**, 164 (1962)
23. H. Tasaki, preprint cond-mat/9512169
24. D.C. Mattis and R.E. Peña, Phys. Rev. B **10**, 1006 (1974)
25. M. Sigrist, H. Tsunetsugu, K. Ueda and T.M. Rice, Phys. Rev. B **46**, 13838 (1992)
26. E. Müller-Hartmann, J. Low Temp. Phys. **99**, 349 (1995)
27. J. Wahle, N. Blümer, J. Schlipf, K. Held and D. Vollhardt, Phys. Rev. B **58**, 12 749 (1998); D. Vollhardt *et al.*, preprint cond-mat/9804112
28. S.R. White and I. Affleck, Phys. Rev. B **54**, 9862 (1996) and references therein.

1.2 Soliton Bound-States in Dimerized Spin Chains

Erik S. Sørensen[1], Ian Affleck[2], David Augier[1], and Didier Poilblanc[1]

[1] Laboratoire de Physique Quantique and UMR CNRS 5626
Université Paul Sabatier
F-31062 Toulouse, France
[2] Department of Physics and Astronomy and
Canadian Institute for Advanced Research
University of British Columbia
Vancouver, BC, V6T 1Z1, Canada

In the beginning of the eighties Faddeev and Takhtajan [1] suggested that the fundamental excitations of the one-dimensional spin one-half Heisenberg model should be viewed as massless kink-like $s = 1/2$ objects, the so called spinons. Although this is not the only way to think of the excitations in the Heisenberg model it is often a quite useful picture. A natural generalization of the Heisenberg model is to include a next-nearest neighbor term $\mathbf{S}_i \cdot \mathbf{S}_{i+2}$ of strength J_2, resulting in the following Hamiltonian:

$$H = \sum_i [J\mathbf{S}_i \cdot \mathbf{S}_{i+1} + J_2 \mathbf{S}_i \cdot \mathbf{S}_{i+2}], \tag{1}$$

Several physical compounds exists that are fairly well described by such a Hamiltonian. Perhaps most notably $CuGeO_3$ [2]. It is in this case useful to think of the spins as being arranged in a zig-zag manner as shown in Fig. 1. At the critical coupling $J_{2c} \simeq 0.241167$ [3–5] systems described

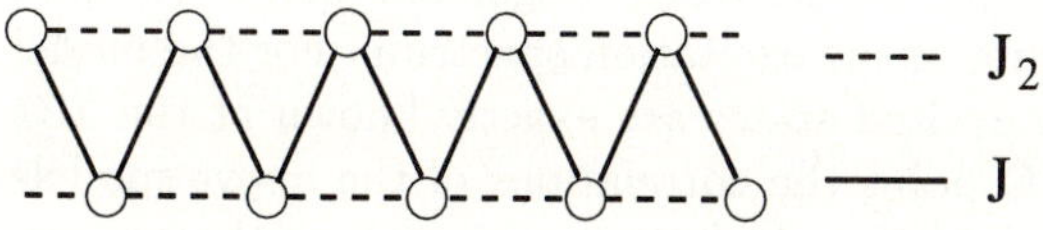

Fig. 1. The zig-zag chain

by the Hamiltonian (1) enters a spontaneously dimerized phase and a gap opens up in the excitation spectrum. The Majumdar-Ghosh [6] (MG) point, $J_2 = J/2$, constitutes a disorder point [7] beyond which the short-range correlations become incommensurate. The associated Lifschitz point, where the peak in the structure factor moves away from π, has been determined to be $J_2/J = 0.52063(6)$ [8] using DMRG methods and the ensuing short-range incommensurate correlations have been studied in detail by White and Affleck [9]. In the present contribution we focus on how the behavior of the spinons is changed once the dimerized phase is entered. The fundamental excitations are now real kinks separating two different dimerization patterns and we therefore refer to them as solitons. These solitons become massive in the dimerized phase.

In order to take into account the three-dimensional lattice in a one-dimensional model one often introduces an explicit dimerization δ yielding a staggered nearest neighbor coupling:

$$H = \sum_i [J(1 + \delta(-1)^i)\mathbf{S}_i \cdot \mathbf{S}_{i+1} + J_2 \mathbf{S}_i \cdot \mathbf{S}_{i+2}]. \tag{2}$$

The phase diagram of this model as well as the correlations have previously been investigated using DMRG techniques by Pati et al. [10] and Chitra et al. [11]. Schönfeld et al. [12] have investigated the magnetic field phase diagram using DMRG methods and the thermodynamics of related models have been investigated using transfer-matrix DMRG by Maisinger et al. [13]. Our interest here is to study the spectrum of low-lying excitations in the dimerized phase, $J_2 > J_{2c}$, to show that solitons and anti-solitons form a ladder of triplet-singlet bound-states [14].

A solitonic picture of the excitations in frustrated spin chains was first introduced by Shastry and Sutherland [15] and more recently by Khomskii et al. [16] as well as others [17–20]. In the context of the so called saw-tooth lattice or delta-chain, corresponding to Fig. 1 with every second J_2 coupling set to zero, the solitonic excitations have been investigated by Nakamura et al. [21] and Sen et al. [22]. In this case the symmetry between solitons and anti-solitons is absent.

1 Numerical Considerations

In order to study the excitations in the dimerized phase we have performed extensive DMRG calculations. See the contribution by Noack and White in Chap. 2 (I). However, where needed we use exact diagonalization techniques to obtain a more complete picture of the excitation spectrum. For the model (1) the ground state and a few excited states are exactly known at the MG point. In the vicinity of the MG point the correlations of the above models are extremely short-ranged and DMRG calculations are exact for the matrix-product ground state of the Hamiltonian (1) at the MG point. For a discussion of matrix-product states see Chap. 3(I). The short correlation length is very helpful when performing numerical work since finite-size corrections will be very small and for the exact diagonalization results we are essentially limited to the neighborhood of the MG point if we want to study the excitations in detail. The short correlation length also improves the precision of the DMRG calculations tremendously and we usually never have to keep more than $m = 128$ states. In order to study the excitations it is however crucial that we can distinguish odd and even multiplets. This is done by performing the calculations in the $S^z_T = \sum_i S^z_i = 0$ subspace and using spin-inversion as a symmetry along with real-space parity. Both of these symmetries reduce the size of the Hilbert space by roughly a factor of two and hence combining them reduces the overall computational over-head by a factor of four. Spin-inversion has the added benefit of being an 'on-site' symmetry and can therefore also be used in combination with the finite-system method. Under spin-inversion even and odd total spin multiplets transform differently and can be distinguished. Symmetries are extensively used in DMRG calculations and for more details we refer the reader to Chaps. 2.2(II) and 2.3(II) A simple way of implementing parity in the infinite-system method is discussed in [23], see also [24]. In [25] the implementation of more complicated symmetries is discussed.

2 The Soliton Gap

We begin with a discussion of systems without any explicit dimerization, described by (1). For a system with an even number of spins the ground state is always a singlet. However, for an odd number of spins the lowest energy state has spin $\frac{1}{2}$. We therefore say that the odd-site system contains one soliton. One way of defining the energy, Δ_{sol}, required to create a soliton is therefore to say that it is equal to $\Delta_{sol} = \lim_{n\to\infty} E_0(L = 2n+1) - (2n+1)e_0$, where $E_0(2n+1)$ is the ground-state energy of an odd-length system with $L = 2n+1$ sites, and e_0 is some measure of the ground-state energy per spin of the corresponding even-length system. Here it is crucial that periodic boundary conditions are used. Similar ideas have been used by Malék et al. [19]. At the MG point the ground-state energy for even-length systems is $-3LJ/8$, hence $e_0 = -3J/8$ does not depend on L. Thus, using exact diagonalization techniques, we have obtained $\Delta_{sol}/J = 0.1168$, $J_2 = J/2$. If

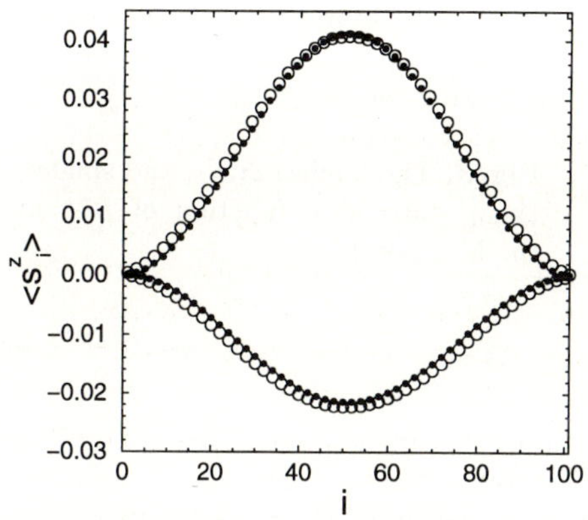

Fig. 2. $\langle S_i^z \rangle$ as a function of i. DMRG results (open circles) are shown for L=101, with $S^z_{tot} = 1/2$, $m = 128$. Clearly the soliton is repelled by the open ends and enters approximately a particle in a box state $\langle S_i^z \rangle \simeq [\text{const} + (-1)^i \sin^2(\pi i/(L+1))]$ indicated by the solid circles.

the soliton does not interact with the chain boundaries we can check this result using DMRG calculations on systems with open boundary conditions. This turns out to be the case and DMRG results for the soliton gap are in complete agreement with the above result. However, the fact that the soliton remains a 'bulk' excitation is non-trivial. In Fig. 2 the ground-state spin-density is shown for an odd-length system. Clearly, the soliton is repelled by the open boundaries and enters roughly a particle in a box state as indicated by the solid circles. If the density of solitons can be increased sufficiently the spin-density should be experimentally observable in NMR experiments. Away from the MG point finite-size corrections to Δ_{sol} become much more important and a reliable determination is only possible using DMRG calculations. As a function of J_2 we have calculated the soliton gap (Ref. [14]) for $J_2 < J/2$, observing an exponential decrease as J_{2c} is approached. For $J_2 \rangle J/2$ short-range correlations become incommensurate [8] rendering a reliable determination of $E_0(L = 2n+1)$ very difficult. Using exact diagonalizations techniques the complete dispersion of the soliton state for odd-length systems can be determined. We have done this at the MG point and a well defined single mode is observable around $q_0 = \pi/2$ [14] with a minimum corresponding to the soliton gap. The soliton s always moves two sites when it moves and hence lives exclusively on one sublattice with the anti-soliton $\bar{s}$ confined to the other. Hence, it is natural to expect a minimum in the dispersion at $q_0 = \pi/2$. Assuming a relativistic dispersion of the soliton, we find for q', the wave-vector closest to q_0:

$$E_{sol}(q') \simeq \Delta_{sol} + \frac{(vq_0)^2}{2\Delta_{sol}} \frac{1}{L^2} + \mathcal{O}(L^{-3}). \tag{3}$$

q' is also the lowest energy state for the odd-length system. Hence, exploiting the finite-size corrections for systems with open boundaries we can determine the soliton velocity, v. At the MG point we find $v/J \simeq 0.43 - 0.45a$.

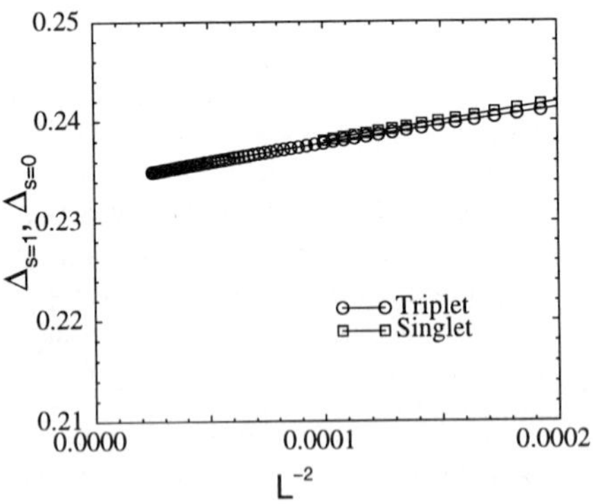

Fig. 3. The triplet, $\Delta_{s=1}$, and singlet, $\Delta_{s=0}$, gaps as a function of L^{-2} at the MG point.

Perhaps the most interesting question to ask is whether $s\bar{s}$ bound-states form below the continuum. If such bound-states were to form one would expect them to come in triplet-singlet pairs since the soliton and anti-soliton each are $s = 1/2$ particles. Furthermore, if the soliton and anti-soliton interact we would expect a splitting between the associated singlet and triplet, whereas in the absence of any interaction they should be degenerate and both have an energy of precisely $2\Delta_{sol}$. Using spin-inversion to separate the triplet and singlets we have calculated the two gaps for $J_2 < J/2$. The results are shown in Fig. 3 for the MG point. Clearly, the singlet and triplet become degenerate in the thermodynamic limit and performing an extrapolation we find $\Delta_{s=0} = \Delta_{s=1} \simeq 0.2340(1)J = 2\Delta_{sol}$. One finds similar results for any $J_2 < J/2$ and we conclude that for these values of J_2 bound-states do not form close to $q = 0, \pi$. However, it remains a possibility that $s\bar{s}$ bound-states form around $q = \pi/2$, where exact bound-states are known [26], or around $q = 0, \pi$ for $J_2 > J/2$ [14]. Thus, the soliton gap, Δ_{sol}, can actually be inferred directly from the triplet gap, $\Delta_{s=1}$, as calculated by White and Affleck [9]. For $J_2 \leq J_{2c}$ the picture is quite different since the system is initially gapless. However, in the presence of a small δ, it is expected from field theoretical arguments that $\Delta_{s=0}/\Delta_{s=1} = \sqrt{3}$ [27,3,28,29] in the limit of $\delta \to 0$ for any $J_2 \leq J_{2c}$.

3 Explicitly Dimerized Systems - $s\bar{s}$ Bound States

We next consider systems described by (2), where an explicit dimerization δ is included. In this case, it has been proposed [18] that a ladder of triplet-singlet bound-states should occur. The introduction of the explicit dimerization will favor one of the two otherwise almost degenerate ground states, the other one being higher in energy by a factor $\sim L\delta$. Hence if the soliton and anti-soliton are separated by only a small distance one can imagine that between them the 'wrong' ground state is found, and it then seems likely that for small δ a linear potential will be created between the soliton and anti-soliton. This is

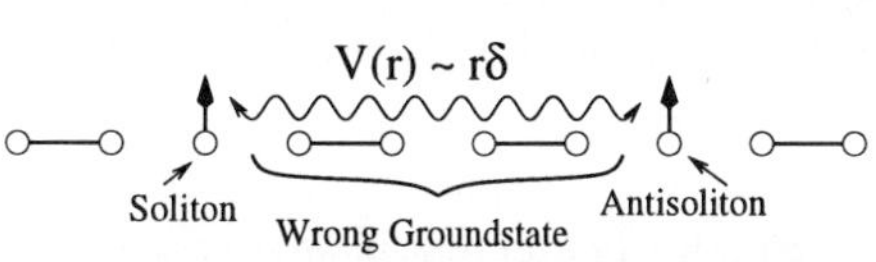

Fig. 4. The creation of a linear potential between a soliton and an anti-soliton at a distance r.

tentatively indicated in Fig. 4. The lowest lying triplet state in the dimerized phase is within this picture best described as a $s\bar{s}$ bound-states and not as a magnon although the difference between the two is largely one of nomenclature. As δ grows the soliton anti-soliton pair becomes tighter and tighter bound and will eventually resemble a single well-defined magnon. The triplet bound-state is always accompanied by a singlet state at a slightly higher energy and at still higher energies a number of other triplet-singlet bound-states appear as indicated in Fig. 5 in the rest-frame of the particle. In real systems these bound-states can acquire a momentum and disperse. Eventually one reaches energies where it becomes favorable to break up the $s\bar{s}$ bound-state in two low-lying $s\bar{s}$ pairs with energy $2\Delta_{s=1}$. This point constitutes the onset of the $s\bar{s}$-$s\bar{s}$ continuum and can be numerically determined as the lowest lying $S = 2$ state. The natural question to ask is now how many bound-states

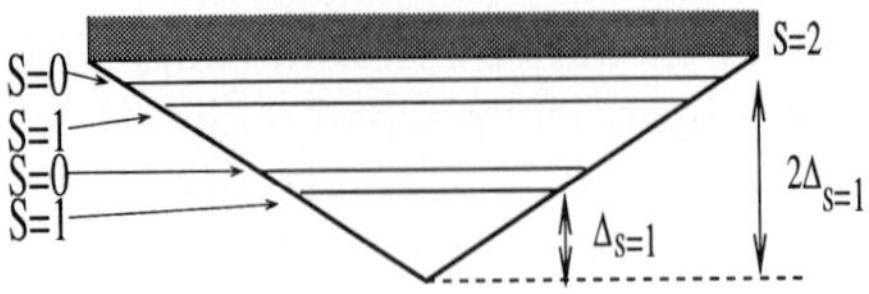

Fig. 5. Singlet and triplet states in the linear potential. The continuum begins at the lowest $S = 2$ state, $E = 2\Delta_{s=1} \sim 4\Delta_{sol}$.

occur. As δ decreases the linear potential becomes more shallow and can accommodate more bound-states. The continuum, defined as the energy where we can 'pair-produce' will also decrease but not as fast. Hence we expect the number of such bound-state to increase with decreasing δ. At the MG point, Casper and Magnus [26] have found two exact $s\bar{s}$ bound-states, a triplet and singlet both with energy J in our units. One can show that these states remain exact eigenstates along the disorder line, $\delta = 1 - 2J_2/J$, with energy $(1+\delta)J$ (triplet) and $(1+3\delta)J$ (singlet), and momentum $q = \pi/2$. These states are therefore rather high-lying and could possible form a well-defined single mode at this rather high energy around $q = \pi/2$ even with $\delta = 0$, as suggested by Shastry and Sutherland [15]. It is interesting to note that the splitting between the first triplet and singlet bound-state is always $2\delta J$ along the disorder line at $q = \pi/2$. Numerically, it turns out that this splitting varies only slightly with q. Hence, even at $q = 0$ do we find a splitting close to $2\delta J$. Using DMRG techniques with spin inversion we can follow the lowest lying triplet and singlet states in most of the parameter space. However, to get a complete picture of the dispersion of these bound-states we are limited to exact diagonalization studies in regions where the correlation length is very small in order to limit finite-size effects. This is the case along the disorder line and in particular in the vicinity of the MG point. Some of our results are shown in Fig. 7 which clearly show three triplet bound-states and at least two singlet bound-state below the continuum for $J_2 = J/2$, $\delta = 0.05$.

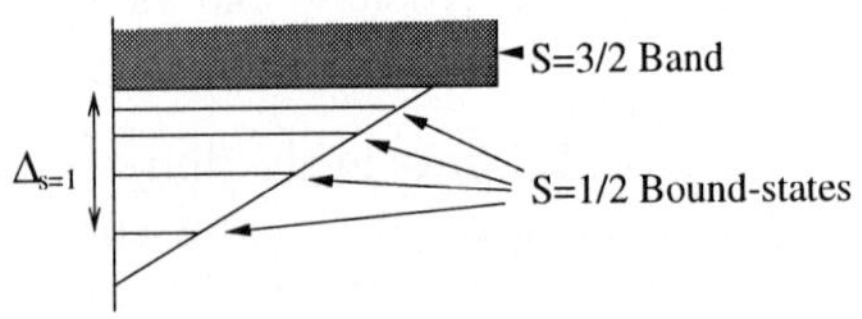

Fig. 6. The potential well created by an open boundary. For the case of an odd site system the continuum will begin at $\Delta_{s=1}$, the lowest state with $S = 3/2$.

Another important effect is the interaction between impurities and the solitons. A strong impurity will completely break the chains and naively one can model them simply as open chains. The question now becomes if the solitons somehow will bind to the ends of an open chain. In order to investigate this, we imagine an open chain with an odd number of sites and hence a single soliton in the ground state. At the end of the chain that starts with a weak link a localizing potential will now be created as indicated in Fig. 6. Due to the non-zero δ the soliton will bind to the chain-end and a number of $S = 1/2$

soliton-impurity bound-states will be possible in this potential well. As was the case for the $s\bar{s}$ states the soliton-impurity complex can thus be excited into a number of higher lying bound-states before one reaches the continuum defined by the energy where we can excite bulk $s\bar{s}$ states, i. e. $\Delta_{s=1}$. Experimentally transitions between such bound-states can be investigated using Raman, NMR or ESR techniques. Numerically, these bound-states are easily accessible since they are localized. Around the MG point a number of them occur [30].

4 Experimental Evidence

Although several compounds exists that might approximately be described using the simplified Hamiltonians, (1) and (2), we shall here concentrate on $CuGeO_3$. Following the initial discovery of a spin-Peierls transition in $CuGeO_3$ by Hase et al. [2] a large number of experimental investigations ensued. The experimental situation has been reviewed by Boucher and Regnault [31], and we concentrate on the more recent results of relevance to the presence of soliton bound-states. Couplings to the three-dimensional lattice play an important role in $CuGeO_3$ [32]. These three-dimensional couplings are taken into account in (2) only at the mean-field level through δ, leading to estimates of J=160 K, $J_2/J = 0.36$, $\delta = 0.014$ [33], and any dynamical effects stemming from phonons are therefore neglected [34]. Inelastic neutron

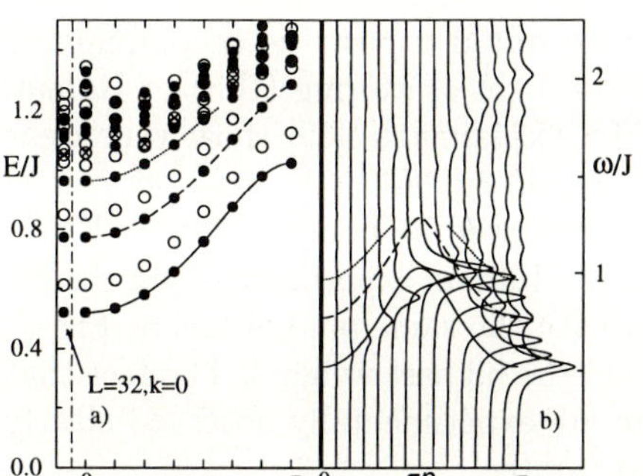

Fig. 7. (**a**) The lowest-lying triplet (solid circles), singlets (open circles) and quintuplets (crosses) for $J_2/J = 0.5$, $\delta = 0.05$, $L = 28$, as a function of $k = 2q$. Results for $L = 32$, $k = 0$ are shown to the left. (**b**) The dynamical structure factor $S^{zz}(q, \omega)$ for $J_2/J = 0.5$, $\delta = 0.05$, $L = 28$ and a broadening of $\varepsilon = 0.04J$. The solid, dashed and dotted lines indicate the 3 triplet branches.

scattering (INS) experiments [35–37] on $CuGeO_3$ have shown rather clear evidence for a well-defined dispersive mode around $q = \pi$, at 2.1meV=16.8cm^{-1}, with a second gap to a continuum. INS is sensitive to triplet excitations and it seems reasonable to interpret this mode as a $s\bar{s}$ triplet bound-state. Additional triplet bound-states appear to be absent in the data. Unfortunately, it is not possible to calculate the number of bound-states at the physical relevant parameters since the correlation length is too big. Secondly, using DMRG techniques we can so far only access the lowest lying singlet and triplet states as well as the continuum at $S = 2$. We therefore study the neighborhood of the MG point ($J_2 = J/2$) where finite-size corrections are very small. With a small dimerization, $\delta = 0.05$, our results are shown in Fig. 7. As shown in the first panel a number of $s\bar{s}$ bound-states clearly occur in agreement with previous results [38]. In the second panel is shown the associated spectral weight. It is interesting to note that the higher-lying triplet bound-states yield an almost negligible weight and hence would be difficult to observe experimentally. Secondly, the spectral weight stemming from the $s\bar{s}$ bound-states is concentrated almost exclusively around $q = \pi$ in agreement with the experiments. Even though no clear evidence is seen

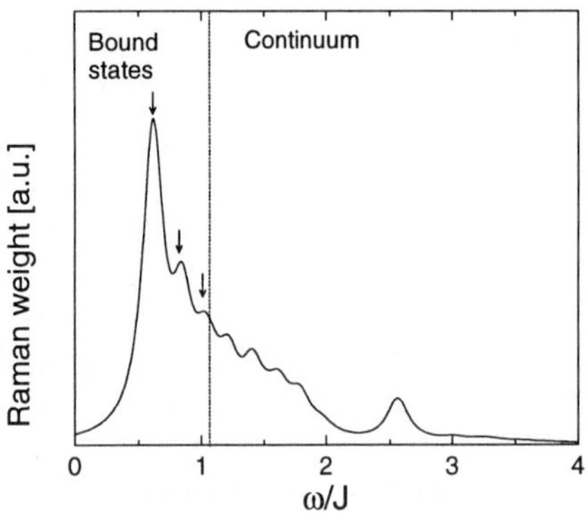

Fig. 8. Raman intensity calculated for a $L = 28$ size system with $J_2/J = 0.5$, $\delta = 0.05$ and a linebroadening $\varepsilon = 0.1J$. The Raman operator $\sum_i \mathbf{S}_i \cdot \mathbf{S}_{i+1}$ was used. The dotted line indicates the lowest-lying $S = 2$ state ($L = 28$), indicating the onset of the continuum. The position of the three singlet $s\bar{s}$ bound-states below the continuum are indicated by arrows. See Fig. 7.

for higher-lying $s\bar{s}$ triplet bound-states we would expect the accompanying singlet excitations to be visible in inelastic light scattering (ILS) or Raman experiments. This is indeed the case. ILS experiments [39] show very clear evidence for a well defined bound-state around 30 cm^{-1}. We have performed numerical calculations to check that the observed $s\bar{s}$ bound-states in Fig. 7 indeed do yield a sizeable non-zero weight in Raman experiments. Our results are shown in Fig. 8 for a 28 site system with the same parameters as Fig. 7. The singlet $s\bar{s}$ are clearly visible and yield the highest weights. The fact that both a triplet and a singlet bound-state are experimentally observed clearly favors a description in terms of $s\bar{s}$ bound-states.

The formation of soliton-impurity bound-states can also be studied with ILS techniques. Experimentally [40] one observes a line at 15 cm^{-1} in addition to the 30cm^{-1} line present in the pure spectrum in $Cu_{1-x}Zn_xGeO_3$. As explained in the previous seciton the additional line at 15cm^{-1} is interpreted as a transition between *two* soliton-impurity bound-states [40]. The triplet gap is $\Delta_{s=1} = 16.8\text{cm}^{-1}$ (see Fig. 6), and the fact that the observed line occurs at energies slightly below this implies that at least *two* $S = 1/2$ bound-states occur.

It is a pleasure to acknowledge discussions with J. Riera and P. Hansen. We thank IDRIS (Orsay) for allocation of CPU time on the C94 and C98

CRAY supercomputers and RS6000 workstations. The research of IA is supported in part by NSERC of Canada.

References

1. L.D. Faddeev and L.A. Takhtajan, Phys. Lett. A **85**, 375 (1981)
2. M. Hase, I. Terasaki and K. Uchinokura, Phys. Rev. Lett. **70**, 3651 (1993)
3. F.D.M. Haldane, Phys. Rev. B **25**, 4925 (1982)
4. K. Okamoto and K. Nomura, Phys. Lett. A **169**, 433 (1992)
5. S. Eggert, Phys. Rev. B **54**, R9612 (1996)
6. C.K. Majumdar and D.K. Ghosh, J. Math. Phys. **10**, 1399 (1969)
7. U. Schollwöck, Th. Jolicœur and T. Garel, Phys. Rev. B **53**, 3304 (1996)
8. R. Bursill, G.A. Gehrig, D.J.J. Farnell, J.B. Parkinson, T. Xiang and C. Zeng, J. Phys. C **7**, 8605 (1995)
9. S.R. White and I. Affleck, Phys. Rev. B **54**, 9682 (1996)
10. S. Pati, R. Chitra, D. Sen, H.R. Krishnamurthy and S. Ramasesha, Euro. Phys. Lett. **33**, 707 (1996)
11. R. Chitra, S. Pati, H.R. Krishnamurthy, D. Sen and S. Ramasesha, Phys. Rev. B **52**, 6581, (1995)
12. F. Schönfeld, G. Bouzerar, G.S. Uhrig and E. Müller-Hartmann, cond-mat/9803084
13. K. Maisinger and U. Schollwöck, condmat/9803122
14. E. Sørensen, I. Affleck, D. Augier and D. Poilblanc, Phys. Rev. B **58**, R14701 (1998)
15. B. Shastry and B. Sutherland, Phys. Rev. Lett **47**, 964 (1981)
16. D. Khomskii, W. Geerstma and M. Mostovoy, Czech. J. of Phys. **46**, Suppl S6, 32 (1996); W. Geerstma and D. Khomskii, Phys. Rev. B **54**, 3011 (1996)
17. J.P. Goff, D.A.Tennant and SE. Nagler, Phys. Rev. B **52** 15992 (1995)
18. I. Affleck, in Dynamical properties of Unconventional Magnetic Systems, Eds. A.T. Skjeltorp and D. Sherrington, Kluwer Academic (1998)
19. J. Málek, S.-L. Drechsler, G. Paasch and K. Hallberg, Phys. Rev. B **56**, R8467 (1997)
20. G.S. Uhrig, F. Schönfeld, M. Laukamp and E. Dagotto, condmat/9805245
21. T. Nakamura and K. Kubo, Phys. Rev. B **53**, 6393 (1996)
22. D. Sen, B.S. Shastry, R.E. Walstedt, and R. Cava Phys. Rev. B **53**, 6401 (1996)
23. E. Sørensen, to appear in J. Phys. Cond. Matt.
24. Ö. Legeza, J. Solyom, Phys. Rev. B **56** 14449 (1997)
25. S. Ramasesha, Swapan K. Pati, H. R. Krishnamurthy, Z. Shuai and J. L.Brédas, Phys. Rev. B **54**, 7598 (1996)
26. W.J. Caspers and W. Magnus, Phys. Lett. **88A**, 103 (1982)
27. M.C. Cross and D.S. Fisher, Phys. Rev. B **19**, 402 (1979)
28. I. Affleck, Nucl. Phys. B. **265** [FS15], 448 (1986)
29. G. Uhrig and H. Schulz, Phys. Rev. B **54**, R9624 (1996)
30. D. Augier, P. Hansen, D. Poilblanc, J. Riera and E. Sørensen, condmat/9807265
31. J. P. Boucher and L. P. Regnault, J. Phys. I (Paris) **6**, 1939 (1996)
32. G. Uhrig, Phys. Rev. Lett. **79**, 163 (1997)
33. J. Riera and A. Dobry, Phys. Rev. B **51**, 16 098 (1995)

34. D. Augier, D. Poilblanc, E. Sørensen and I. Affleck, Phys. Rev. B **58**, 9110 (1998)
35. M. Arai, M. Fujita, M. Motokawa, J. Akimitsu and S.M. Bennington, Phys. Rev. Lett. **77**, 3649 (1996)
36. L. P. Regnault, M. Aïn, B. Hennion, G. Dhalenne and A. Revcolevschi Phys. Rev. B **53**, 5579 (1996)
37. M. Aïn, J.E. Lorenzo, G. Dhalenne, A. Revcolevschi, B. Hennion and Th. Jolicœur, Phys. Rev. Lett. **78**, 1560 (1997)
38. G. Bouzerar, A. P. Kampf and F. Schönfeld, condmat/9701176
39. P.H. M. van Loosdrecht, J. P. Boucher, G. Martinez, G. Dhalenne and A. Revcolevschi, Phys. Rev. Lett. **76**, 311 (1996); G. Els, P.H.M. van Loosdrecht, P. Lemmens, H. Vonberg, G. Güntherodt, G.S. Uhrig, O. Fujita, J. Akimitsu, G. Dhalenne and A. Revcolevschi, Phys. Rev. Lett. **79**, 5138 (1998)
40. G. Els, G.S. Uhrig, P. Lemmens, H. Vonberg, P.H.M. van Loosdrecht, G. Güntherodt, O. Fujita, J. Akimitsu, G. Dhalenne and A. Revcolevschi, cond-mat/9808064, to appear in Europhys. Lett.

1.3 Haldane Phase, Impurity Effects and Spin Ladders

Xiaoqun Wang

Institut Romand de Recherche Numérique en Physique des Matériaux
PPH-EPFL, CH-1015 Lausanne, Switzerland
Max-Planck-Institut für Physik komplexer Systeme
Nöthnitzer Str. 38, D-01187 Dresden, Germany

In the last fifteen years, a lot of effort has been devoted to the understanding of a disordered spin liquid with a spin gap for low dimensions. On one hand, Haldane predicted that a spin gap exits for integer as opposed to half-integer antiferromagnetic (AF) quantum spin chains [1]. On the other hand, a spin gap has been important for the understanding of fundamental properties of high-T_c superconducting materials [2] and recently discovered ladder materials [3,4].

For quantum spin chains, the Haldane conjecture has been examined very carefully by both numerical calculations [5,6] and experiments [7]. In particular, the low-energy physics is now well established for spin one and two using the DMRG techniques. The values of the Haldane gap Δ are $0.4105J$ for $S = 1$ [8] and $0.087J$ for $S = 2$ [9], respectively. Moreover, when the translation symmetry is broken one can find important impurity effects as shown by experiments on non-magnetic and magnetic doping [10,11]. As pointed out before [12], the impurity effects allow to characterize the low-energy behaviour of these systems as a Haldane phase.

As arrays of coupled chains, antiferromagnetic spin-$\frac{1}{2}$ ladders which interpolate between one and two dimensions have attracted much attention in recent years [3,4,13–17]. The ladders are now realized in materials such as $Sr_{n-1}Cu_{n+1}O_{2n}$ [3]. Numerically, White, Noack and Scalapino found that the ladders can display significantly different physical properties according to their number of legs [13]. For instance, the spin correlation function in the ground state shows an algebraic decay if the number of legs is odd and an exponential decay if it is even [13,14]. In this sense, the ladders are analogous to antiferromagnetic spin-S chains where a half-integer spin corresponds to an odd number of legs and an integer spin to an even number of legs [1,6].

In this chapter, we show that the low-energy properties for (quasi) one-dimensional (1D) spin gapped systems are classified into a Haldane phase and a singlet phase. In Sect. 1, we first discuss how to treat impurity problems within DMRG framework, and then demonstrate the characterization of the Haldane phase solving a bond impurity problem. In Sect. 2, we show that the Haldane phase and the singlet phase exist (i) for the standard ladder depending on the sign of the interchain coupling and (ii) for a frustrated spin AF ladder depending on the usual interchain coupling for a given frustration coupling. Finally, we discuss the relevance of the frustration for materials and the impurity effects for other spin gapped systems.

1 Impurity Effects Characterize the Haldane Phase

The open boundary chain is obtained by doping a closed chain with a non-magnetic impurity. It is thus via a kind of impurity effect that the non-trivial degeneracy of the ground state shows up for the open chain. This degeneracy was numerically found by Kennedy for a valence bond solid model (VBS) [18,19] and confirmed by White for the $S = 1$ chain [8]. According to the VBS theory [19], an open boundary condition implies that one of the valence bonds is broken. Two effective $S = 1/2$-spins thus become asymptotically free at both ends of the chain. This effect has been observed for the materials NENP and Y_2BaNiO_5 with non-magnetic doping [10].

On the other hand, DiTusa et al found a new state appearing below the Haldane gap when Y_2BaNiO_5 is doped into $Y_{2-x}Ca_xBaNiO_5$ [11]. According to the experiment, the added hole is localized on the oxygen site with $S = \frac{1}{2}$, which *either* directly breaks one exchange coupling J and thus couples to two nearest neighbour $S = 1$ Ni^{+2} ions *or* otherwise changes J into J'. These two impurity configurations are illustrated in Fig. 1. Both the bond-impurity and

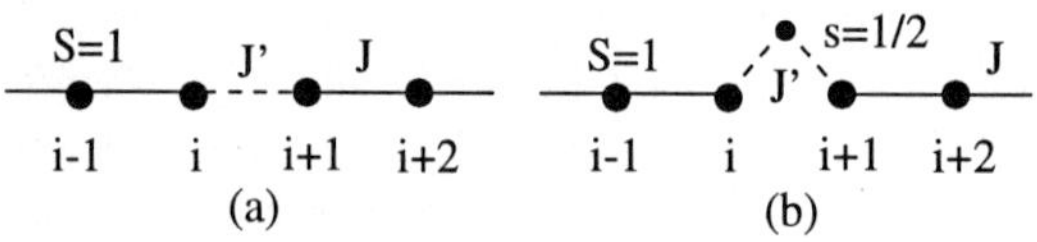

Fig. 1. Two different configurations: (**a**) bond impurity and (**b**) site impurity

the site-impurity problems have been studied extensively by several groups [20,21]. The main concern was to see *whether and how* impurity states appear below the Haldane gap in each case. In fact, the impurity effects involved in both cases essentially reveal the same intrinsic feature, i.e. the Haldane phase. Below we demonstrate this feature by treating the bond-impurity problem numerically.

Impurity DMRG schemes

The DMRG technique discussed in Chap. 2(I) has provided highly accurate results for some homogeneous systems. However, a straight application to impurity problems leads to additional systematic errors because of the DMRG truncation. Therefore proper impurity schemes are required to *reliably* describe the full properties of impurity problems. In general, impurity schemes are expected to reduce the extra systematic errors. Below we present three different DMRG impurity schemes shown in Fig. 1(a).

Closed boundary scheme: The superblock in this case consists of two renormalized blocks B (solid boxes), two added sites (black dots) and an impurity part (inside dashed box) as shown in Fig. 2 for the bond impurity problem. For the site-impurity problem shown in Fig. 1(b), one just replaces

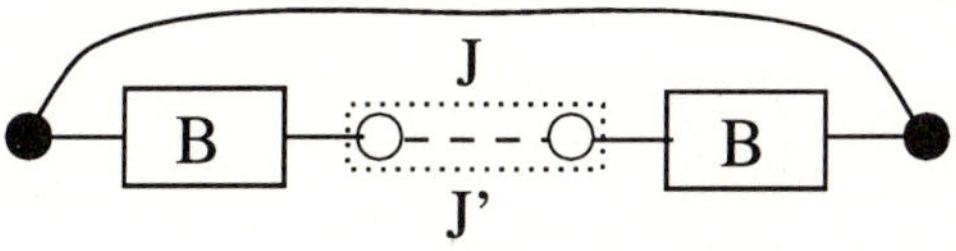

Fig. 2. Closed boundary scheme

the two $S = 1$ spins (open circles) inside the dashed box by one $S = \frac{1}{2}$ spin which is thus coupled to two nearest $S = 1$ spins in the blocks B with J'.

Open boundary schemes: Scheme I: (1) Use the infinite-system DMRG algorithm to build up a $L-2$ site superblock; (2) Add the impurity part in the middle instead of two normal sites; (3) Make sweeps as in the finite-system algorithm to reduce systematic errors.

Scheme II [22]: (1) Initialize a superblock, which can be exactly diagonalized, to contain the impurity part in the middle; (2) In order to increase the size of the superblock, the superblock is sweeped. Two sites are added whenever the environment is decreased to a size at which its Hamiltonian can be exactly diagonalized. Then the environment block and the systems are interchanged and the sweep is proceeded towards an opposite direction. (3) Make more sweeps whenever physical quantities are evaluated at a given length.

The closed boundary scheme has two peculiarities. (i) Using the infinite-system algorithm a truncation is carried out only for the block B not including the impurity part; (ii) Added sites depart further and further from the impurity part in the spirit of Wilson's renormalization group as applied to the Kondo impurity problem. On the other hand, since a closed boundary condition is usually unfavourable for DMRG calculations, some analytic scaling theory and accurate raw date are requested. For the open boundary schemes, the finite-system algorithm rather than analytic scaling theory are requested. All these impurity schemes allow to calculate of dynamical properties of the impurity directly, since local commutation relations are not violated.

We have used the closed boundary scheme, since the Shanks extrapolation is valid also for this impurity system [5]. In order to utilize this extrapolation, we have to obtain sufficiently accurate raw data by keeping sufficient number of states. For this purpose, a mapping technique was designed to generally optimize DMRG calculations. In our calculation, we kept up to 1100 states. The truncation error is typically of the order of 10^{-10}. The errors of the final results for the impurity states are at most in the fourth digit [20].

Impurity state arising from degenerate ground states

We now turn to magnetic impurity effects in the Haldane systems. Let us consider the bond-impurity problem. The corresponding Hamiltonian reads:

$$\hat{H} = J \sum_{i=1}^{L-1} \mathbf{S}_i \cdot \mathbf{S}_{i+1} + J' \mathbf{S}_1 \cdot \mathbf{S}_L, \qquad (1)$$

where $J > 0$ and $0 \le J' \le \infty$.

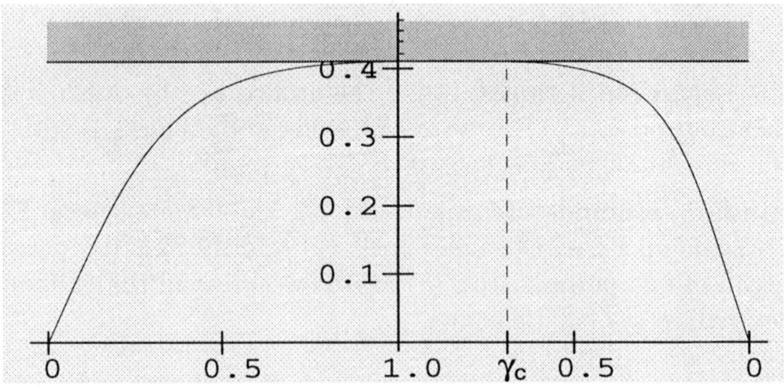

Fig. 3. Excitation energies relative to the ground-state energy as functions of J'/J on the left side and of J/J' on the right side

In Fig. 3, we show the energy of the low-lying states with respect to impurity coupling J'. The left and right sides of the vertical axis correspond to the weak coupling regime ($0 \le J' < J$) and the strong-coupling regime($J < J' \le \infty$), respectively. The horizontal axis indicates the ground-state energy. The darker shaded area represents the continuum spectrum and the straight line at its bottom marks the Haldane gap which is independent of the impurity coupling J'. Solid curves represent the impurity-state energy relative to the corresponding ground-state energy. The results presented here are for thermodynamic limit. The numerical determination of these results and discussions for the other aspects can be found in [20]. Below we will focus on the intrinsic feature of the Haldane systems to characterize the Haldane phase.

- For $J' = J$, i.e. the periodic boundary case, the ground state is a singlet. The continuum starts at the Haldane gap and the lowest one is a triplet state with $S = 1$ (for convenience the chain length L is assumed to be even).
- For $J' = 0$, i.e. the open boundary case, the lowest energy state having $S = 1$ becomes degenerate with the singlet ground state in the thermodynamic limit. The Haldane gap is now the energy difference between the ground states and the lowest $S = 2$ state.

- As $J' \to \infty$, the spins linked by the impurity bond J' form a singlet, which leaves the problem of a chain with $L-2$ sites, open boundary and the usual properties.
- As $J' \neq 0, 1, \infty$, the ground state is a singlet and the bottom excitation has $S = 1$. The impurity state always appears below the Haldane gap except $J < J' \leq \gamma_c J$ where the system is nonperturbative and $1/\gamma_c = 0.708(2)$ [20].

From these results, it is clear that the impurity state and one of the ground states of an open chain and the bottom excitation of a periodic chain are *all the* same state with $S = 1$, which evolves with J'. Kennedy and Tasaki argued that the degeneracy of the ground states is a consequence of the $Z_2 \times Z_2$ hidden symmetry breaking [23]. In fact, the open boundary condition is just a special case where the impurity strength J' equals to zero. Therefore, the impurity effects reveal an intrinsic feature of the Haldane systems. This is exhibited if and only if the translational symmetry is broken. Moreover we believe that this feature is generally valid for the integer spin chains. For instance, one can find a non-trivially degenerate ground state and corresponding impurity effects for a $S = 2$ AF chain. Furthermore we note that impurity states appear not only below the Haldane gap but also above the Haldane gap such as in the nonperturbative regime $J \leq J' < \gamma_c J$. We conclude that the Haldane phase *for spin gapped systems* is characterized by the fact that one can find non-trivially degenerate ground states or corresponding impurity effects which can be experimentally detected.

2 A Frustrated Spin Ladder

In this section, we want to contrast the standard spin ladders and single chains using a two-leg frustrated AF spin-$\frac{1}{2}$ ladder as described by [24–26,12]:

$$\begin{aligned}\hat{H} = \sum_{i=1,N} & [J_{\|}(\hat{\mathbf{S}}_{1,i} \cdot \hat{\mathbf{S}}_{1,i+1} + \hat{\mathbf{S}}_{2,i} \cdot \hat{\mathbf{S}}_{2,i+1}) + J_{\perp}\hat{\mathbf{S}}_{1,i} \cdot \hat{\mathbf{S}}_{2,i} \\ + & \quad J_{\times}(\hat{\mathbf{S}}_{1,i} \cdot \hat{\mathbf{S}}_{2,i+1} + \hat{\mathbf{S}}_{1,i+1} \cdot \hat{\mathbf{S}}_{2,i})], \end{aligned} \tag{2}$$

where $\hat{\mathbf{S}}_{n,i}$ is a spin-$\frac{1}{2}$ operator at site i of the nth chain. $J_{\|} = 1$ and $J_{\perp}$ are usual intrachain and interchain couplings, respectively. $J_{\times}$ is a diagonal coupling leading to frustration. Owing to the exchange symmetry of $J_{\times}$ and $J_{\|}$ we consider $J_{\times} \leq J_{\|}$ only. The standard ladder correspond to $J_{\times} = 0$.

Antiferromagnetic $J_{\perp}$ *vs* ferromagnetic $J_{\perp}$

For the standard ladder, a gap exists for all non-zero $J_{\perp}$. If $J_{\perp} \to -\infty$, the two spins of each rung form a triplet. The low-energy properties are effectively governed by a $S = 1$ Hamiltonian with $J_{\text{eff}} = \frac{1}{2}J_{\|}$, i.e. Δ is just half of that for the spin one chain. If $J_{\perp} \to \infty$, the two spins of each rung form a singlet.

The gap is caused when a rung triplet is generated and then $\Delta \sim J_\perp$. In Fig. 4, Δ is shown as a function of $J_\perp$. The dependence of Δ on $J_\perp$ is strikingly different for $J_\perp > 0$ and $J_\perp < 0$.

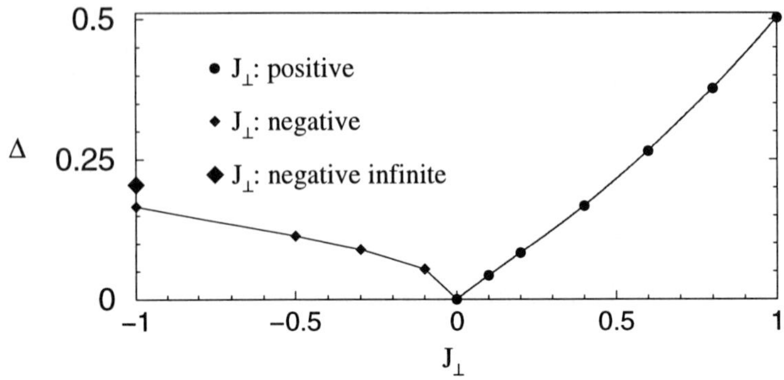

Fig. 4. Gap as a function of $J_\perp$ for both ferromagnetic and antiferromagnetic cases.

Moreover, for $J_\perp < 0$, we found the $S = 0, 1$ degenerate ground states, the $S = 2$ bottom excitation, and associated impurity effects. Therefore, it is justified to consider the low-energy properties of the two-leg ferromagnetic ladder as the Haldane phase. However, for $J_\perp > 0$, the ground state is always a singlet and the bottom excitation has $S = 1$. The averaged rung singlet density is always more than $1/4$. This distinguishes the low-energy properties of $J_\perp > 0$ from those of $J_\perp < 0$. We call this a *singlet phase*. A second order transition between two phases takes place at $J_\perp = 0$ where the gap closes.

Phase diagram

Now we generalise the above description to the frustrated AF ladder of (3). Shown in Fig. 5 is the phase diagram which has been numerically determined by the calculation of the ground-state energy, singlet density per rung and low-lying excitations (see [12] for the details). It consists of three parts: The Haldane phase, the singlet phase and the phase boundary. The phase transition is of first order if $J_\times > \lambda = 0.278$ and of second order otherwise.

Haldane phase: The system has the intrinsic property of showing impurity effects. Two typical spectra are shown in Fig. 6(a_o) and (a_p) for open and periodic boundary conditions, respectively. In fact, in the limiting case $J_\times = 1$ and $J_\perp = 0$, all states for $E \in [-1.4015, 0]$ are *identical* to those of a $S = 1$ chain. On the other hand, for open boundary conditions only, the region C_1 appears due to the presence of midgap states. The number of levels is one as we enter the C_1 region from below (Fig. 6(b)) and grows larger and larger as

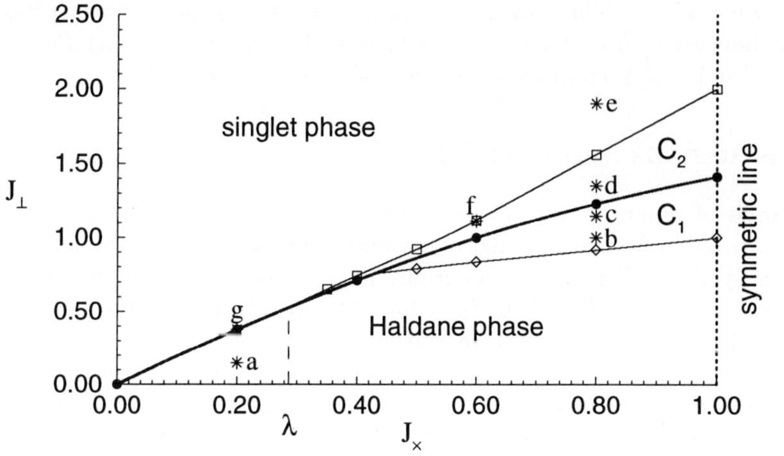

Fig. 5. Phase diagram: the thick solid curve fitting dots indicates the phase boundary; dotted line the symmetric line; C_1 (C_2) is surrounded by the phase boundary, a thin curve fitting diamonds (squares) and the symmetric line, respectively. Schematic spectra of points a–g are shown in Fig. 6.

we approach the phase boundary (Fig. 6(c)). This effect is a peculiar feature of this frustrated spin ladder in contrast to the $S = 1$ chain.

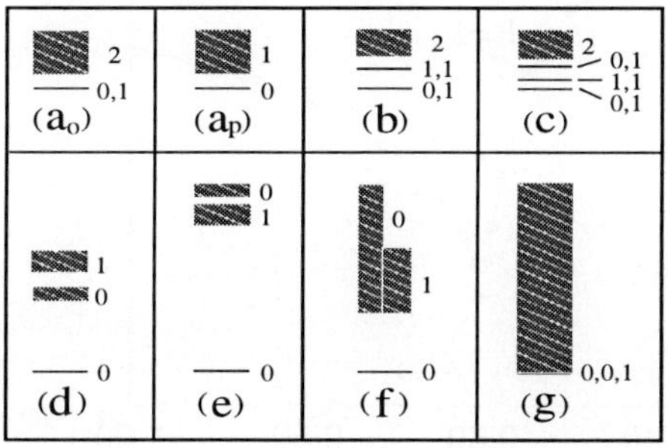

Fig. 6. Schematic spectra for the typical points a-g of Fig. 5. Line, shaded rectangles and numbers denote levels, the continuous bands and corresponding spins, respectively. a_o and a_p indicate the spectra for the point a with open and period boundary cases, respectively

Singlet phase: Independently of the boundary conditions, the ground state is a singlet and the continuum starts from a state with $S_t = 1$ outside the region C_2 but $S_t = 0$ inside the region C_2 (Fig. 6(d)-(f)). Moreover, we find neither midgap states nor impurity effects.

Phase boundary: There are at least three non-trivial degenerate states. Two of them have $S_t = 0$ and one has $S_t = 1$. The gap vanishes as $J_\times \leq \lambda$ (Fig. 6(g))and is a discontinuous function of $J_\perp$ otherwise.

3 Discussions and Conclusion

Relevance of frustration: Frustration effects are less well-known but expected to result in dramatic change of low-energy properties in low-dimensional quantum systems. For ladder materials, it is unclear to what extent the frustration can be neglected. In fact, inter-ladder coupling in the material $SrCu_2O_3$ effectively induces the frustration.

In Fig. 7, it is shown how the gap depends on the frustration coupling $J_\times$ specially for $J_\perp = 0.555$ and $J_\perp = 1$. These are two possible values for $SrCu_2O_3$ according to the spin susceptibility measurements. It is remarkable that at $J_\perp = 1$ the gap changes little for $J_\times \leq 0.4$. At $J_\times = 0.25$, it has a maximum value of $\Delta = 0.519$. At $J_\perp = 0.555$, the values of Δ for $J_\times \leq 0.1$ *and* $J_\times \approx 0.5$ are also not really distinguishable from experimental point of view. These results indicate that the frustration cannot be excluded by only considering the values of Δ for ladder materials. To examine whether or not the ladder materials can be described only by $J_\perp$, one should fit the experimental data up to an intermediate temperature regime, because at low temperatures the thermodynamic quantities are determined only by the value of Δ.

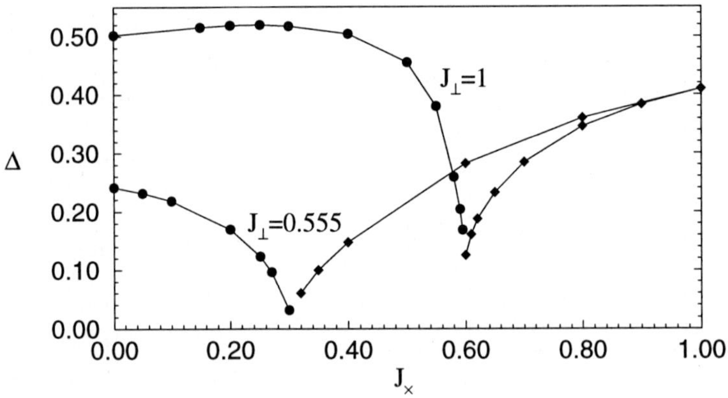

Fig. 7. Gap as a function of $J_\perp$ at the frustration couplings $J_\times = 0.555$, and $J_\times = 1$

Other quasi-1D spin gapped systems: Here are several interesting (quasi-)1D spin gapped systems: a on-site anisotropic S=1 AF chain, 1D attractive Hubbard model, 1D t-J model, Zigzag ladder, half-filled Kondo lattice, half-filled Hubbard ladder, and t-J ladder, etc. So far we have examined the boundary effects and associated impurity effects for some of them.

We found that the large-D phase of the on-site anisotropic S=1 AF chain, 1D attractive Hubbard model, Zigzag ladder, half-filled Kondo lattice do not involve the Haldane phase. However it will be further interesting to investigate others, the Hubbard and t-J ladder with frustration.

In conclusion, the impurity effects, which are the intrinsic feature of the integer-spin AF chain, characterize the low-energy properties of spin gapped systems as the Haldane phase. The standard spin AF ladder involves only the singlet phase where the rung singlet is favourable and the ground state is always a singlet.

The author is very grateful to E. Dagotto, P. Fulde, A. Honecker, A.K. Kolezhuk, M. Long, J.Z. Lou, S. Mallwitz, R.M. Noack, I. Peschel, T.M. Rice, Z.B. Su, A.M. Tsvelik, Lu Yu, Y. Zhao, and X. Zotos for fruitful discussions and helpful communications.

References

1. F.D.M. Haldane, Phys. Lett. A **93**, 464(1983)
2. P.W. Anderson, Science **235**, 1196 (1987)
3. M. Azuma *et al*, Phys. Rev. Lett. **73**, 3463 (1994)
4. E. Dagotto and T.M. Rice, Science, **271**, 618 (1996) and references therein
5. O. Golinelli, Th. Jolicoeur and R. Lacaze, Phys. Rev. B **50**, 3037 (1994) and references therein
6. S.R. White and D.A. Huse, Phys. Rev. B **48**, 3844 (1993)
7. See e.g. Y. Ajiro, et al, Phys. Rev. Lett. **63**, 1424 (1989); L.P. Regnault et al., Phys. Rev. B **50**, 9174 (1994); M.Takigawa, et al., Phys. Rev. Lett. **76**, 2173 (1996); G.E.Granroth, et al. Phys. Rev. Lett. **77**, 1616 (1996); references therein
8. S.R. White, Phys. Rev. Lett. **69**, 2863 (1992)
9. S.J. Qin, X. Wang and Lu Yu Phys. Rev. B. **56**, R14251 (1997) and to be published
10. M. Hagawara, et al., Phys. Rev. Lett. **65**, 3181 (1990); S.H. Glarum et al., ibid, **67**, 1614 (1991)
11. J.F. DiTusa et al, ibid **73**, 1857 (1994)
12. X. Wang, cond-mat/ 9803299 and to appear in Phys. Rev. B
13. S.R. White, R.M. Noack and D.J. Scalapino Phys. Rev. Lett. **73**, 886 (1994)
14. M. Greven, R.J. Birgeneau and U.-J. Wiese, Phys. Rev. Lett. **77**, 1865 (1996)
15. E. Dagotto, J.Riera and D.J. Scalapino Phys. Rev. B **45**, 5744 (1992)
16. S.P. Strong and A.J. Millis, Phys. Rev. Lett. **69**, 2419 (1992)
17. M. Sigrist, T.M. Rice and F.C. Zhang, Phys. Rev. B **49**, 12058 (1994)
18. T. Kennedy, J. Phys. Condens. Matter **2**, 5737 (1990)
19. I. Affleck, T. Kennedy, E.H. Lieb and H. Tasaki, Phys. Rev. Lett. **59**, 799 (1987)
20. X. Wang and S. Mallwitz, Phys. Rev. B **48**, R492 (1996)
21. M. Kaburagi, I. Harda and T. Tonegawa, J. Phys. Soc. Jpn, **62**, 1848 (1993); ibid, **63**, 420 (1993); Z.Y. Lu, Z.-B. Su and Lu Yu, Phys. Rev. Lett. **74**, 4297

(1995), E.S. Sorensen and I. Affleck, Phys. Rev. B. **51**, 16115 (1995); K. Penc and H. Shiba, ibid, **52**, R715 (1995); W. Wang, S.J. Qin, Z.Y. Lu, Z.-B. Su and Lu, Yu, ibid, **53**, 40 (1996); E. Dagotto, J.Riera, A. Sandvik and A. Moreo, Phys. Rev. Lett. **76**, 1731 (1996); Q. Yuan, Y. Chen, H. Chen and Y. Zhang, ibid, **57**, 1324 (1998); J.Z. Lou, S.J. Qin, Z.-B. Su and Lu Yu, to appear in Phys. Rev. B.; V, Barzykin and I. Affleck, cond-mat/9810075
22. X. Wang, Mod. Phys. Lett. B, **12**, 667 (1998)
23. T. Kennedy and H. Tasaki, Phys. Rev. B **45**, 304 (1992)
24. W.H. Zheng, V. Kotov and J. Oitmaa, Phys. Rev. B. **57**, 11439 (1998)
25. H.Q. Lin and J.L. Shen, cond-mat/9805269
26. F. Mila, cond-mat/9805029

1.4 Spin Chain Properties

Karen Hallberg

Centro Atómico Bariloche and Instituto Balseiro
8400 Bariloche, Argentina

The first applications of DMRG were for spin chains because of the simplicity in programming and its excellent performance. We will very briefly discuss this topic for the case of zero temperature and in the second part specialize to an application to spin-one chains where very accurate results can be obtained.

1 Ground-State Calculations

Here we will concentrate on Heisenberg spin chains. The properties of these chains differ if the spins are integer (having a spin gap) or half-odd integer (no spin gap). We will consider each group separately.

Concerning $S = 1/2$ systems, DMRG has been crucial for obtaining the logarithmic corrections to the $1/r$-dependence of the spin-spin correlation functions in the isotropic Heisenberg model [1]. For this, very accurate values for the energy and correlation functions were needed. For $N = 100$ sites, using $m = 150$ and the infinite-system DMRG algorithm, a relative error (with respect to the exact Bethe Ansatz value) of 10^{-5} is achieved. Much more precise results can be obtained with DMRG by using the finite-system algorithm or by increasing m. However, the precision is worse than for the spin one case because, although the latter has more degrees of freedom per site, it possesses a gap in the excitation spectrum. For this model it was found that the data for the correlation function has a very accurate scaling behaviour and this could be used to obtain the logarithmic corrections in the thermodynamic limit. Other calculations of the spin correlations have been performed for the anisotropic case [2].

We have also performed similar calculations for the $S = 3/2$ Heisenberg chain [3]. In this case we found a larger logarithmic correction to the spin correlation function in short chains (comparing with the $S = 1/2$ case). For this model we were interested in obtaining the central charge c to elucidate whether it corresponds to the same universality class as the $S = 1/2$ case. This central charge can be obtained from the finite-size scaling of the ground-state energy. This involves also a term $\sim 1/\ln^3 N$ and previous attempts using exact diagonalization which had reached only $N = 12$ sites [4] had failed. With the DMRG, the value $c = 1$ could be unambiguously obtained. While for the previous cases standard infinite-size DMRG ($m \leq 200$) was used, we were forced to keep many more states ($m \sim 1000$) and several finite-size sweeps for the $S = 3/2$ case in order to have sufficient accuracy to account for the logarithmic corrections.

Considering integer spin systems, a very impressive result with unprecedented acccuracy was obtained by White and Huse [5] when calculating the spin gap in a $S = 1$ Heisenberg chain obtaining $\Delta = 0.41050J$. They also calculated very accurate spin correlation functions and excitation energies for one-magnon and several-magnon states and performed a very detailed analysis of the excitations for different momenta. They obtain a spin correlation length of 6.03 lattice spacings. Simultaneously Sørensen and Affleck[6] also calculated the structure factor and the spin gap for this system up to length 100 with very high accuracy, comparing their results with the nonlinear σ-model. In a subsequent paper [7] they applied the DMRG to the anisotropic $S = 1$ chain obtaining very precise values for the Haldane gaps and also performed a detailed study of the $S = 1/2$ end excitations in an open chain. We will come back to this point later.

For larger integer spins there have also been some studies. Nishiyama and coworker [8] calculated the low-energy spectrum and correlation functions of the $S = 2$ antiferromagnetic Heisenberg open chain. They found spin-one end excitations in agreement with the Valence Bond Theory theory (see below). Almost at the same time Schollwöck and Jolicoeur [9] calculated the spin gap in the same system with up to 350 sites obtaining $\Delta = 0.085J$. They also calculated correlation functions that showed topological order and a spin correlation length of 49 lattice spacings. Note that for this $S = 2$ spin system the gap is much smaller than for the $S = 1$ case implying a larger correlation length.

2 Thermodynamic Properties of Finite-Length Spin-One Systems

In this section we will present specific heat calculations for the doped antiferromagnetic spin-one chain compound $Y_2BaNi_{1-x}Zn_xO_5$ in the presence of a magnetic field [10]. Y_2BaNiO_5 has an orthorhombic crystal structure with the Ni^{2+} ($S = 1$) ions arranged in linear chains with a nearest neighbour AF superexchange coupling J (three orders of magnitude larger than the perpendicular interchain coupling). Each Ni is surrounded by six O atoms in near octahedron coordination. In doping with Zn atoms ($S = 0$), these replace Ni atoms and cut the chains, generating a random distribution of finite chains with different lengths. At low enough temperatures the specific heat in this compund is dominated by the effect of the impurities in the chains. By using DMRG we obtain very accurate energies for large enough systems, allowing for an excellent agreement with experiments.

Integer spin chains continue to be a very active research field, both theoretical and experimentally. Affleck *et al.* [11] have shown that the valence-bond-solid state is a very good approximation to the exact ground state of the nearest neighbour $S = 1$ Heisenberg Hamiltonian. In this picture, an open chain has two unpaired $S = 1/2$ spins, one at each end. This has been confirmed numerically [5,12]. This picture is also supported by electron paramagnetic resonance (EPR) measurements of NENP doped with non-magnetic ions [13], where resonances corresponding to the fractional spin $S = 1/2$ states at the "open" ends of the Ni chains were observed. Similar measurements for doping with magnetic ions are also consistent with $S = 1/2$ end states [14].

In spite of these positive results, no such states were found by Ramirez *et al.* [15] who also tested the presence of free $S = 1/2$ states by studying the specific heat of defects in NENP and Y_2 $BaNiO_5$, with magnetic fields up to 9T and low temperatures (between 0.2 and 7K). On the basis of a comparison between two very simplified pictures, they found that the shape and magnitude of the Schottky anomaly associated with the chain-cutting Zn defects in $Y_2BaNi_{1-x}Zn_xO_5$ are well described by a simple model involving spin-1 excitations, instead of the $S = 1/2$ excitations of the VBS.

As Y_2BaNiO_5 has a Haldane gap $\sim 100K$ [16], the spin wave contribution to the specific heat is negligible below $7K$ so we restrict our Hamiltonian to the subspace generated by the four lowest-lying states. The open ends in a finite $S = 1$ chain introduce new states within the spin gap (there are a singlet and a triplet). These four states become degenerate when the system size N goes to infinity.

We were able to reconcile all the experiments by solving the low-energy spectrum of the following anisotropic Hamiltonian which describes the Ni chains segments of Y_2BaNiO_5 with N spin-one Ni atoms between two spin-zero Zn impurities [16,17]:

$$H = \sum_i \{J\mathbf{S}_i\mathbf{S}_{i+1} + D(S_i^z)^2 + E[(S_i^x)^2 - (S_i^y)^2]\} - g\mu_B\mathbf{B}\mathbf{S}_t \tag{1}$$

where z is along the chain axis and $\mathbf{S}_t$ is the total spin.

By means of the standard infinite-size ($m = 200$) DMRG method, we have calculated the lowest singlet and triplet states, *i. e.* the two lowest eigenenergies in the $S_t^z = 0$ subspace and the lower one with $S_t^z = 1$ (this state is degenerated with the $S_t^z = -1$ due to the time-reversal symmetry of H) for all $N \leq 40$ and $E = 0$. The energy difference between the excited states and the ground state decays exponentially to zero with increasing N. This behavior, shown in Fig. 1 for an arbitrary value of anisotropy $D = -0.1$, allows us to extrapolate the energies to all values of $N > 40$, and demonstrate that the free S =1/2 spins at the end of the chain persist in presence of anisotropy. At this level, it is important to remark that the two S =1/2 spin excitations, predicted by VBS, have a finite localization length $l \sim 6$ sites [5,12]. Therefore, while they are nearly free for large ($N >> l$) open chains, the interaction between them is considerable when the length of the chain N is comparable to $2l$. This interaction splits the two S =1/2 states into a singlet (ground state for even N) and a triplet (ground state for odd N) [18].

The difference between any two energies of the above mentioned low-energy states is linear in D, and the quadratic corrections are negligible [17]. This widely justifies the validity of perturbation theory to first order in D.

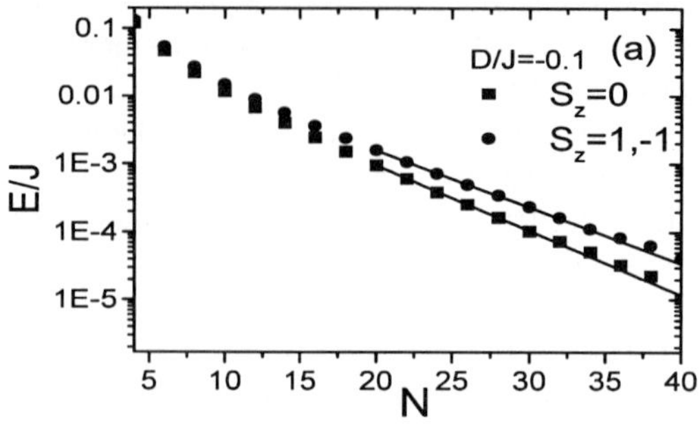

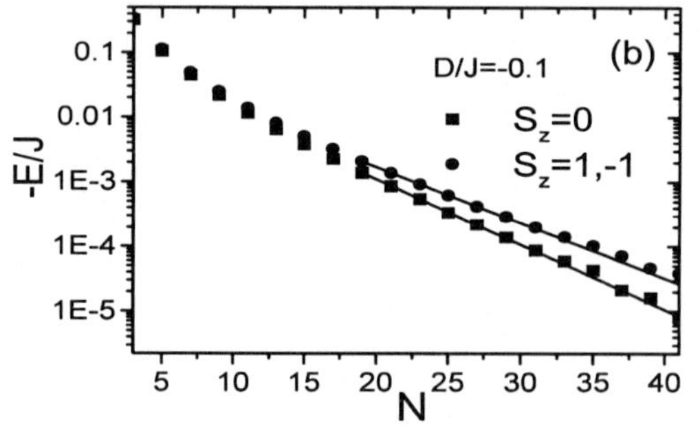

Fig. 1. Difference between the energy of the triplet $S_z = \pm 1$ ($S_z = 0$) and the singlet states in the presence of anisotropy D are denoted with circles (squares) . a) N even, b) N odd. The full line is an exponential fit of these differences for $N \geq 19$.

Then, by symmetry we can include the term $\sum_i E[(S_i^x)^2 - (S_i^y)^2]$ to first order.

For a random distribution of defects, the specific heat per elementary cell of Zn-doped Y_2BaNiO_5 is $C(B,T) = \sum_{N=1}^{\infty} x^2(1-x)^N \; C_N(B,T)$, where $C_N(B,T)$ is the specific heat of a segment of length N and x is the concentration of missing Ni atoms.

In order to take into account other low-energy contributions, we have included the interaction between neighbouring chains in an effective way (J'). We model the effect of J' by considering a collection of chain segments described by H (restricted to the lowest singlet-triplet subspace, as before) with effective 1D topology having exchange interaction J' between two consecutive segments, with a random and uniform distribution between $-J'_{\max}$ and $J'_{\max}$ [10]. Each segment has a random length and the corresponding low-lying energies calculated previously using DMRG. This effective chain (up to 8 sites long) is then treated exactly.

In Fig. 2 we represent the experimental data for $\delta C/T = [C(B=3\mathrm{T},T) - C(B=6\mathrm{T},T)]/T$ with $x = 0.04$, scanned from [15], and compare it with different theories where we take the experimental values for the parameters ($J \sim 280K$, $D \sim -.039J$, and $E \sim -.014$ [16] and $g_z = 2.25$, $g_x = g_y = 1.17$ [19]). The difference between two different values of B, eliminates the contribution of phonons, two-level systems, or any field independent contribution to $C(B,T)$. The Heisenberg model without anisotropy and interchain interactions is unable to reproduce the data. Including realistic anisotropy, there is already a considerable improvement, and the remaining discrepancy is removed by our treatment of the interchain interactions with $J'_{\rm max} = 4$ K. When $g\mu_B B > J'$ the effect of interchain interaction decreases for larger B. In Fig. 3 we have separated the contributions of different chain lengths to $\delta C/T$, neglecting J'. The full line corresponds to the result for non-interacting $S = 1/2$ end spins. This behavior is obtained for $N > 50$ where the singlet-triplet gap is much smaller than $g\mu_B B$.

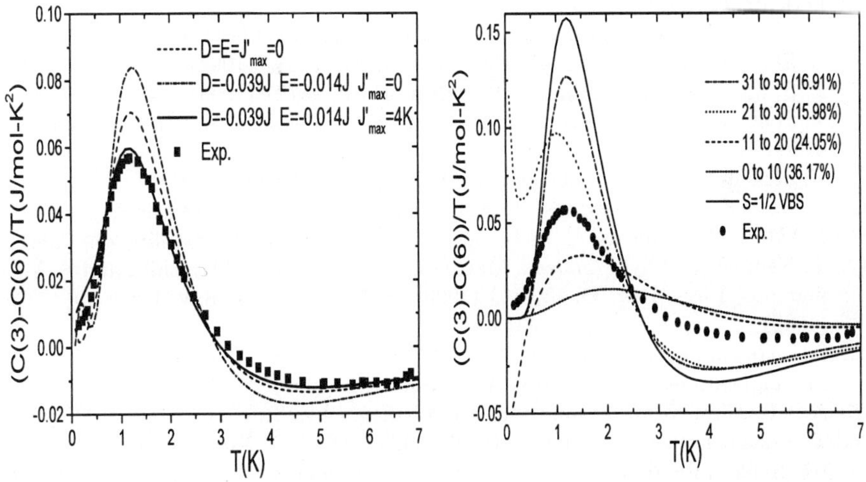

Fig. 2. Left: Specific heat difference $\delta C/T$ for $Y_2BaNi_{0.96}Zn_{0.04}O_5$ for several parameters. Right: Contributions of chains in different length intervals to the specific heat difference $\delta C/T$. The full line corresponds to the prediction of VBS theory for non-interacting $S = 1/2$ moments. The final result (Fig. 2) is obtained adding each curve with the weight denoted inside the figure.

In conclusion, by solving using DMRG the low-energy spectrum of a Heisenberg Hamiltonian H which includes experimental axial and planar anisotropy, and (approximately) the effect of interchain exchange through the only fitting parameter J'_{max}, we have reproduced the low-temperature specific-heat data measured in $Y_2BaNi_{1-x}Zn_xO_5$ for different magnetic fields [15]. Our results are consistent with valence-bond-solid predictions of $S = 1/2$ chain end excitations, which are asymptotically free for large chain segments. However, the intrachain interactions for short segments are critical for the understanding of the specific heat behavior. These results remove the apparent discrepancy between the specific heat data for Y_2BaNiO_5, interpreted in [15] with a singlet-triplet model, and electron paramagnetic resonance data for NENP.

I would like to thank Dr. Cristian Batista for fruitful discussions and his collaboration in preparing this manuscript.

References

1. K. Hallberg, P. Horsch and G. Martínez, Phys. Rev. B, **52**, (1995) R719
2. T. Hikihara and A. Furusaki, cond-mat/9803159
3. K. Hallberg, X. Wang, P. Horsch and A. Moreo, Phys. Rev. Lett. **76**, 4955 (1996)
4. A. Moreo, Phys. Rev. B **35**, 8562 (1987); T. Ziman and H. Schulz, Phys. Rev. Lett **59**, 140 (1987)
5. Steven White and David Huse, Phys. Rev. B **48**, 3844 (1993)
6. E. S. Sørensen and I. Affleck, Phys. Rev. B **49**, 13235 (1994)
7. E. S. Sørensen and I. Affleck, Phys. Rev. B **49**, 15771 (1994)
8. Y. Nishiyama, K. Totsuka, N. Hatano and M. Suzuki, J. Phys. Soc. Jpn **64**, 414 (1995)
9. U. Schollwöck and T. Jolicoeur, Europhys. Lett. **30**, 493 (1995)
10. C. Batista, K. Hallberg and A. Aligia, Phys. Rev. B **58**, 9248 (1998)
11. I. Affleck, T. Kennedy, E. Lieb and H. Tasaki, Phys. Rev. Lett. **59**, 799 (1987)
12. T. Kennedy, J.Phys.Condens. Matter **2**, 5737 (1990); S. Miyashita and S. Yamamoto, Phys. Rev. **B 48**, 913 (1993); S. White, Phys. Rev. Lett. **69**, 2863 (1992)
13. S.H. Glarum et al., Phys. Rev. Lett. **67**, 1614 (1991)
14. M. Hagiwara et al., Phys. Rev. Lett. **65**, 3181 (1990)
15. A.P. Ramirez, S-W. Cheong and M.L. Kaplan, Phys. Rev. Lett. **72**, 3108 (1994)
16. T. Sakaguchi et al., J. Phys. Soc. **65**, 3025 (1996); G. Xu et al., Phys. Rev **B 54**, R6827 (1996)
17. O. Golinelli, Th. Jolicoeur and R. Lacaze, Phys. Rev **B 45** , 9798 (1992)
18. E.H. Lieb and D.C. Mattis, J. Math. Phys. **3**, 749 (1962)
19. C.D. Batista, K. Hallberg and A.A.Aligia, preprint

2 Molecules and Polymers

2.1 Electronic Structure using DMRG

Steven R. White

Department of Physics and Astronomy, University of California
Irvine, CA 92697–4575, USA

The success of DMRG for studying ground states of strongly correlated models suggests that the application of DMRG to electronic structure could be quite fruitful. What I have in mind here is the extraction of accurate *ab initio* ground states for atoms, molecules, and possibly solids. As I discussed in the first chapter of this book, I have been interested in this problem since graduate school. In this chapter, I will discuss some work in this area. This work supports the idea that DMRG is very promising, but also indicates that more work needs to be done to make DMRG as useful in this context as it is in strongly correlated models.

In Sect. 1, I discuss the conceptually straightforward application of DMRG to a small molecule in a standard basis set from quantum chemistry. In this Gaussian basis set, after any orthogonalization procedure, such as the transformation to the Hartree-Fock orbitals, the system has a second quantized Hamiltonian which resembles an extended Hubbard model, but with many more long ranged interaction terms, different local potentials on each site, long range hopping, etc. Despite the large number of terms, finite-system DMRG can be applied without modification. What one finds is that DMRG can work very well in this context. However, because of the many terms in H, even after various tricks are used to treat interaction terms more efficiently, the computational effort is still rather large. Further progress to make DMRG more efficient will require some modification of the DMRG algorithms or modifications of the basis set.

In Sect. 2, I discuss a novel type of basis set for electronic structure which I believe will be well-suited for DMRG. This basis is local and orthogonal from the start, which is a great asset for DMRG. Unlike other local basis sets with properties of completeness and orthogononality, these *orthlet* basis sets do not need to have enormous numbers of functions to represent both the valence and core regions of a molecule. The reduced number of functions is possible because of a technique borrowed from DMRG: the functions are adapted to the problem at hand using a local density matrix.

Eventually, I hope to combine DMRG calculations and orthlet basis sets, but at this stage these two topics are separate.

1 DMRG and Quantum Chemistry

Here I consider the application of DMRG to the fully *ab initio* determination of the electronic structure of atoms and molecules. The successful adaptation of DMRG to this field could potentially open up a wide range of improved calculational techniques, characterized by high accuracy and improved scaling of calculation time with system size. As a first step in this direction, I show here that DMRG can be successfully used to obtain very accurate many-body solutions for small molecules.

I will use DMRG within the conventional quantum chemical framework of a finite basis set with non-orthogonal basis functions made from products of gaussian radial functions and Cartesian harmonics centered on each atom. The initial step of the calculation is a standard Hartree-Fock (HF) calculation in which a Hamiltonian is produced within the orthogonal HF basis. DMRG is then used as a procedure for including correlations beyond HF, much as the configuration interaction (CI) or coupled cluster methods are used.

Within the HF basis, the Hamiltonian is in principle no different from other model Hamiltonians which have been studied using DMRG. It can be written as

$$H = \sum_{ij\sigma} T_{ij} c^\dagger_{i\sigma} c_{j\sigma} + \frac{1}{2} \sum_{ijkl\sigma\sigma'} V_{ijkl} c^\dagger_{i\sigma} c^\dagger_{j\sigma'} c_{k\sigma'} c_{l\sigma}. \tag{1}$$

Here T_{ij} contains the electron kinetic energy and the Coulomb interaction between the electrons and the nuclei, while V_{ijkl} describes the electron-electron Coulomb interaction. The most important difference in this Hamiltonian from model Hamiltonians is the large number of interaction terms V_{ijkl}: N^4, where N is the number of basis functions or orbitals. The number of electrons, $N_{\rm el}$, is less important in a DMRG calculation. The large number of terms makes standard DMRG programs very inefficient. There are a number of procedures one can devise for improving the efficiency of the treatment of these terms [1], but I will not discuss these procedures here.

To use DMRG, an ordering of the orbitals is chosen, and each orbital is treated as a "site" in a one-dimensional lattice. Since this arrangement is artificial, the Hamiltonian is long-ranged. The orbitals can be sorted according to various criteria. Sorting them in order to minimize strong interactions between widely separated orbitals is probably best, but they can also be arranged by HF orbital energy.

Once this ordering is chosen, a standard DMRG finite-system algorithm can be used [2]. The most significant difference from the usual case is in the many operators which must be kept in a block. The operators to be kept are chosen in order to be able to generate the Hamiltonian operator for the system. For example, in order to construct the kinetic energy, one must keep matrix elements for the operators $c^\dagger_{i\sigma}$, for all orbitals i in the block. These operators allow us to construct terms $c^\dagger_{i\sigma} c_{j\sigma}$ where j is not in the block. In addition, one must keep matrices for $c^\dagger_{i\sigma}c_{j\sigma}$ if both i and j are in the block. Note, as is always the case for DMRG, that one cannot avoid storing a matrix for AB simply because one has stored matrices for A and B: the incomplete nature of the basis means that the matrix for AB is not the product of the matrices for A and B. In order to describe the Coulomb interaction, it appears that $o(N^4)$ operators of the form $c^\dagger_{i\sigma}c^\dagger_{j\sigma'}c_{k\sigma'}c_{l\sigma}$, must be kept, where i, j, k, and l are all in the block. In addition, $o(N^3)$ additional operators are needed to construct terms when some of the $ijkl$ are not in the block.

Although it appears that $o(N^4)$ operator matrices must be stored per block, a completely standard DMRG "trick" reduces this number to $o(N^3)$. The trick is to sum terms together into a single block Hamiltonian matrix once all of the parts of the term are in the block. Hence there is no need to store terms of the form $c^\dagger_{i\sigma}c^\dagger_{j\sigma'}c_{k\sigma'}c_{l\sigma}$; these terms are multiplied by V_{ijkl} and summed into H. In a typical model Hamiltonian, this trick reduces the number of operators stored per block from $o(N)$ to $o(1)$. Here, we still have $o(N^3)$ operators. Since $o(N)$ blocks must be stored, the storage is $o(N^4m^2)$.

Additional improvements can be made by combining other operators, as was first done by Xiang in adapting DMRG to momentum space calculations [3]. These improvements reduce the total storage to $o(N^3m^2)$, of which only $o(N^2m^2)$ needs to be in RAM; the rest can be on disk with little cost in calculation time. The overall calculation time can be reduced to scale as $o(N^3m^3) + o(N^4m^2)$. This computational load is manageable for small molecules. In the test calculations below, the number of states kept per block m is typically a few hundred, and $N = 25$.

As a test case [1], we have studied a water molecule in a standard basis, comparing with the benchmark full configuration interaction calculations of Bauschlicher and Taylor [4]. In this work, exact results for the H_2O molecule within a particular basis were compared with various approximate approaches. We used the same basis for the calculations shown here. In Fig. 1(**a**), DMRG results are compared with Hartree-Fock and singles and doubles configuration interaction (SDCI), for the molecule in its equilibrium geometry. Both DMRG and SDCI results are variational. The DMRG results become more accurate than the SDCI for $m \approx 70$.

In Fig. 1(**b**), results on an expanded scale are shown for the same system. Also shown are multireference configuration interaction calculations (MRCI), and MRCI plus an estimated correction (the Davidson correction) (MRCI+Q). MRCI is variational, but MRCI+Q is not. We see that DMRG becomes more accurate than MRCI for $m \approx 110$, and more accurate than MRCI+Q for $m \approx 200$. The most accurate DMRG result is off by 0.00024 Hartrees. The MRCI+Q results were the most accurate of the approximate results reported by Bauschlicher and Taylor, with an error of 0.0014.

The numerical effort expended for these DMRG calculations was probably more than for the other approximate methods, but less than for the full CI calculations. The point of our calculations was not to present a fully devel-

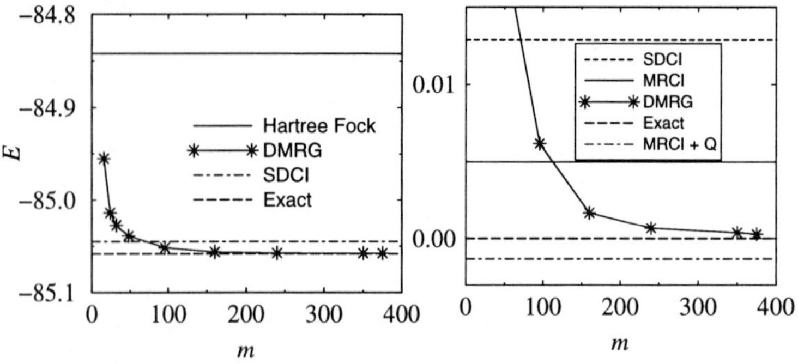

Fig. 1. (**a**) Ground state energy of a water molecule in a 25 orbital basis using various methods. DMRG results are as a function of the number of states kept per block m; other results, which have no m dependence, are plotted as horizontal lines. Energies are in Hartrees. (**b**) Errors in the energy.

oped technique, but to demonstrate the potential of a new type of approach. DMRG is a versatile technique, and I believe substantial improvements can be made over the calculations described here. I will now discuss two general directions I believe could be successful in improving the method.

An area for potential progress is in the choice of a single particle basis. The Hartee Fock wavefunctions used here are reasonable choices for a small molecule: this basis gives a good answer even if only one state per block is kept, $m = 1$. Within the HF basis, the occupancy of orbitals tends to be either almost 0 (if the orbital is above the Fermi level) or almost 2 (if the orbital is below), which helps the convergence of DMRG as a function of m. One could also try natural orbitals, which are the eigenvectors of the single particle function $\langle c_i^\dagger c_j \rangle$. The occupancy of the natural orbitals is even closer to 0 or 2. However, both HF and natural orbitals have a significant flaw: they are delocalized on a large system. Experience on lattice models suggests that DMRG tends to be much more accurate with localized bases. An ideal basis would be one that has occupancies close to 0 or 2, but would also be as local as possible. A basis developed with these goals is described in the next section.

Another area for potential progress is in the grouping of similar orbitals into clusters. In the standard version of DMRG, blocks are formed for either the left or right half of the system. However, it is also possible to form blocks built out of clusters of orbitals which are strongly coupled to each other. As discussed in Chapter 5.1(II), this sort of procedure was found to be quite effective in an electron-phonon model, where a local density matrix was used to reduce the size of the local phonon space for each oscillator from up to 128 states to only 2 or 3 [5]. Here, one could form a cluster out of one "occupied" orbital plus a group "virtual" orbitals which are used to correlate the pair of electrons in the occupied orbital. One would want to organize the basis set into a set of such clusters, with each cluster having one occupied and a set of closely coupled virtual orbitals. A density matrix would be used to form an accurate many body basis for each of these clusters. Our preliminary calculations using this approach suggest that fewer than 50 states would be sufficient to describe such a cluster to millihartree accuracy. A standard DMRG calculation could then procede using these clusters as "sites", with the expectation that much higher accuracy for a given m would be obtained than in the usual approach. Alternatively, one might group these clusters into superclusters, describing shells, atoms, or even molecules. Note that efficient clustering would require the use of localized orbitals, so that progress in these two areas may be coupled

2 Orthlet Basis Sets

Basis sets for electronic structure have a long history. A major difficulty in finding an optimal basis is that various desirable properties are mutually exclusive. Among the desirable qualities are orthogonality, locality, compactness (i.e. compact support), the ability to represent space uniformly, the ability to represent singular regions with higher resolution, the ability to incorporate prior knowledge about singular regions, the ability to ignore empty regions, and the availability of specialized efficient algorithms (such as the fast Fourier transforms (FFTs) or wavelet transforms) for solving differential equations.

For example, in electronic structure calculations for solids using density functional methods, plane waves are widely and successfully used [6]. These are orthogonal, have uniform resolution, and the FFT allows rapid switching between real and Fourier space. Pseudopotentials are normally used to represent atomic cores. However, for more accurate treatment of the cores, and for nonperiodic systems (including molecules and surfaces), the plane wave basis is inconvenient. Furthermore, the delocalized nature of the basis is completely inappropriate for DMRG calculations.

In quantum chemistry, the standard choice for basis functions is the product of a radial function centered on an atom times a cartesian or spherical harmonic. Because the radial functions which solve the Hartree-Fock equations for atoms are known, remarkably small numbers of basis functions are needed—often only about twice as many basis functions as there are electrons. However, as discussed in the previous section, the nonlocality of this basis, particularly after orthogonalization, is quite inconvenient for DMRG, and it gives a Hamiltonian with $o(N^4)$ terms.

The orthlet basis sets discussed in this section are orthogonal, very localized and compact, allow variable resolution, and allow prior knowledge about singularities to be incorporated into the basis. Furthermore, these goals are achieved while keeping the number of basis functions to a minimum. This approach is most closely related to the finite element basis using orthogonal shape functions developed by White, Wilkins, and Teter (WWT) [7]. The major problem with the approach of WWT was the difficulty in obtaining adequate resolution for the cores. The new approach overcomes that difficulty.

Consider a set of localized *shape* functions $S_i(\boldsymbol{r})$, and a lattice $\{\boldsymbol{R}_j\}$. We can generate a set of functions for each lattice site by translation, $S_{ij} = S_i(\boldsymbol{r} - \boldsymbol{R}_j)$. Let the functions have the following properties: 1) the set of functions $S_i(\boldsymbol{r})$ is orthonormal; 2) each function is also orthogonal to the functions on all other lattice sites; and 3) the total set of functions on all lattice sites is *complete*. (Wannier functions also have these properties.) We define a projection operator for site j by

$$P_j = \sum_i |S_{ij}\rangle\langle S_{ij}|. \tag{2}$$

In coordinate notation this operator is

$$P_j(\boldsymbol{r}, \boldsymbol{r}') = \sum_i S_i(\boldsymbol{r} - \boldsymbol{R}_j) S_i(\boldsymbol{r}' - \boldsymbol{R}_j), \tag{3}$$

where the S_i are assumed real. Completeness implies that $\sum_j P_j = 1$. Now consider the application of P_j on an arbitrary function $f(\boldsymbol{r})$: $f_j(\boldsymbol{r}) = P_j f(\boldsymbol{r})$. Then

$$f(\boldsymbol{r}) = \sum_j f_j(\boldsymbol{r}), \tag{4}$$

and

$$\langle f_j(\boldsymbol{r}) | f_{j'}(\boldsymbol{r}) \rangle = 0, \qquad j \neq j'. \tag{5}$$

We will call functions such as $f_j(\boldsymbol{r})$, which are local, orthogonal, and specifically adapted to a function or to a set of functions, *orthlets*. The basis formed by the orthlets $f_j(\boldsymbol{r})$ is ideal for representing $f(\boldsymbol{r})$: it is orthogonal, represents the function exactly, and has the minimum number of functions given the scale set by the lattice spacing. We describe below how to form an orthlet basis describing an arbitrary *set* of functions $f^\alpha(\boldsymbol{r})$ to a specified accuracy.

In order for the orthlets to be useful, the shape functions $S_i(\boldsymbol{r})$ should be smooth and local, and preferably compact. In two or more dimensions, shape functions can be written as products of one dimensional shape functions [7], so that we need only consider the 1D case. WWT developed a set of four shape functions with continuous derivatives up to third order, which were able to represent exactly polynomials up to third order [7]. These shape functions were compact, with a total width of two, where we assume a lattice spacing of unity henceforth. The compactness means that orthogonality must only be specifically arranged for nearest neighbor functions. Here we give six orthogonal shape functions which are smooth, i.e. all derivatives are continuous, which also are compact with width two, and which are able to represent polynomials exactly up to order five. We believe these shape functions are sufficiently complete for most uses, because we give alternative ways of generating the orthlets in the vicinity of singularities and also for changing lattice spacings. The smoothness allows additional functions to be added to increase completeness without adjusting the functions one already has.

All shape functions $S_n(x)$ are defined for $x \geq 0$; $S_n(x)$ is even (odd) if n is. We construct $S_0(x)$ so that $S_0(x)$ and $S_0(x-1)$ together are able to reproduce a constant function between $0 \leq x \leq 1$, and hence the set of functions $S_0(x-i)$ can represent a constant for all x. By adding additional functions $S_j(x)$, we will be able to generate an exact fit to higher order polynomial functions. Then, given an arbitrary function, the set of shape functions can generate fits to it that are similar to the piece-wise polynomial fits used by spline functions. We first define a smooth "splicing function" $p(x)$ which divides a function to be fit to into pieces. We require that $p(x) = 0$ for $|x| \geq 1$, $p(x) = p(-x)$, and that

$$p(x) + p(x-1) = 1 \qquad 0 \leq x \leq 1. \tag{6}$$

We choose the function

$$p(x) = \frac{1}{2} - \frac{1}{2}\tanh(\frac{3}{2}\,\frac{x-1/2}{[x(1-x)]^{1/2}}) \qquad 0 < x < 1. \tag{7}$$

This bell-shaped function has essential singularities at $x = 0, \pm 1$, allowing it to have compact support yet be smooth.

The shape function is obtained by multiplying the function to fit to, in this case unity, by $p(x)$, and then adding a smooth oscillating function $o(x)$ to induce orthogonality. The fit will not be spoiled if

$$o(x) + o(x-1) = 0 \qquad 0 \leq x \leq 1. \tag{8}$$

We choose

$$o(x) = \sum_{m=1}^{M} a_{0m} o_m(x) \tag{9}$$

where

$$o_m(x) = p(2x-1)\sin(2m\pi(x-1/2)). \tag{10}$$

For the first shape function, $S_0(x)$, is suffices to take $M = 1$, with $a_{01} = -0.507021142747521$ producing the required orthogonality between $S_0(x)$ and $S_0(x-1)$:

$$S_0(x) = p(x) + a_{01} o_1(x) \qquad 0 \leq x \leq 1. \tag{11}$$

Note that $S_0(x) + S_0(x-1) = 1$ for $0 \leq x \leq 1$, so that $S_0(x)$ is already normalized. We show $S_0(x)$ in Fig. 2.

The second shape function is obtained similarly, by requiring an exact fit to the function x. First, we attempt to fit the function x using only $S_0(x)$. We then make $S_1(x)$ out of the error or residual of this fit,

$$r_1(x) = x - S_0(x-1) \qquad 0 \leq x \leq 1. \tag{12}$$

Since $S_1(x)$ is odd, the extra orthogonalizing functions must be made out of cosine functions rather than sine functions. For $0 \leq x \leq 1$

$$S_1(x) = N_1 \left(p(x) r_1(x) + \sum_{m=1}^{2} a_{1m} e_m(x) \right), \tag{13}$$

with $S_1(-x) = -S_1(x)$,

$$e_m(x) = p(2x-1)\cos((2m-1)\pi(x-1/2)), \tag{14}$$

$a_{11} = 0.13240379\ldots$, $a_{12} = -0.04884462\ldots$, and $N_1 = 3.78750743\ldots$.

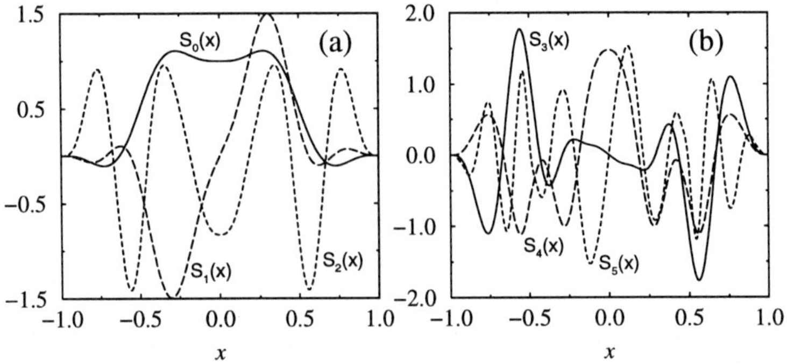

Fig. 2. (**a**) and (**b**) shape functions $S_n(x)$. These functions have completeness properties only when combined with their neighbors $S_n(x \pm 1)$. They are orthonormal, including orthogonality between neighboring functions.

To determine the coefficients a_{1m}, and later coefficients for higher order shape functions, two different linear combinations of the $e_m(x)$ or $o_m(x)$ were found which induced orthogonality to all the lower order shape functions centered at $x = 1$. These two different combinations were then combined in order to induce orthogonality of $S_n(x)$ and $S_n(x-1)$. This last procedure required solving a quadratic equation, which might not have real roots. If there were no real roots, more orthogonalizing functions were included. A solution in this case required finding both positive and negative eigenvalues of a symmetric matrix, which also sometimes did not occur. There does not appear to be any guarantee of a real solution; in fact, we did not find a satisfactory $S_6(x)$ able to fit x^6 exactly [8].

In Fig. 3, we use these shape functions to generate an orthlet basis for a function with a slope discontinuity. For the lattice sites away from the singularity, the overlap integral of the function with each shape function was computed numerically. Resumming the shape functions on a site with these integrals as coefficients gives the orthlet. In the case of the orthlet at $x = 0$, where the slope discontinuity is, an expansion would converge too slowly. Instead, the orthlet was obtained by subtracting from $f(x)$ the orthlets on the adjacent sites. Another procedure for dealing with singularities, which is not shown here, is to add a set of localized functions near the singularity, which are chosen for their ability to represent the singularity.

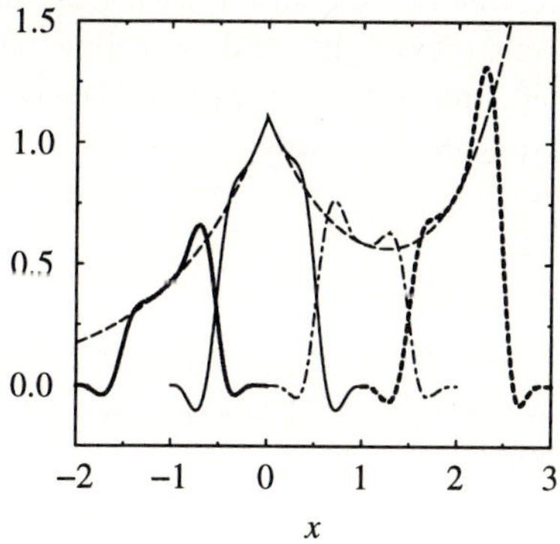

Fig. 3. Orthlets used to fit the function $f(x) = \exp(-|x|) + 1/((x-3)^2 + 1/2)$. The sum of these four orthlets is equal to $f(x)$ within the region $-1 \leq x \leq 2$.

In three dimensions, we form shape functions as cartesian products of the 1D shape functions, $S_{\boldsymbol{n}}(\boldsymbol{r}) = S_{n_x}(x)S_{n_y}(y)S_{n_z}(z)$. In preliminary calculations to represent the hydrogen atom ground state, we have found that adding a set of narrow gaussians, multiplied by a windowing function similar to $p(x)$, is a convenient and effective for representing the cusp singularity at the nucleus. Part of the convenience is that 3D gaussians are products of 1D gaussians, like the 3D shape functions.

Now, suppose one wishes to generate orthlets to represent a set of functions $\{f^\alpha(\boldsymbol{r})\}$. For example, one might want to build an orthlet basis which is able to represent a standard radial basis set from quantum chemistry, since these basis sets are known to represent Hartree-Fock orbitals well. Note that the f_α need not be orthogonal, but the orthlet basis generated from them is. The orthlet basis will automatically have additional degrees of freedom allowing improved treatment of correlations. For simplicity, we will let $S_{\boldsymbol{n}}(\boldsymbol{r})$ represent both shape functions and any extra singularity basis functions. To adapt the basis to represent the f_α, we apply the procedure in the density matrix renormalization group for targetting more than one state [2]. For each lattice site j, we find the coefficients

$$c_{\alpha\boldsymbol{n}} = \langle S^j_{\boldsymbol{n}}(\boldsymbol{r}) | f^\alpha(\boldsymbol{r}) \rangle. \tag{15}$$

We now form the positive semidefinite density matrix

$$\rho^j_{\boldsymbol{n}'\boldsymbol{n}} = \sum_\alpha c_{\alpha\boldsymbol{n}'} c_{\alpha\boldsymbol{n}}. \tag{16}$$

Note that positive weighting factors a_α can also be included in the sum if some of the functions are considered more important than others. The eigenvectors of ρ^j, $v^\beta_{\boldsymbol{n}}$, define orthlet functions $v^\beta(\boldsymbol{r}) = \sum_{\boldsymbol{n}} v^\beta_{\boldsymbol{n}} S^j_{\boldsymbol{n}}(\boldsymbol{r})$ which optimally represent the functions $\{f^\alpha(\boldsymbol{r})\}$ in the basis for site j. Each density matrix eigenvalue w_β gives the weight associated with that orthlet in representing the f_α. By choosing a cut off weight, and retaining all orthlets with weight greater than the cut off, one obtains a systematically improvable basis for the f_α. In particular, one can show that

$$w_\beta = \sum_\alpha (\langle v^\beta | f^\alpha \rangle)^2. \tag{17}$$

Thus, if a density matrix eigenvalue w_β is very small, then none of the f^α have significant overlap with the corresponding function v^β, and v^β need not be included in the basis. The basis set becomes exact if the number of orthlets kept per site is equal to the number of functions f_α. If only one function is in the set f_α, then there is only one nonzero eigenvalue and the orthlet is simply the normalized projected function $P_i f_\alpha$.

In electronic structure calculations, one would use a lattice spacing appropriate for the valence electrons, say 0.3-1.0 angstroms. Cores would be treated using orthlets derived from localized cusp functions, which would be tied to each nucleus. Lattice sites near cores would have a dozen or more orthlets; in other areas, we expect only a few might be needed, making perhaps 100-300 basis functions per atom. Although this is an order of magnitude more than with the radial functions used in quantum chemistry, it is much less than for other localized basis sets. The localized nature of the basis immediately reduces the number of terms in the Hamiltonian from $o(N^4)$ to $o(N^2)$. These basis sets appear to be ideal for DMRG calculatons for electronic structure, but one will not really know for sure until they have been tried. Work is currently underway to generate an orthlet basis for a small molecule, in order to test the effectiveness of orthlets for DMRG as well as other approaches to electronic structure.

The calculations in Sect. 2 were done in collaboration with R. L. Martin. I acknowledge support from the NSF under Grant No. DMR-9870930.

References

1. S.R. White and R.L. Martin, cond-mat/9808118
2. S.R. White, Phys. Rev. Lett. **69**, 2863 (1992), Phys. Rev. B **48**, 10345 (1993)
3. T. Xiang, Phys. Rev. B **53**, 10445 (1996)
4. C. W. Bauschlicher and P.R. Taylor, J. Chem. Phys. **85**, 2779 (1986)
5. C. Zhang, E. Jeckelmann and S.R. White, Phys. Rev. Lett. **80**, 2661 (1998)
6. M.C. Payne, M.P. Teter, D.C. Allan, T.A. Arias and J.D. Joannopoulos, Rev. Mod. Phys. **64**, 1045 (1992)
7. S.R. White, J.W. Wilkins, and M.P. Teter, Phys. Rev. B **39**, 5819 (1989). For a more recent approach, see E. Tsuchida and M. Tsukada, Phys. Rev. B **54**, 7602 (1996)
8. C programs to generate S_0 to S_5 are available at `http://hedrock.ps.uci.edu`

2.2 Symmetrized DMRG Method for Conjugated Polymers

S. Ramasesha and Kunj Tandon

Solid State and Structural Chemistry Unit, Indian Institute of Science
Bangalore 560 012, India

Conjugated polymers and molecules have been of considerable interest to chemists for over a century, because of their high reactivity, long wavelength *uv* absorption and magnetic anisotropy [1]. In recent years, these polymers have been recognized as providing real life examples of quasi low-dimensional systems. In long polymers, the σ - backbone of the framework formed by the sp^2 - hybrid orbitals of carbon is well separated, on the energy scale, from the $2p_z$ orbitals involved in π - conjugation. The low energy electronic excitations of the polymers can therefore be thought of as arising from the itinerant π - system. While the early picture of conjugated polymers was extensively derived from a study of the noninteracting π - system within the framework of the Hückel theory, even almost half-a-century ago it was recognized that electron-electron interactions were of crucial importance in resolving the discrepancy between experiments and Hückel predictions. The interacting π-electron models were independently proposed by Pariser and Parr [2] and by Pople [3] in 1953 and the resulting Pariser-Parr- Pople (PPP) model has formed the corner stone for studying low-energy excitations in large conjugated systems.

The Hückel model in one dimension is nothing but the nearest neighbour single-band tight-binding model whose Hamiltonian is given by,

$$H_0 = \sum_{i=1}^{N} t_i(\hat{a}^\dagger_{i,\sigma}\hat{a}_{i+1,\sigma} + h.c.) \tag{1}$$

where, t_i is the transfer or resonance integral for the ith bond and $\hat{a}^\dagger_{i,\sigma}$ ($\hat{a}_{i,\sigma}$) creates (annihilates) an electron of spin σ at site i and the last term in the summation is modified according to the chosen boundary condition. Introducing electron-electron interactions leads to the full π-electron Hamiltonian given by,

$$H_\pi = H_0 + \frac{1}{2}\sum_{ijkl}[ij|kl](\hat{E}_{ij}\hat{E}_{kl} - \delta_{jk}\hat{E}_{il}) \tag{2}$$

where the two electron integral $[ij|kl] = \int \phi_i^*(1)\phi_j(1)\frac{1}{r_{12}}\phi_k^*(2)\phi_l(2)d^3r_1d^3r_2$ and $\hat{E}_{ij} = \sum_\sigma \hat{a}^\dagger_{i,\sigma}\hat{a}_{j,\sigma}$.

The PPP model is obtained from the above many-body Hamiltonian by making Zero Differential Overlap (ZDO) approximation; $[ij|kl] = [ij|kl]\delta_{ij}\delta_{kl}$. This results only in one and two-center two-electron integrals, in the interaction part. The two-center two-electron integrals $[ii|jj]$ $(i \neq j)$ are further approximated by interpolating between $[ii|ii]$ and $\frac{e^2}{r_{ij}}$ in the limit $r_{ij} \longrightarrow \infty$ [4,5]. The PPP model Hamiltonian can now be written as,

$$H_{PPP} = H_0 + \sum_i \frac{U_i}{2}\hat{n}_i(\hat{n}_i - 1) + \sum_{i>j} V_{ij}(\hat{n}_i - z_i)(\hat{n}_j - z_j) \tag{3}$$

where $\hat{n}_i = \hat{a}_i^\dagger \hat{a}_i$, $U_i = [ii|ii]$, $V_{ij} = [ii|jj]$ and z_i are the local chemical potentials.

The PPP model reduces to the Hubbard model if the range of electron repulsions is mercilessly truncated to include only the on-site repulsion, U, an approximation which is valid in metals. However, in polymers which are semiconducting, the minimal model includes the nearest neighbour electron repulsion integral V and is known as the "$U - V$" model.

The PPP model has been very widely studied in the chemistry literature using a variety of quantum chemical techniques [6]. In the context of polymers, with several atoms per unit cell, no reliable technique existed until the invention of the DMRG technique [7]. However, the DMRG technique in its original form is well suited for studying only the ground states and a few low-lying states lying immediately above the ground state and it cannot access the excited states of interest in polymers. This is particularly so since, if the electronic states are numbered in the increasing order of energy, the serial number at which the interesting states occur changes unpredictably both with chain length and model parameters. This problem can be overcome if the crucial symmetries of the model Hamiltonian are exploited in the DMRG calculations. The states of experimental interest then turn out to be either the lowest states or one of the few low-lying states in each symmetry subspace.

1 Symmetrized DMRG Technique

The PPP Hamiltonian is spin independent and hence conserves both the total spin and the z-component of the total spin. Besides this symmetry, most polymers of interest also possess alternancy symmetry, whereby the carbon sites in the polymer backbone can be divided into two sublattices, with the transfer part of the Hamiltonian operating only between sites belonging to different sublattices. This is only an approximate symmetry since all the carbon sites in a polymer are not in truly identical environments. This symmetry is also known variously as the electron-hole symmetry and charge-conjugation symmetry. Most simple polymers also possess inversion or reflection symmetry. The linear optical spectra of the polymers are almost entirely governed by the low-lying states in the subspace which has opposite electron-hole and inversion symmetries relative to the subspace in which the ground state lies. However, nonlinear optic (NLO) properties of the polymer are governed by the low-lying states in both of the above subspaces. The phosphorescence properties of the polymer are governed by the low-lying states in the subspace which has opposite spin parity to that of the ground-state subspace.

While in exact diagonalization techniques it is easy to exploit the total spin invariance of the spin-independent Hamiltonians [8], within the DMRG scheme, it appears to be problematic since the truncation of the Fock space of the system block at each iteration may not strictly preserve the total spin invariance. However, in the study of polymers, one is rarely interested in electronic states with a total spin higher than one (triplets). At intermediate correlation strengths, most often, the triplet states intrude between the ground state and excited singlet states. Thus, if the Hilbert space can be divided into space of even and odd total spins, the singlets and the triplets could be effectively separated. This could be achieved by exploiting the spin parity which exists in the total $M_s = 0$ subspace. The spin parity operation involves flipping all the spins and the subspace with even spin parity consists of basis functions with even integer total spin, S, while the subspace of odd spin parity is spanned by basis function with odd integer S.

The symmetrized DMRG (SDMRG) method involves block-diagonalizing the Hamiltonian matrix of the total system at each iteration and targetting the desired states in the blocks. Implementing symmetries in such a scheme appears to be difficult at first glance since the physical system changes from iteration to iteration. However, the alternancy symmetry and the parity symmetry are essentially associated with individual sites and the symmetry of the total system can be written as a direct product of the site symmetry operators. The matrix of the site symmetry operators in the Fock space of a single site for parity $\hat{P}_i$ and alternancy $\hat{J}_i$ can be easily obtained since $\hat{P}_i|0\rangle = 0; \hat{P}_i|\uparrow\rangle = |\downarrow\rangle; P_i|\downarrow\rangle = |\uparrow\rangle; \hat{P}_i|\uparrow\downarrow\rangle = -|\downarrow\uparrow\rangle$ and similarly $\hat{J}_i|0\rangle = |\uparrow\downarrow\rangle; \hat{J}_i|\uparrow\rangle = (-)^{\eta_i}|\uparrow\rangle; J_i|\downarrow\rangle = (-)^{\eta_i}|\downarrow\rangle; \hat{J}_i|\uparrow\downarrow\rangle = -|0\rangle$ where i is the site index and η_i is 1 for i belonging to one sublattice and 0 for the other. The $\hat{C}_2$ symmetry is defined over the entire system. The matrix representation of this operation in the direct product DMRG basis $\{\mu, \sigma, \sigma', \mu'\}$ is easily obtained from the effect of $\hat{C}_2$ on the basis states given by,

$$\hat{C}_2|\mu\sigma\sigma'\mu'\rangle = (-1)^{\gamma}|\mu'\sigma'\sigma\mu\rangle \tag{4}$$

where $\mu(\mu')$ is the density matrix eigenvector (DMEV) of the left (right) block, $\sigma(\sigma')$ is the Fock-space state of the new single site adjacent to the left (right) block and the phase factor $\gamma = (n_\mu + n_\sigma)(n_{\mu'} + n_{\sigma'})$ where the n's are the occupations of the respective blocks/sites.

The symmetry operators $\hat{C}_2, \hat{J}$, and $\hat{P}$ commute amongst each other and hence there are eight one-dimensional irreducible representations of this group. We use the label A (B) to represent the subspace which transforms as even (odd) under $\hat{C}_2$; similarly, $+$ $(-)$ for the subspace even (odd) under

$\hat{J}$; and e (o) for the subspace even (odd) under $\hat{P}$. The ground state of most systems usually lies in the subspace ${}^eA^+$ while the dipole allowed excited state lies in the subspace ${}^eB^-$. The lowest energy spin excitation is to the lowest state in the subspace ${}^oB^+$ and the lowest two-photon excitation is to the second state in the subspace ${}^eA^+$.

The projection operator for a given irreducible representation Γ is given by,

$$\hat{P}_\Gamma = \frac{1}{h}\sum_{\hat{R}} \chi_\Gamma(\hat{R})\hat{R} \tag{5}$$

where $\hat{R}$ is a member of the group; h is the order of the group and $\chi_\Gamma(\hat{R})$ is the character under the irreducible representation Γ for $\hat{R}$. Matrix representation of the projection operator in the DMRG basis for any given representation can now be obtained from the matrix representations of the individual symmetry operations. The symmetrized combinations of the DMRG basis vectors follow from the operation of the matrix $\mathbf{P_\Gamma}$ on the basis. The resulting SDMRG basis is however overcomplete. To avoid the problem of over completeness, we recognize that the dimensionality of the symmetrized basis in the representation Γ is given by,

$$n_\Gamma = \frac{1}{h}\sum_{\hat{R}} \chi_\Gamma(\hat{R})\chi_{red.}(\hat{R}) \tag{6}$$

where $\chi_{red.}$ is easily obtained as the sum of the diagonal elements in the matrix of $\hat{R}$ in the DMRG basis. We can now obtain a rectangular matrix $\mathbf{S_\Gamma}$ with n_Γ rows and $4m$ columns by Gram-Schmidt orthonormalization of the rows of P_Γ. The orthonormalization process is stopped when the number of rows obtained equals n_Γ.

The Hamiltonian matrix $\mathbf{H_\Gamma}$ in the SDMRG basis can be obtained from the nonsymmetrized Hamiltonian matrix $\mathbf{H}$ by the following transformation,

$$\mathbf{H}_\Gamma = \mathbf{S}_\Gamma \mathbf{H} \mathbf{S}_\Gamma^\dagger. \tag{7}$$

Once the desired symmetrized eigenstate is obtained in the symmetrized basis, the density matrix for the next step is obtained from the unsymmetrized eigenstate obtained from a back transformation.

The eigenvectors of the density matrix are not simultaneously eigenvectors of the symmetry operations. Thus, in the straightforward implementation of the truncation process, it is likely that not all the symmetry partners of the DMEV are included. This problem can be avoided by marginally increasing the cut-off from the chosen value whenever such a situation occurs.

As test of the above technique, we present results of DMRG calculations for uniform Hubbard chains at half-filling, for U/t of 4.0 and 6.0, with chain lengths of up to 50 sites [9]. We have obtained the lowest energy states in all the eight subspaces (of $\hat{C}_2$, $\hat{J}$ and $\hat{P}$), keeping 70 to 150 eigenvectors of the density matrix.

The energy of the one-photon transition from the ground state (lying in the

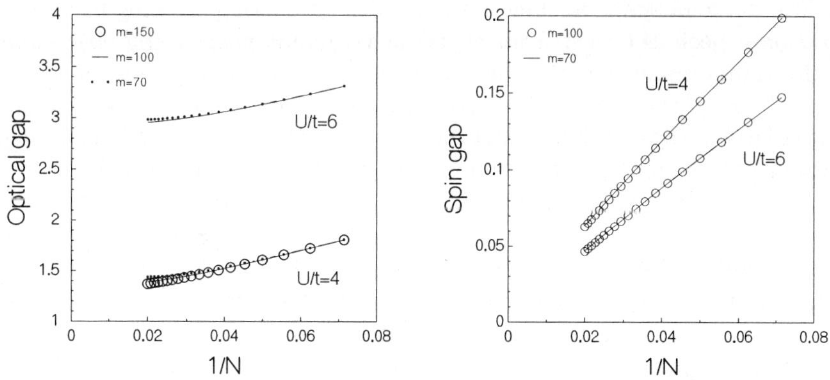

Fig. 1. Optical Gap (left) and Spin Gap (right) *vs* $1/N$ for Hubbard chains.

${}^{e}A^{+}$ subspace) to the lowest energy state in the ${}^{e}B^{-}$ subspace defines the optical gap of the chain. In Fig. 1, we show a plot of optical gap *vs.* inverse chain length (1/N) for $U/t = 4.0$ and 6.0 for different values of the cut-off m. The optical gaps from the Bethe *ansatz* solution for the infinite chain, are $1.2867t$ for $U/t = 4.0$ and $2.8926t$ for $U/t = 6.0$ [10]. The corresponding extrapolated values from a polynomial fit in powers of $1/N$ from DMRG are $1.278t$ and $2.895t$, obtained with a cut-off of $m = 150$ for $U/t = 4.0$ and m=100 for $U/t = 6.0$. We also find that the DMRG optical gap tends to saturate at shorter chain lengths as we decrease m. The fit of the optical gap to $1/N$ is a reasonably good straight line although the best fit is to a polynomial in this variable. The spin gap (the energy gap between the lowest triplet state and the ground-state singlet) as a function of $1/N$ is also shown in Fig. 1. The Bethe *ansatz* solution yields a vanishing spin gap in the thermodynamic limit of the uniform Hubbard chain. Polynomial fits to our DMRG data are consistent with this.

The above technique has proved successful in studying polymers such as poly-para-phenylenes (PPP) and poly-para-phenylene-vinylenes (PPV) [11,12] which are of great current interest.

2 Dynamics from the Correction-Vector Method

The DMRG method [7], as it was first introduced, is essentially a static technique for a few low-lying eigenstates. However, one of the main properties of interest in the field of conjugated polymers is the study of their dynamic nonlinear optical (NLO) response [13,14]. It is a major challenge to obtain reliable dynamics of interacting electron systems. While for short oligomers there exist reasonable approximations for computing these properties [15], longer chains had proved elusive. Yet most interest lies in the longer chains since the NLO properties exhibit dominant finite-size effects.

The dynamic response functions of finite interacting systems have most commonly been obtained from an explicit computation of the eigenstates of the Hamiltonian and the matrix elements of the appropriate operators in the basis of these eigenstates [16]. This has been a widely used method particularly in the computation of the dynamic NLO coefficients of molecular systems and is known as the sum-over-states (SOS) method. In the case of model Hamiltonians, the technique that has been widely exploited to study dynamics is the Lanczos method [17]. The spectral intensity corresponding to an operator $\hat{O}$ is given by,

$$I(\omega) = -\frac{1}{\pi}\mathrm{Im}[\langle G|\hat{O}^{\dagger}\frac{1}{(\omega + E_0 + i\epsilon - \hat{H})}\hat{O}|G\rangle] \tag{8}$$

where $|G\rangle$ is the ground state eigenvector of the Hamiltonian $\hat{H}$, with eigenvalue E_0, ω is the frequency at which the response is sought, ϵ is the mean life time parameter. In the Lanczos method, $I(\omega)$ is computed as a continued fraction by employing a truncated tridiagonal form of the Hamiltonian matrix obtained in the Lanczos scheme [17].

In the context of NLO properties of interacting models such as the Hubbard and extended Hubbard models, it was shown by Soos and Ramasesha that the model exact dynamical NLO coefficients could be obtained by solving for correction vectors [18]. If we define the correction vector $\phi^{(1)}(\omega)$ by the equation,

$$(\hat{H} - E_0 - \omega - i\epsilon)\phi^{(1)}(\omega) = -\hat{O}|G\rangle \tag{9}$$

then the spectral function, $I(\omega)$, can be expressed as,

$$I(\omega) = -\frac{1}{\pi}\mathrm{Im}\langle G|\hat{O}^{\dagger}|\phi^{(1)}(\omega)\rangle. \tag{10}$$

The correction vector is solved for in the basis of the configuration functions, which is also the basis in which the Hamiltonian matrix is set-up for obtaining the ground state. Given the ground state and the correction vector, it is straightforward to compute the spectral function. This method is quite general and has been employed in the computations of dynamic NLO coefficients of a wide variety of Hamiltonians [19]. The inhomogeneous linear algebraic equations encountered in this method involve large sparse matrices. An iterative small matrix algorithm, which runs parallel to the Davidson algorithm for eigenvalue problems, gives rapid convergence for the solution of the system of equations [20].

The correction-vector method is not limited to tridiagonal matrices. Furthermore, by defining higher-order correction vectors, it is possible to compute nonlinear spectral functions as are encountered in the study of nonlinear optical properties. This technique provides a way of computing the dynamics which is exact for the chosen restricted configuration space, unlike other techniques which involve truncations over and above that imposed by the choice of the space of configurations. Although the dynamics of a spin chain [21] and also that of a spinless fermion system [22] have been reported in the literature using the DMRG technique, the first one employed the Lanczos algorithm which amounts to further truncation of the Hilbert space over and above the truncation involved in the DMRG procedure and the latter used the maximum entropy method which relies on analytic continuation and involves large uncertainties. There is also a discussion of dynamics from the DMRG method by Wang, et al. in Chap. 7(I)

3 Computation of Dynamic Optical Coefficients

The computation of the dynamic NLO coefficients by the correction vector method requires the ground-state eigenfunction, the Hamiltonian matrix and the dipole displacement matrices [19]. The DMRG method as implemented, readily provides us with the ground state and the Hamiltonian matrix. The matrices of the dipole operators are constructed in the DMRG scheme by renormalizing the matrix representations of these operators corresponding to the system blocks in way analogous to the Hamiltonian operators for the fragments. The matrix representation of the dipole operators for the full system are obtained as direct products of the fragment matrices, the *dipole displacement* matrices are then obtained by subtracting the corresponding components of the ground state dipole moments from the diagonal elements of the full system dipole matrices.

The two correction vectors $|\phi_i^{(1)}(\omega_1)\rangle$ and $|\phi_{ij}^{(2)}(\omega_1,\omega_2)\rangle$ encountered in the computation of first-order and third-order polarizabilities are obtained by solving the following linear equations:

$$(H - E_0 + \omega_1 + i\epsilon)|\phi_i^{(1)}(\omega_1)\rangle = \tilde{\mu}_i|G\rangle \tag{11}$$

$$(H - E_0 + \omega_2 + i\epsilon)|\phi_{ij}^{(2)}(\omega_1)\rangle = \tilde{\mu}_j|\phi_i^{(1)}(\omega_1)\rangle \tag{12}$$

where $\tilde{\mu}_i$'s are the dipole displacement operators, the indices i and j correspond to the Cartesian coordinates (x, y, z) of the physical system and other quantities are as defined in (8). In terms of these correction vectors, the components of the first polarizability, α_{ij}, and third-order polarizability, γ_{ijkl}, can be written as:

$$\alpha_{ij}(\omega) = \langle\phi_i^{(1)}(\omega)|\tilde{\mu}_j|G\rangle + \langle\phi_j^{(1)}(-\omega)|\tilde{\mu}_i|G\rangle \tag{13}$$

$$\gamma_{ijkl}(\omega_1,\omega_2,\omega_3) = \hat{P}\langle\phi_i^{(1)}(-\omega_1-\omega_2-\omega_3)|\tilde{\mu}_j|\phi_{kl}^{(2)}(-\omega_1-\omega_2,-\omega_1)\rangle \quad (14)$$

where the operator $\hat{P}$ generates all the permutations: $(-\omega_\sigma, i), (\omega_1, j), (\omega_2, k)$ and (ω_3, l) leading to 24 terms for γ_{ijkl} with $\omega_\sigma = -\omega_1 - \omega_2 - \omega_3$. The tumbling averaged $\bar{\alpha} = \frac{1}{3}\sum\alpha_{ii}$ and $\bar{\gamma} = \frac{1}{15}\sum(2\gamma_{iijj} + \gamma_{ijji})$ are usually computed to allow comparison of the calculated NLO response with experiments on systems containing molecules in random orientations [13].

While the extension of the DMRG scheme for obtaining the correction vectors appears to be straightforward, given the eigenstate and the matrices corresponding to the relevant operators, in the actual implementation there exist some severe problems. The correction vector lies in the symmetry subspace which is connected to the ground state by the electric dipole operator. For ω values corresponding to resonance between the ground state and eigenstates in that particluar symmetry subspace, *lhs* of (11) and (12) become singular for $\epsilon = 0$ and present numerical difficulties even when solving them for reasonable nonzero ϵ values. However, if we do not exploit the symmetries of the Hamiltonian, we encounter singularities in (11) and (12) even for those ω values corresponding to eigenstates of the Hamiltonian found in other symmetry subspaces which are not connected to the ground state by the dipole operator. Therefore, numerically it would be impossible to obtain the correction vectors at these frequencies and thereby the associated response of the system. For example, in Hubbard chains at intermediate correlation strengths, a triplet excited state lies below the lowest singlet state in the *ionic* B subspace [23]. The resonances in polarizability are expected only at frequencies corresponding to the energy levels in the *ionic* B space, relative to the ground state. However, we can not solve for the correction vector using (11) at an excitation energy corresponding to the energy of the lowest triplet state. Thus, the technique of correction vectors will not be able to give the complete dispersion of the polarizabilities up to the first one-photon resonance, unless interferences due to spurious intruders such as the triplet states are eliminated by suitably block diagonalizing the Hamiltonian matrix. This problem of intruders becomes more severe with increasing system size due, as mentioned previously, to increasing number of intruder states lying below the frequency corresponding to the first *true* resonance.

We have therefore exploited the electron-hole symmetry, the reflection symmetry of the polymers and spin parity to block-diagonalize the Hamiltonian [9]. In the course of our test calculations, we have considered a single excitation frequency ($\omega = \omega_1 = \omega_2 = \omega_3$). We have also exploited sparseness of all the matrices to improve upon the computational efficiency and have employed the finite-size DMRG algorithm in some cases to check the convergence of the DMRG results. In these cases, the spatial symmetry is introduced only at the end of the DMRG procedure when the left and the right density matrices correspond to fragments of the same size, *i.e.*, at the end of each finite-size iteration.

We have compared our results with the *model exact* α and γ values obtained from the correction-vector method in the valence-bond basis, for chains of up to 12 sites. The results compare well with a cut-off of $m = 100$, *i.e.*, by retaining the dominant 100 DMEVs in the DMRG scheme. For the 8-site problem, DMRG with $m = 100$ is exact and the DMRG results compare with exact results to numerical accuracy. In the case of the 12-site chain, the largest relative error is 1% for $U/t = 4$ in the uniform Hubbard model. The exact $\bar{\alpha}$ is 5.3433×10^{-24} *esu* while $\bar{\alpha}$ from DMRG is 5.1848×10^{-24} *esu*. The corresponding exact $\bar{\gamma}$ and DMRG $\bar{\gamma}$ values are 5.9832×10^{-34} *esu* and 5.8439×10^{-34} *esu*. The dominant α and γ components (namely the longitudinal α_{xx} and γ_{xxxx} components) are in relative error of 0.1% and 0.2%, respectively.

4 Lattice Relaxation in Polyenes

As an interesting example of the application of SDMRG method to conjugated polymers, we present the relaxed geometries of excited states of conjugated polymers [24]. Barford and Bursill also discusses some applications of the DMRG method to polymers in Chap. 2.3(II). One of the predictions of the Su-Schrieffer-Heeger model [25,26] is that the electronic excitations are closely related to the topological distortions of the conjugated polyene chain. It is predicted that solitons are the elementary excitations both for optical excitation and for charge doping. Su and Schrieffer further proposed photoexcitation-induced soliton creation, which is now known as the Su-Schrieffer mechanism [27]. According to this mechanism, photoexcited states can relax to a pair of separated positively and negatively charged solitons; unlike the doping process, this mechanism will generate solitons of both charges. This mechanism, if correct, implies a similarity between the spectra of photogenerated states obtained from a pump-probe experiment and the absorption spectra of doped polyacetylene. Unfortunately, a straightforward comparison of the photogenerated gap states and states of the doped systems is not possible. This is because doping introduces counter-ions which provide an electrostatic potential for the charged solitons leading to a shift in the solitonic absorption energy [28–30].

The state that is expected to decay into two neutral solitons is the lowest triplet state. The spins constituting the triplet state can dissociate to give rise to two neutral solitons each with spin-$\frac{1}{2}$ [31]. There have been studies of the triplet state [32–34], with this idea in focus employing the optically detected magnetic resonance technique [35]. Another excited state that has been discussed in the context of solitons is the 2^1A_g state. It was shown by Tavan and Schulten in the context of finite polyenes that this state could be regarded as composed of two triplets [36]. This was also later studied by others with a view to find the relaxation products of the 2^1A_g state. It is worth mentioning that among these studies, the work of Su dealt with longer polyene chains. Su [37] who employed a degenerate perturbation approach in both the weak correlation limit (U as perturbation) and the strong correlation limit (spin-lattice model), found that in both limits the $2A_g$ state relaxed to four solitons. However, it should be pointed out that the perturbation theory employed by Su is not size-extensive. Thus, his result for N=80 is not reliable in the weak correlation limit. In the strong correlation limit, it is now well established that employing spin models is quite inappropriate [38,39].

In so far as energetics of low-lying excited states of odd and even carbon systems and doped systems are concerned, earlier studies [40–44] have shown that once electron correlations are introduced, the notion of solitons as elementary electronic excitations becomes somewhat blurred [28]. However, it is important to understand whether the associated topological features of the low-lying excitations survive when we introduce realistic electron-electron correlations. We therefore employed the SDMRG method to study the equilibrium geometries of excitations in the Peierls-PPP model to discover if there is a relation between the relaxed geometries and the topological lattice distortions associated with solitons in the presence of realistic electron correlations.

Our study employs the PPP model (3) with independent bond distortions for a fixed electron-phonon coupling constant. The distortions of the bond lead to change in the corresponding transfer integrals. The bond distortions are introduced such that the total bond-order of the chain remains a constant. In the absence of this constraint, the purely π-electron Hamiltonian would lead to a collapse of the chain to a point since the total energy of the system tends to decrease with decrease in chain length; equivalently, the π-electron Hamiltonian does not have an in-built repulsive term which keeps the atoms apart since this term comes from the σ-framework and the internuclear potentials. Imposition of the constraint of constant total bond-order serves the purpose of the σ-framework and the inter-nuclear potentials.

The Hamiltonian including the electron-phonon coupling is given by,

$$H = H_{PPP} + \frac{1}{\pi\lambda}\sum_i \delta_i^2 \tag{15}$$

where λ is the dimensionless coupling constant, δ_i represents the deviation of the transfer integral for the $i, i+1$ bond from uniform transfer integral. All the interaction integrals appearing in the Hamiltonian are scaled in units of the uniform transfer integral.

The constraint of constant total bond-order is introduced via the method of Lagrange undetermined multiplier, so that the functional that is minimized with respect to δ_i is $F = H + L$, where $L = \zeta \sum_i \delta_i$, ζ being the undetermined multiplier. The Lagrange multiplier ζ turns out to be the average bond order $\bar{p}$ and we obtain,

$$\delta_i = \pi\lambda(p_i - \bar{p}) \; ; \quad p_i = \frac{1}{2}\langle(\hat{E}_{i,i+1} + \hat{E}_{i+1,i})\rangle \tag{16}$$

In the actual calculations, we start with either $\delta_i = 0$ or the δ_i value corresponding to the ground state and iterate on the δ_i values obtained from the desired eigenstate until the bond order profile δ_i vs i converges.

The π coherence length for a reasonable value of the electron-phonon coupling constant of $\lambda = 0.1$ is rather large, implying the need to study fairly long chains in the range $40 \leq N \leq 60$, where the usual quantum chemical or related many-body methods for obtaining the ground and excited states to reliable accuracy fail. However, the SDMRG method can be applied to this problem since the distortion does not break any of the crucial symmetries of the undistorted model. We have used the finite SDMRG algorithm in all the calculations to enhance the accuracy of our computations over and above what is inherent in the infinite SDMRG method for these systems. Besides, in the finite DMRG iteration, we obtain the bond orders of the newly introduced bonds at each stage of the iteration to modify the transfer integrals corresponding to those bonds at the next stage of the iteration. The finite DMRG iterations are continued until the bond order profile converges. Our calculations for the excited states start with the relaxed ground-state geometry and hence can give the vertical excitation energies. Furthermore, the calculations for the self-consistently relaxed geometries give us the relaxation energies for various states.

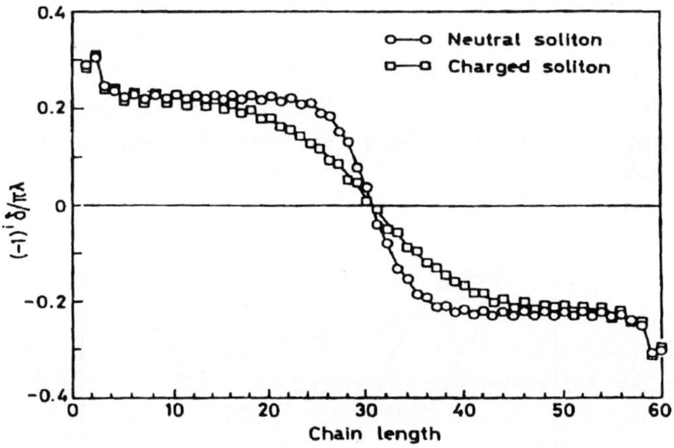

Fig. 2. Bond order profile for neutral(square) and charged(circle) solitons as a function of chain length for a polyene chain of 61 carbon atoms

We have studied the geometries of four states, namely, the ground state, the lowest one-photon state, the lowest two-photon state, and the lowest triplet state. We have studied the excited states in both the unrelaxed geometry (corresponding to the geometry of the ground state) and the relaxed geometry, for various chain lengths. We have also studied the relaxed geometries of the neutral and charged solitonic configurations, corresponding to the ground state of an odd-polyene chain and its ion, respectively. For a value of the coupling constant $\lambda = 0.1$, we find that the ground state of the polyene has a bond alternation of $\delta = 0.07$, which is in agreement with experiments. The magnitude of bond alternation scales linearly with λ for $0 \leq \lambda \leq 0.2$. We have carried out all our studies fixing the value of λ to be 0.1.

For both neutral and charged solitons we find the typical tangent hyperbolic dependence of the bond order as a function of bond number (Fig. 2). What is interesting to find is the fact that the soliton width of the neutral soliton is less than that of the charged soliton. This is purely a correlation effect and the bond lengths are rearranged over a longer length of the chain to accommodate the excess charge in the charged soliton. We also find that the relaxation energy associated with the neutral soliton is smaller than that associated with the charged soliton. The energy of the charged soliton state is now dependent upon the nature of the charge and thus lifts the degeneracy of the positively and negatively charged solitons. We also find that the $2A_g$ relaxation energy is 0.8 ± 0.06 eV for all chain lengths studied and is in fact the largest in magnitude of all the states considered in this study. The relaxation energy of the $1B_u$ state is 0.2 ± 0.01 eV while the polaron relaxation energy is about 0.04 eV in the long chain limit, which is about the same as in original SSH model. Note that in SSH model, the relaxation energy of $1B_u$ is about 0.56 eV, thus correlation which induces eletron hole binding, tends to reduce the relaxation energy of this excitation.

In order to understand these relaxation energies, we need to look at the bond-order profiles of these states (Fig. 3). The ground-state equilibrium geometry corresponds to an almost ideal bond alternation pattern along the entire length of the chain, except for some chain-end effects. The $1B_u$

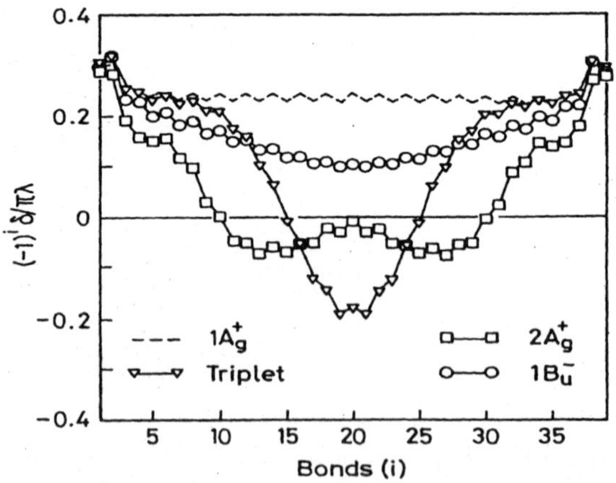

Fig. 3. Bond order profile for $1A_g, 2A_g, 1B_u$ and lowest triplet states as a function of bonds in the chain.

excitation shows an almost gradual distortion of the profile with the largest change in bond order for the central bonds is about 0.15 (while at the ends the difference in the bond order profiles of the $1A_g$ state and the $1B_u$ state is much less). In contrast, the $2A_g$ profile is very different from the ground state bond order profile. In fact, the bond order profile of this two-photon state shows the signature of *two* soliton-antisoliton pairs; the bond profile of the triplet state shows *one* soliton-antisoliton pair. The rather large deviation of the bond order profiles is consistent with the fact that the relaxation energies associated with these states are considerably larger than the relaxation energy of the $1B_u$ state.

In the original SSH model, the relaxation of the $1B_u$ state goes to a charged (S^+/S^-) soliton pair. However, once electron correlation effect is taken into account, the negative and positive solitons appear to attract one another to form an exciton. Nevertheless, experimentally, photo-excited charged solitons have been observed. We can reconcile these contradicting results by recalling that the exciton binding energy for polyacetylene has been found to be small (less than 0.1 eV). Thus, the appearance of photo-induced charged solitons can be due to the relaxation of higher B_u states which form a continuum band and where the electron-hole separation is large so that correlation is very small. Thus, the main conclusions from the SSH model does hold for these band states.

It is a pleasure to thank S.K. Pati, Z. Shuai, Y. Anusooya, J.-L. Brèdas, and H.R. Krishnamurthy for various collaborations and the organisers of DMRG98 for their kind hospitality.

References

1. L. Salem, Molecular Orbital Theory of Conjugated Systems, Benjamin (1966)
2. R. Pariser and R.G. Parr, J. Chem. Phys. **21**, 466 (1953)
3. J. A. Pople, Trans. Farad. Soc. **49**, 1375 (1953)
4. K. Ohno, Theor. Chem. Acta **2**, 219 (1964); G. Klopman, J. Am. Chem. Soc. **86**, 4550 (1964)
5. N. Mataga and K. Nishimoto, Z. Physik. Chem. **12**, 335 (1957); **13**, 140 (1957)
6. For a review see, Z.G. Soos and S. Ramasesha, in Valence Bond Theory and Chemical Structure, eds., D. J. Klein and N. Trinajstic, Elsevier, p 81 (1990)
7. S.R. White, Phys. Rev. Lett. **69**, 2863 (1992); Phys. Rev. B **48**, 10345 (1993)
8. S. Ramasesha and Z.G. Soos, Int. J. Quant. Chem. **25**, 1003 (1984); J. Chem. Phys. **80**, 3278 (1984)
9. S. Ramasesha, Swapan K. Pati, H.R. Krishnamurthy, Z.Shuai and J.L. Brédas, Phys. Rev. B **54**, 7598 (1996); Synth. Metals **85**, 1019 (1997)
10. E.H. Lieb and F.Y. Wu, Phys. Rev. Lett. **25**, 1445 (1968)
11. Y. Anusooya, Swapan K. Pati and S. Ramasesha, J. Chem. Phys. **106**, 1 (1997)
12. S. Ramasesha, Kunj S. Tandon, Y. Anusooya and Swapan K. Pati, Proceedings of SPIE **3145**, 282 (1997)

13. Nonlinear Optical Properties of Organic Molecules and Crystals, ed. D.S. Chemla and J. Zyss, Academic (1987)
14. Materials for Nonlinear Optics: Chemical Perspectives, edited by S.R. Marder, J.E. Sohn, and G.D. Stucky, American Chemical Society, Washington DC, **455**, 1991
15. P.W. Langhoff, S.T. Epstein and M. Karplus, Rev. Mod. Phys. **44**, 602 (1972); B.J. Orr and J.F. Ward, Mol. Phys. **20**, 513 (1971)
16. e.g. The Recursion Method and Its Applications, eds. D.G. Pettifor and D. L. Weaire, Springer (1985)
17. E. Dagotto, Rev. Mod. Phys. **66**, 763 (1994)
18. Z.G. Soos and S. Ramasesha, Phys. Rev. B **29**, 5410 (1984)
19. S. Ramasesha, Z. Shuai and J. L. Brédas, Chem. Phys. Lett. **245**, 224 (1995)
20. S. Ramasesha, J. Comput. Chem. **11**, 545 (1990)
21. K. Hallberg, Phys. Rev. B **52**, 9827 (1995)
22. H.B. Pang, H. Akhlaghpour and M. Jarrel, Phys. Rev. B **53**, 5086 (1996)
23. Z.G. Soos, S. Ramasesha and D.S. Galvao, Phys. Rev. Lett. **71**, 1609 (1993)
24. Swapan K. Pati, Ph.D. Thesis, Indian Institute of Science, Bangalore, India, February 1998.; Swapan K. Pati, S. Ramasesha, Zhigang Shuai and J.-L. Brédas *unpublished results*
25. W.P. Su, J.R. Schrieffer and A. J. Heeger, Phys. Rev. B **22**, 2099 (1980); Phys. Rev. B **28**, 1138 (1983) (erratum)
26. A. J. Heeger, S. Kivelson, J.R. Schrieffer and W.P. Su, Rev. Mod. Phys. **60**, 781 (1988)
27. W.P. Su and J.R. Schrieffer, Proc. Natl. Acad. Sci. U.S.A. **77**, 5626 (1980)
28. C.Q. Wu, X. Sun and K. Nasu, Phys. Rev. Lett. **59**, 831 (1987); S. Mazumdar and D.K. Campbell, *ibid.*, **55**, 2067 (1985); J.L. Brédas, and A.J. Heeger, *ibid.* **63**, 2534 (1989)
29. B. Horovitz, Solid State Commun. **41**, 729 (1982)
30. A. Terai and Y. Ono, J. Phys. Soc. Jpn. **55**, 213 (1986)
31. W.P. Su, Phys. Rev. B **34**, 2988 (1986)
32. M.J. Rice and I.A. Howard, Phys. Rev. B **28**, 6089 (1983)
33. R.H. Friend, D.D.C. Bradley, and P.D. Townsend, J. Phys. D: Appl. Phys. **20**, 1367 (1987)
34. B.S. Hudson and B.E. Kohler, Annu. Rev. Phys. Chem. **25**, 437 (1974)
35. L. Robins, J. Orenstein and R. Superfine, Phys. Rev. Lett. **56**, 1850 (1986)
36. P. Tavan and K. Schulten, J. Chem. Phys. **85**, 6602 (1986)
37. W.P. Su, Phys. Rev. Lett. **74**, 1167 (1995)
38. Z. Shuai, S.K. Pati, W.P. Su, J.L. Brédas, and S. Ramasesha, Phys. Rev. B. **55**, 15368 (1997)
39. Z. Shuai, J.L. Brédas, S.K. Pati, and S. Ramasesha, Phys. Rev. B **56**, 9298 (1997)
40. L. Rothberg, T.M. Jedju, P.D. Townsend, S. Etamad and G.L. Baker, Phys. Rev. Lett. **65**, 100 (1990)
41. X. Wei, B.C. Hess, Z.V. Vardeny and F. Wudl, Phys. Rev. Lett. **68**, 666 (1992)
42. Z.G. Soos and S. Ramasesha, Phys. Rev. Lett. **51**, 2374 (1983)
43. S. Mazumdar and S.N. Dixit, Synth. Met. **28** D463 (1989)
44. J. Orenstein and G.L. Baker, Phys. Rev. Lett. **49**, 1043 (1982)

2.3 Conjugated One-Dimensional Semiconductors

William Barford[1] and Robert J. Bursill[2]

[1] Department of Physics and Astronomy, The University of Sheffield Sheffield, S3 7RH, United Kingdom.
[2] School of Physics, The University of New South Wales Sydney, NSW 2052, Australia.

Conjugated oligomers (finite-length chains) and polymers exhibit a variety of fascinating low-energy excitations. These arise from the inter-play of electron-electron correlations and electron-lattice coupling. $(CH)_n$ polyene oligomers, and the associated polymer (polyacetylene (PA)), have long been the subject of interest due to their soliton-like excitations [1]. Recently, however, the phenylene based semiconductors have also attracted wide attention because of their light-emmitting properties, and possible applications in non-linear optical devices [2].

1 The Pariser-Parr-Pople Model

In a conjugated molecule the carbon atom undergoes sp^2 hybridisation, resulting in a convenient separation of energy scales between the bonding σ orbitals and the higher energy p_z or π orbitals. The π orbitals hybridise, leading to electron delocalisation (or conjugation) throughout the carbon back bone. A convenient, semi-empirical, model of the π-electron system is provided by the Pariser-Parr-Pople (P-P-P) model,

$$\mathcal{H} = \mathcal{H}_{1p} + \mathcal{H}_{2p}, \tag{1}$$

where

$$\mathcal{H}_{1p} = - \sum_{<ij>\,\sigma} t_{ij} \left[c^\dagger_{i\sigma} c_{j\sigma} + c^\dagger_{j\sigma} c_{i\sigma} \right] \tag{2}$$

and

$$\mathcal{H}_{2p} = U \sum_i \left(n_{i\uparrow} - \frac{1}{2} \right) \left(n_{i\downarrow} - \frac{1}{2} \right) + \frac{1}{2} \sum_{i \neq j} V_{ij} (n_i - 1)(n_j - 1). \tag{3}$$

$t_{ij} = t_0(1 + \delta(r_0 - r_{ij}))$ and $<>$ represents nearest neighbours. A standard interpolation for the Coulomb interaction is the Ohno function: $V_{ij} = U/(1 + \beta r_{ij}^2)^{1/2}$, where $\beta = (U/14.397)^2$ and the bond lengths are in Å. The parameters are chosen to fit the optical spectra of small molecules. In practice, benzene serves as a good candidate. An optimal parametrisation results in $U = 10.06$ eV, $t = 2.539$ eV and $\delta = 1.142 Å^{-1}$ [3,4].

The one dimensional nature of conjugated oligomers and polymers imply that electron correlations will play a key role. Until recently, the only rigorous way to deal with correlations was via exact diagonalisation calculations, which is limited to approximately 14 atoms. The configuration interaction technique, which generates a Hilbert space by creating one, two or mulitple excitations from the Hartree-Fock ground state, can study somewhat larger systems if less accuracy is required. However, it suffers from so-called finite-size inconsistency [5]. The DMRG method, which builds up a finite lattice by a successive addition of repeat units, is ideal for the study of finite-size oligomers. As will be discussed in Sect. 4, the standard DMRG approach can also be extended to complex one-dimensional oligomers, where the fundamental repeat unit is not just one or two carbon atoms, but a complex molecular building block.

Table 1. The classification of the relevant low-lying states.

State	$^1A_g^+$	$^1B_u^-$	$^3B_u^+$
Spatial Inversion	+	−	−
Spin-flip	+	+	−
Spin	0	0	1
Particle-hole	+	−	+

Since the energies and wavefunctions of the excited states are required, particularly for the calculation of optical properties, the symmetries of the Hamiltonian (1) must be exploited to target these states. These are, spatial inversion, particle-hole and spin-flip symmetries. The group theoretic labelling of the states discussed in the following sections are shown in Table 1. The symmetry operators can be constructed for the superblock and used to generate a trial state of definite symmetry, $|\Psi\rangle$, from a random superblock state, $|\Phi\rangle$, [6] (see also [7] and Chap. 2.2(II)) i.e.,

$$|\Psi\rangle = \frac{1}{2}\left(|\Phi\rangle \pm \hat{O}|\Phi\rangle\right) . \tag{4}$$

$|\Psi\rangle$ can then be used in the diagonalisation of the superblock Hamiltonian. This procedure is numerically stable, because the block symmetry operators commute with the density matrix. It is also possible to calculate excited states in the same symmetry sector by projecting new trial states from the lower-lying states. The electric dipole operator, $\hat{\mu}$, connects states with the same total spin and states of opposite spatial and particle-hole symmetries. Since the ground state is $^1A_g^+$, the optically allowed transitions are to $^1B_u^-$ states. In order to calculate the matrix elements of $\hat{\mu}$, the density matrix must be constructed from states from both of these symmetry sectors.

In the non-interacting limit of the P-P-P model (1) the lowest-lying singlet ($1^1B_u^-$) and triplet ($1^3B_u^+$) excitations are degenerate, and lie below the lowest excited even-parity singlet ($2^1A_g^+$) state. However, electron correlations can lead to a reversal of the energetic ordering of the $1^1B_u^-$ and $2^1A_g^+$ states. Some of the key issues in organic semiconductors, which can be addressed by DMRG calculations, are:

1. What is the relative ordering of the $1^1B_u^-$ and $2^1A_g^+$ states? What are the roles of electron-electron correlations and electron-lattice coupling in determining this ordering?
2. In the case of degenerate systems, such as PA, what are the solitonic strutures? How do electron-electron correlations change the non-interacting picture?
3. Are there excitonic states, and if so, what are their binding energies?
4. What is the connection between theory and experiment?

The first two issues have been addressed for polyene oligomers by Bursill and Barford, and will be discussed in Sect. 2. The last two issues were considered by Barford, Bursill and Lavrentiev for the phenylene-based semiconductors, and is the subject of Sect. 3.

The DMRG method has recently been used by Yaron et al. [8] and Fano et al. [9] to solve the P-P-P model for linear and cyclic polyenes, respectively. Jeckelmann [10] studied the metal-insulator transition in doped PA by solving the U-V Hubbard-Peierls model. He found that a metal-insulator transition occurs when $V = 0$, in contrast to an earlier DMRG study by Wen and Su [11], but not when $V \neq 0$. Wen and Su [11] also considered the relative stability of bipolarons with respect to two polarons with the Hubbard-Peierls model supplemented by a Brazovskii-Kirova symmetry breaking term. This was extended by Kuwabara et al. [12] for non-zero V. Shuai et al. [13], and Boman and Bursill [14] have examined exciton binding energies in the U-V Hubbard model, while Ramasesha et al. have studied the phenylene-based systems (see Chap. 2.2(II)).

2 Polyene Oligomers

Bursill and Barford [15] performed a DMRG calculation on long polyene oligomers where both long range Coulomb interactions and electron-lattice coupling in the adiabatic limit were considered on an equal footing. They solved the Pariser-Parr-Pople-Peierls model and utilised the Hellmann - Feynman (H-F) theorem to calculate the low-lying excited states and the lattice relaxation associated with them.

The Hamiltonian is defined as

$$\mathcal{H} = -2\sum_{i=1}^{N-1} t_i \hat{T}_i + \frac{1}{4\pi t_0 \lambda}\sum_{i=1}^{N-1} \Delta_i^2 + \Gamma \sum_{i=1}^{N-1} \Delta_i + \mathcal{H}_{2p}, \tag{5}$$

where $t_i = \left(t_0 + \frac{\Delta_i}{2}\right)$, $\hat{T}_i = \frac{1}{2}\sum_\sigma (c_{i+1\sigma}^\dagger c_{i\sigma} + h.c.)$ is the bond order operator of the ith. bond and $\mathcal{H}_{2p}$ is defined in (3). The single and double bond lengths used in the evaluation of V_{ij} are 1.46 Å and 1.35 Å, respectively, and the bond angle is 120^0. The dimensionless electron-phonon coupling constant, λ, is defined by $\lambda = 2\alpha^2/\pi K t_0$, where K is the elastic spring constant (estimated to be 46 eV $\AA^{-2}$ [1]), and α relates the actual distortion of the ith. bond from equilibrium, y_i, to Δ_i: $y_i = \Delta_i/2\alpha$. Γ is defined such that the relaxed ground state of an infinite polymer has the same chain length as the unrelaxed state, i.e., $\sum_{i=1}^{N-1} \Delta_i = 0$. This ensures that the average hopping integral is t_0, which is applicable to C-C bond lengths of 1.40 Å. However, the chain length is permitted to change for excited states, and for all the states of finite oligomers. The remaining parameter, λ, is chosen to fit the *vertical* energies of the $1^1B_u^-$ and $2^1A_g^+$ states of hexatriene, giving $\lambda = 0.115$ and $\Gamma = 0.602$.

The equilibrium values of the bond length distortion are determined by the H-F condition that,

$$\frac{\partial E(\{\Delta_i\})}{\partial \Delta_i} = 0 \Rightarrow \Delta_i = 2\pi t_0 \lambda \left[\langle \hat{T}_i \rangle - \Gamma \right] . \tag{6}$$

The calculation of the relaxed energy of a given state for a given chain length proceeds as follows:

1. The eigenstate is calculated for an initial choice of $\{t_i\}$ by building up the lattice to the target chain size using the infinite-lattice algorithm of the DMRG method.
2. At the target chain size the H-F condition (6) is repeatedly applied until the $\{t_i\}$ have converged (typically less than 10 times).
3. Using the new values of $\{t_i\}$, steps 1 and 2 are repeated. The procedure is succesfully terminated when the energies have converged after successive lattice and H-F iterations.

It is necessary to sweep through the lattice after each set of H-F iterations to ensure that the electronic states and the lattice distortion have converged simultaneously [16]. The Hamiltonian (5) possesses the symmetries discussed above, so the states are labelled as in Table 1.

Table 2 shows the convergence of the low-lying states as a function of the superblock Hilbert space size, showing that the energy gaps have converged to considerably better than 1%.

Table 2. Convergence of the ground-state ($1^1A_g^+$) energy and vertical and 0-0 transition energies of the $2^1A_g^+$ and $1^1B_u^-$ states for 102 sites as a function of the superblock Hilbert space size (SBHS).

SBHS	$1^1A_g^+$	$2^1A_g^+$	$2^1A_g^+$(0-0)	$1^1B_u^-$	$1^1B_u^-$(0-0)
15844	−509.6330807	2.8927	1.8051	2.7719	2.6785
25492	−509.6330971	2.8795	1.8008	2.7650	2.6483
36312	−509.6331002	2.8764	1.7972	2.7617	2.6392
54916	−509.6331009	2.8744	1.7963	2.7605	2.6345
67240	−509.6331010	2.8737	1.7959	2.7601	2.6336

The vertical energy, E^{v}, (i.e. the energy of the excited state with the ground-state geometry) and the relaxed energy, E^{0-0}, (i.e. the energy of the state determined by (6)) of the $1^3B_u^+$, $1^1B_u^-$ and $2^1A_g^+$ states are shown in Fig. 1 as a function of $1/N$. Notice that the vertical energies of the $1^1B_u^-$

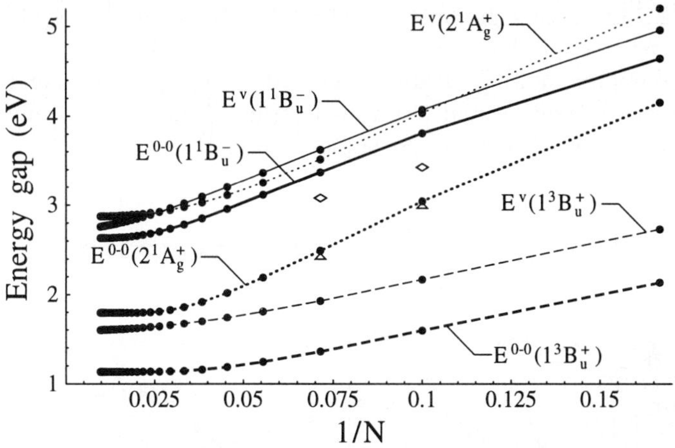

Fig. 1. The low-lying electronic spectrum of polyene oligomers as a function of $1/N$. $1^1B_u^+$ (solid lines) $2^1A_g^+$ (dotted lines) and $1^3B_u^+$ (dashed lines) states. Vertical/0-0 transitions are indicated by thin/thick lines. Experimental 0-0 energies of the $1^1B_u^-$ (diamonds) and $2^1A_g^+$ (triangles) states for polyenes in hydrocarbon solution [1].

and $2^1A_g^+$ states are very close, with a crossing at short chains, and again for long chains. In the thermodynamic limit $E^{\mathrm{v}}(1^1B_u^-) < E^{\mathrm{v}}(2^1A_g^+)$. This large N crossing has also been observed in the U-V dimerised Hubbard model [13] [14]. The relaxation energy of the $1^1B_u^-$ state is modest (ca. 0.3 eV). By contrast, the relaxation energies of the $1^3B_u^+$ and $2^1A_g^+$ states are substantial, being ca. 0.5 eV and 1.0 eV, respectively. Finally, the experimental values of $E^{0-0}(1^1B_u^-)$ and $E^{0-0}(2^1A_g^+)$ for $N = 10$ and 14 are shown. The $2^1A_g^+$ values are in excellent agreement with the calculation. The $1^1B_u^-$ values are ca. 0.3 eV lower than the predictions; a result probably due to solvation effects [21].

Fig. 2 shows a plot, as a function of bond index from the centre of the chain, of the normalised staggered bond dimerisation, defined as, $\delta_i \equiv (-1)^i(t_i - \bar{t})/\bar{t}$, where $\bar{t}$ is the average value of t_i in the middle of the chain. Note that the $1^3B_u^+$ and $2^1A_g^+$ states undergo considerable bond distortion, whereas the $1^1B_u^-$ state and the charged state (denoted E_g) show a weak polaronic distortion of the lattice. The oscillatory behaviour of δ_i in the polaronic distortions indicates a local expansion of the lattice. The $1^3B_u^+$, $1^1B_u^-$ and charged state fit a 2-soliton form [17,18],

$$\delta_i = \bar{\delta}\left\{1 + \tanh\left(\frac{2x_0}{\xi}\right)\left[\tanh\left(\frac{i - x_0}{\xi}\right) - \tanh\left(\frac{i + x_0}{\xi}\right)\right]\right\}, \tag{7}$$

whereas the $2^1A_g^+$ state fits a 4-soliton [17,18] fit of the form,

$$\delta_i = \bar{\delta}\left\{1+\tanh\left(\frac{2x_0}{\xi}\right)\left[\tanh\left(\frac{i-x_d-x_0}{\xi}\right)-\tanh\left(\frac{i-x_d+x_0}{\xi}\right)\right.\right.$$
$$\left.\left.+\tanh\left(\frac{i+x_d-x_0}{\xi}\right)-\tanh\left(\frac{i+x_d+x_0}{\xi}\right)\right]\right\}. \tag{8}$$

x_0, x_d and ξ are fitting parameters. These functions give good fits to the relaxed geometries of the $N = 102$ site system, as shown in Fig. 2. The 4-soliton character of $2^1A_g^+$ state indicates the strong inter-play between electron-lattice relaxation and electron-electron correlations in polyenes as this state has a considerable triplet-triplet character.

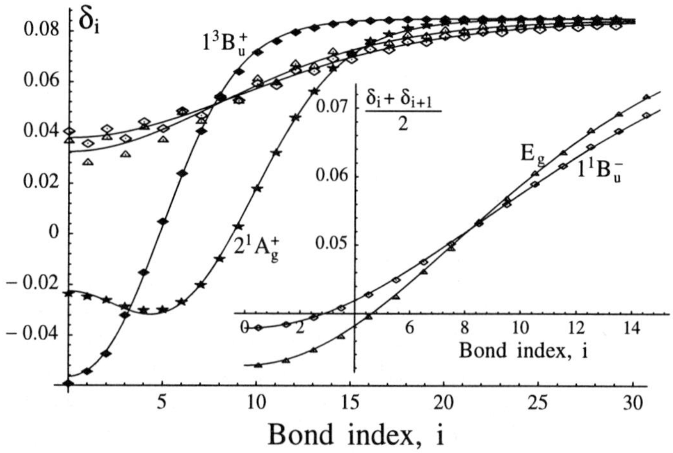

Fig. 2. The staggered bond dimerisation of the excited states near the centre of the 102 site chain: $1^1B_u^-$ (open diamonds), $1^3B_u^+$ (filled diamonds), $2^1A_g^+$ (stars) and the charged state, E_g, (open triangles). The solid lines are fits to the 2-soliton form (7) (and the 4-soliton form (8) for the $2^1A_g^+$). The inset shows the two-point averages $((2i+1)/2, (\delta_i+\delta_{i+1})/2)$ for the polaronic E_g and $1^1B_u^-$ states

Bursill and Barford [15] showed that the fitting parameters, ξ and x_0 (and x_d for the $2^1A_g^+$ state) converge as a function of N. This indicates that the soliton/anti-solitons are bound, because, if they were not, x_0 would increase with N. This result was confirmed by a calculation of the adiabatic potential energy curves [15], and is partly a result of the fact that polyenes with open ends are not degenerate, i.e. the energy of a chain changes when the single and double bonds are reversed.

In conclusion, the DMRG study of (5) has shown that: (i) the relaxed energy of the $2^1A_g^+$ state lies below that of the $1^1B_u^-$, hence PA is not electroluminescent; (ii) the $1^1B_u^-$ is an exciton-polaron, while the $1^3B_u^+$ and $2^1A_g^+$ states show strong lattice distortion and fit a two and four soliton form, respectively; (iii) the soliton/anti-solitons are bound; and (iv) where comparisons to oligomer experimental results are possible, there is good agreement with the calculated energies.

3 The Phenylene Based Semiconductors

Poly(p-phenylene) (PPP) and poly(p-phenylene vinylene) (PPV) are the most widely studied light emmitting polymers. The structure of PPV, a sequence of phenylene and vinylene units capped by phenyl rings, is shown in Fig. 3. PPP is similar, except for the absence of vinylene units. Each phenylene

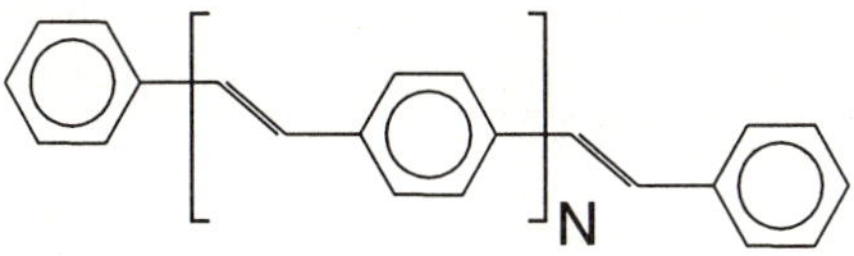

Fig. 3. The structure of PPV

repeat unit has six Hubbard sites with a total Hilbert space of 2^{12} states. This is far too many states to combine with the system block to generate an augmented block at each lattice iteration, and so a truncation of the repeat block Hilbert space is necessary. Barford and Bursill overcame this difficulty by developing the two-state molecular orbital (MO) model [6]. The essential assumption of this model is that for a description of the low energy physics, the six MOs of the phenylene ring, arising from the six atomic orbitals, may be replaced by the bonding highest occupied molecular orbital (HOMO) and the bonding lowest unoccupied molecular orbital (LUMO). This is equivalent to adding two Hubbard sites, so there are now only 2^4 states per repeat unit. The vinylene unit, however, is exactly described by its two HOMO and LUMO states. The 2-MO model is an effective linear chain model which is intended to provide an accurate description of the low energy physics. Since it also possesses spatial inversion, spin-flip and particle-hole symmetries, the low-lying excited states are labelled as in Table 1. Exact P-P-P model (1-3) calculations of the molecular building blocks, i.e. benzene, biphenyl [3] and stilbene [22] are used to parametrise the interactions. The 2-MO model is described in detail and applied to PPP in [19] and PPV in [20].

A controversial issue in the theory of conjugated semiconductors is the identification of bound and unbound states, and hence the determination of the exciton binding energies. Provided that the electronic correlations are not too strong, it was shown in [19] and [20] that a suitable particle-hole correlation function is,

$$C_{ij}(|n\rangle) = \langle n| S^{\dagger}_{ij} |1^1A_g^+\rangle , \tag{9}$$

where

$$S^{\dagger}_{ij} = \frac{1}{\sqrt{2}}(a^{\dagger}_{i2\uparrow}a_{j1\uparrow} + a^{\dagger}_{i2\downarrow}a_{j1\downarrow}) \tag{10}$$

is a singlet exciton creation operator, which creates a hole in the HOMO ($|1, j\rangle$) on repeat-unit j and creates a particle in the LUMO ($|2, i\rangle$) on repeat-unit i. The mean square particle-hole spacing is then defined as,

$$l^2(|n\rangle) = \langle (i-j)^2 \rangle = \frac{\sum_{ij} (i-j)^2 C_{ij}^2}{\sum_{ij} C_{ij}^2}. \tag{11}$$

Using (11), excitonic $1^1B_u^-$ and $2^1A_g^+$ states were identified at 2.7 eV and 2.9 eV, respectively in the model calculations on the 13 phenylene unit PPV oligomer. The lowest unbound states, denoted by $m^1A_g^+$ and $n^1B_u^-$, where $m = 7$ and $n = 4$ for oligomers of 10 to 13 phenylene units, were identified by their mean particle-hole spacing increasing linearly with oligomer size. Their energies of 3.6 eV, indicates that the $1^1B_u^-$ and $2^1A_g^+$ excitons exist below the conduction band threshold.

Having calculated the energies, symmetries and spatial correlation functions of the low-lying states a comparison to and interpretation of experiment can be made via the optical spectroscopies. A number of optical probes have been employed to ascertain the character of the low-lying states. In particular, one-photon absorption identifies the ${}^1B_u^-$ states, two-photon absorption identifies the ${}^1A_g^+$ states, and electro-absorption (EA) and third harmonic generation identify both kinds of states. The lowest non-zero non-linear optical (NLO) spectroscopy for centro-symmetric systems is $\chi^{(3)}$. This has been calculated for PPP [19] and PPV [20] oligomers using the sum-over-states approach. In PPV the $1^1A_g^+$, $1^1B_u^-$, $2^1A_g^+$, $n^1B_u^-$ and $m^1A_g^+$ states dominate the NLO spectra, and are readily identifiable in the EA spectra.

4 Future Developments

The 2-MO model of the phenylene based semiconductors, discussed in Sect. 3, is derived from a truncation of the phenylene repeat unit Hilbert space. It is based on the assumption that the *single particle* MO basis provides a good separation of energy scales. The fact that a direct mapping from the P-P-P model to the 2-MO basis does not provide an accurate prediction of the low-lying excitations indicates that this assumption is not good enough to achieve quantum chemistry accuracy. A more sophisticated approach is to truncate the *many body* states of the repeat unit by constructing a reduced density matrix for the repeat unit when the repeat unit forms part of a suitable superblock. For the phenylene unit, this superblock would be two phenylene rings (i.e. biphenyl, which can be solved exactly via the P-P-P model) with both open and periodic boundary conditions. Such an approach is being developed by Bursill and Barford, and is near completion.

Further progress in calculating the electronic structure of conjugated semiconductors will include going beyond the adiabatic limit to include dynamical phonons (see Chap. 5.1(II)) and the use of *ab initio* Hamiltonians (see Chap.

2.1(II)). Finally, to understand solid state effects (e.g. aggregation and three dimensional screening) coupled chain calculations are necessary.

We thank our co-worker, Dr. M. Y. Lavrentiev, for discussions.

References

1. H. Kiess, Conjugated Conducting Semiconductors, Springer (1992)
2. T.A. Skotheim, R.L. Elsenbaumer and J.R. Reynolds, Handbook of Conducting Polymers, Marcel Dekker (1997)
3. R.J. Bursill, C. Castleton and W. Barford, Chem. Phys. Lett. **294**, 305 (1998)
4. The standard parametrisation is $U = 11.26$ eV, $t_0 = 2.4$ eV and $\delta = 1.22\AA^{-1}$.
5. P. Fulde, Electron Correlations in Molecules and Solids, Springer (1991)
6. W. Barford and R.J. Bursill, Chem. Phys. Lett. **268**, 535 (1997)
7. S. Ramasesha, S.K. Pati, H.R. Krishnamurthy, Z. Shuai and J.L. Bredas, Phys. Rev. B **54**, 7598 (1996)
8. D. Yaron, E.E. Moore, Z. Shuai and J.L. Bredas, J. Chem. Phys. **108**, 7451 (1998)
9. G. Fano, F. Ortolani and L. Ziosi, J. Chem. Phys. **108**, 9246 (1998)
10. J. Jeckelmann, Phys. Rev. B **57**, 11 838 (1998)
11. G. Wen and W-P Su, Synth. Met. **78**, 195 (1996)
12. M. Kuwabara, Y. Shimoi and S. Abe, J. Phys. Soc. Jpn. **67**, 1521 (1998)
13. Z. Shuai, J.L. Bredas, S.K. Pati and S. Ramasesha, Phys. Rev. B **56**, 9298 (1997)
14. M. Boman and R.J. Bursill, Phys. Rev. B **57**, 15167 (1998)
15. R.J. Bursill and W. Barford, Phys. Rev. Lett. **82**, 1514 (1999)
16. The finite-lattice method would avoid the lattice sweeps, however states of definite spatial symmetry cannot be targeted in this method.
17. D.K. Campbell and A.R. Bishop, Nucl. Phys. B **200**, 297 (1982)
18. W. P. Su, Phys. Rev. Lett. **74**, 1167 (1995)
19. W. Barford, R.J. Bursill and M.Y. Lavrentiev, J. Phys.: Condens. Matt., **10** 6429 (1998)
20. A Theoretical investigation of the low-lying electronic structure of poly(p-phenylene vinylene). M.Y. Lavrentiev, W. Barford, S.J. Martin, H. Daly and R.J. Bursill, to appear in Phys. Rev. B
21. E. Moore, B. Gherman and D. Yaron, J. Chem. Phys. **106**, 4216 (1997)
22. C. Castleton and W. Barford, proceedings of ICSM98 (Synth. Met.)

2.4 Strongly Correlated Complex Systems

Shaojin Qin[1], Jizhong Lou[1], Zhaobin Su[1], and Lu Yu[2,1]

[1] Institute of Theoretical Physics, Academia Sinica
Beijing 100080, P R China
[2] International Centre for Theoretical Physics
I-34100 Trieste, Italy

Since it was developed by White, the DMRG has been applied to many one-dimensional (1D) systems to attack a number of long-standing problems. The algorithm itself is discussed in Chap. 2(I). The 1D systems are split into two parts: the subsystem and the environment [1]. The DMRG probes effectively a huge many-body Hilbert space step by step according to the reduced density matrix of the subsystem at each step. However, the situation here is similar, to some extent, to the calculation of the lowest energy states of spin glasses. There is no universal DMRG recipe to calculate low-energy states for various kinds of finite-size strongly correlated systems. The difficulty is how to use the infinite-size DMRG to build up "correctly" the system's subsystems and environments (the environment problem) for finite-size DMRG sweeps. As a rule, specific DMRG recipes have to be found for different types of systems.

We first encounter this difficulty when considering a non-half-filled fermionic system, for example, in the 1D Hubbard model. In some cases we optimize subsystems and environments for a system not at the correct filling. These subsystems optimized for wrong fillings will be used to construct new subsystems and/or new environments. As a consequence, the final results do not have a high precision. Some complex finite systems, *e.g.* the C_{60} molecule, exhibit a similar difficulty. To our knowledge, no efficient algorithms have been proposed for such systems so far.

In this chapter we show a method to keep the correct filling at all steps of DMRG sweeping when preparing the environments, and thus to obtain high enough precision results for the analysis of singularity exponents [2] in the 1D Hubbard model. Then we discuss the environment problem in DMRG calculation for strongly correlated complex molecules like C_{60}, and demonstrate how to avoid it using a momentum-space-like scheme [3].

1 One-Dimensional Non-Half-Filled Fermionic Systems

We use the 1D Hubbard model as an example to describe a high-precision DMRG method for non-half-filled fermionic systems. We have studied a 1/4-filled system, and calculated the singularity exponents of the momentum distribution functions and spin correlations [2]. These exponents have been analytically studied, and some of them have also been numerically calculated. Our studies show that the exponent at $3k_f$ for the momentum distribution is different from the other predictions at finite chain length and raise the question whether there are some new kinds of excitations [4,5].

In Fig. 1, we show the singularity of the momentum distribution for the 1D Hubbard model with different chain lengths. It is clear that the position of the singularity is related to k_f, which is determined by the filling factor. Without keeping the correct filling at all DMRG steps, wrong correlations are built into the system, for instance, the Fermi surface jump of the momentum distribution function at fillings which deviate slightly from the exact one ("close fillings") will contribute to the final correlations and make the singularities rounded off.

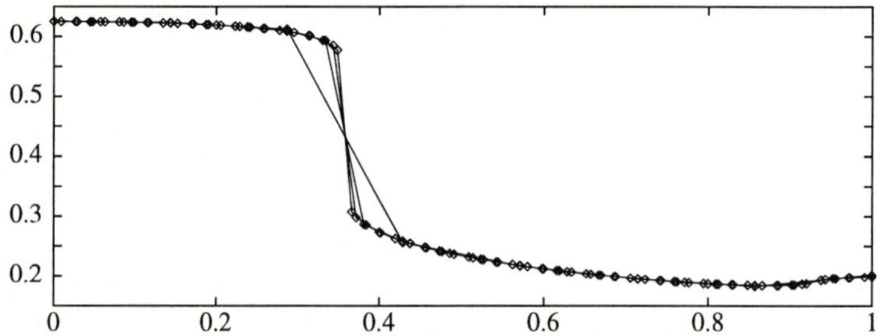

Fig. 1. Momentum distribution function for 1D large U Hubbard model at filling 5/14 calculated for four different values of L. The unit for the momentum k is π

At 1/4 filling, only $L = 4n$ systems have the correct filling. We do not increase the subsystem by one site and use its reflection as the environment at each DMRG step because it forces us to calculate for several close fillings [6] at lengths $L = 4n$ and $4n + 2$. We have instead increased the chain length by 4 sites at each DMRG step to keep the correct filling [2]. We use the infinite-size algorithm to increase the chain length up to certain size, say 60 sites, and then use the finite-size algorithm to compute the correlation functions. For comparison, we have also done the calculation increasing the chain length by two at each DMRG step. The convergence is much slower, and the quality of results is much poorer. The obtained rounded behavior around the singularities can still be illustrative, but is not good for numerical analysis of the singularity exponents.

So far for 1/4 filling. Now we show an example on how to keep the correct filling 3/7 for the Hubbard chain at all DMRG steps. For simplicity we do not split the sites into spin-up electron orbit and spin-down one. By using the finite-size algorithm at chain length $L = 7n$, we can make blocks of length $7n/2$, $7n/2 + 1$, $7n/2 + 2$, and $7n/2 + 3$. Then we can use these blocks to construct the system of length $L = 7(n+1)$, and then construct $L = 7(n+2)$, and so on. Thus DMRG calculations are performed at the correct fillings. Good quantum numbers here are the total particle number and the total spin in z-direction S_z. One more point is that we should keep the ground state unique. We should use periodic boundary conditions for odd number of both spin up and down electrons and antiperiodic boundary conditions for even number cases. This can be easily implemented in the DMRG approach.

The above algorithm of DMRG is also applicable to periodic quasi-1D systems with complex cells, *e.g.* spin and Kondo ladders or lattices. In particular, the algorithm for filling 3/7 systems can be used for systems with 7 orbits or spins in each cell.

2 Environment Problem in C_{60} Real-Space DMRG

In this section we discuss the environment problem in the real-space DMRG calculation for complex molecule C_{60}. Since C_{60} can be readily prepared [7], and has been found to be superconducting when doped with certain alkali metals [8], it has led to considerable interest [9,10] in exploring its intrinsic physical properties. It has been suggested [11] that the electron-electron interaction plays an important role. Experimental evidence [12] shows that the on-site repulsive Hubbard U is big. Many authors have analyzed the electron-electron interaction by perturbation [13–18]. Metal insulator transition [19] and screening effects [20] have also been studied. Numerical calculations considering several energy levels near the lowest unoccupied molecular orbital (LUMO) and the highest occupied molecular orbital (HOMO), Monte Carlo, and exact diagonalization have been performed for the Hubbard model of C_{60} [21–25]. Here we study the effects of the electron-electron interaction by DMRG and try to gain further understanding of some molecular physics issues – such as the perturbative vs. nonperturbative region of interactions, energy scale for single particle state pictures, etc.

We first calculate the ground-state energy for neutral C_{60} using the real-space DMRG. The model Hamiltonian for C_{60} includes only π-electrons and the on-site Hubbard interaction:

$$H = -\sum_{(i,j)\sigma} (c^\dagger_{i\sigma} c_{j\sigma} + h.c.) + U \sum_i n_{i\uparrow} n_{i\downarrow}, \qquad (1)$$

where (i,j) denotes the ninety nearest neighbor bonds on the C_{60} ball in Fig. 2(**a**), and the hopping integral has been set to -1.

We prepare the environment as in the DMRG study for 2D regular systems [1]. In that scheme the energy obtained is not always good. The idea is to choose a path – a 1D chain, to order the sites in the 2D lattice, and then map the hopping in the 2D lattice onto long range hoppings in the 1D chain as we show in Fig 2(**c**). We have tried several paths and different approaches to prepare the environment blocks by keeping 200 to 300 states in DMRG calculations. The energies obtained are higher than -83 for neutral C_{60} at $U = 0$ for the Hamiltonian (1), while the exact ground-state energy is -93.1616038. These results are not good because some environment blocks are obtained from systems rather different from C_{60} during the preparation, and they are optimized for correlations, quite different from those of C_{60}.

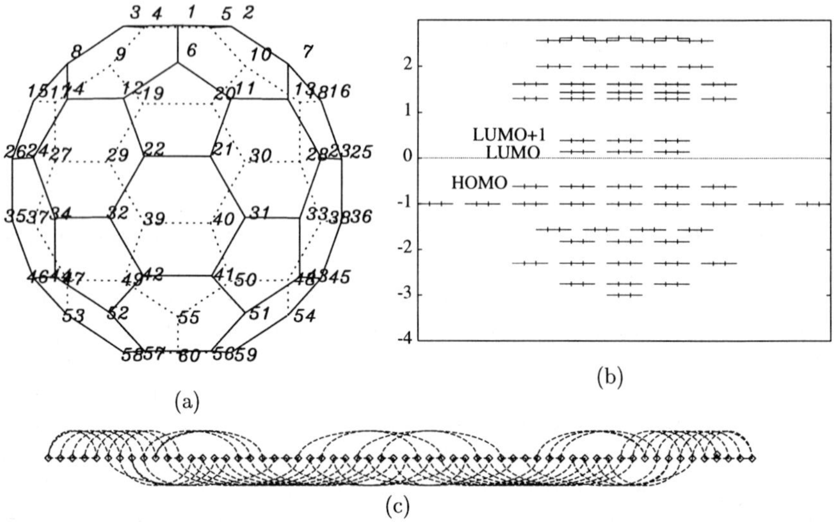

Fig. 2. (**a**) C_{60} molecule: 60 atoms with labels and 90 bonds. (**b**) C_{60} single electron spectrum [28] and related molecular orbitals. (**c**) Atoms viewed as a chain. The atoms are labeled from 1 to 60 from left to right of the chain as in Figure (**a**)

The real-space DMRG must be more accurate for the large-U Hubbard model than for small U. The key argument is that spin up electrons and spin down ones are decoupled at $U = 0$. So when we could have an error of order 10^{-n} by keeping m states in DMRG for the spinless fermion system with two states on each site, we will have to keep m^2 states to obtain the same precision for the Hubbard system with four states on each site – a direct product of the original two state Hilbert space. For large U we have only three or fewer effective states on each site in the low-energy region. The environment problem is still there for the large U system, but increasing the number of states kept in DMRG might be a way to overcome partially the problem due to smaller number of states on each site. In the small U region, the number of states one can possibly keep in DMRG is limited, and the precision of the calculation will not be improved significantly. We will show that a momentum-space-like DMRG [3], which is almost exact in the small U region, overcomes the environment difficulty for the Hubbard model to certain extent.

3 *K*-space DMRG for a Hubbard Model of Molecules

In Chap. 6(I), Xiang and Wang have described the K-space algorithm of DMRG, and White has shown his implementation of the single-particle state

DMRG for the water molecule H_2O in Chap. 2.1(II). We will present our single particle state space calculation for C_{60} in this section. Before these studies, Liang has also used DMRG in the single particle state space [26]. K-space DMRG is the name for the specific algorithm proposed by Xiang for systems with conserved momentum. The term is also used for generic DMRG using the single particle state space. It includes White's quantum chemistry molecular orbital approach, and the atomic and molecular orbital approach in the previous and this section, respectively.

The K-space DMRG deals with general systems which may have long range hoppings and interactions. If we diagonalize the hopping part of the Hamiltonian, then only the interaction part is off-diagonal, and this may improve the DMRG calculation in some cases. In principle, the K-space DMRG can deal with any kind of fermionic systems. By choosing the appropriate single particle states, we may handle the environment problem. How to find the best single particle states is also an interesting problem by itself. An important practical issue is how many interaction terms will be found in this representation. In the following we will show a method to reduce the number of the interaction terms to "order N" – the number of the single particle states in the system. Otherwise, the method is of little practical value.

The Hubbard interaction for molecules are defined in real space (atomic orbitals):

$$H_I = U \sum_i c_{i\uparrow}^\dagger c_{i\uparrow} c_{i\downarrow}^\dagger c_{i\downarrow}. \tag{2}$$

In any other single particle representation defined by $d_l^\dagger$, we use

$$c_i^\dagger = \sum_l \phi(l,i) d_l^\dagger \tag{3}$$

to transform the Hamiltonian. In particular, we will use the molecular orbitals shown in Fig. 2(**b**)as single particle states in the next section. We order them from $l = 1$ to $l = N$ as a 1D chain (Fig. 2(**c**)). *i.e.* these states are "sites" in the standard DMRG sense. The Hubbard interaction is "diagonal" in atomic orbitals, and is "off-diagonal" in other orbitals. One may think each term in the atomic orbital will give rise to N^k (with $k \sim 3$) terms in these orbitals. However, by an appropriate grouping of these terms we will have only 14 terms, instead. We use $[m,n]$ to represent the chain segment from $l = m$ to $l = n$. When combining two blocks $[1,m]$ and $[m+1,n]$ into a block $[1,n]$, we will have 14 terms coming from each site i in interaction (2). For site i, we construct the interaction term for block $[1,n]$ from operators in the left block $[1,m]$ and the right block $[m+1,n]$. The expansions of the operator $c_i^\dagger$ in the full block $[1,n]$, left block $[1,m]$, and right block $[m+1,n]$ are $c_{iF}^\dagger = \sum_{l=1}^{n} \phi(l,i) d_l^\dagger$, $c_{iL}^\dagger = \sum_{l=1}^{m} \phi(l,i) d_l^\dagger$, and $c_{iR}^\dagger = \sum_{l=m+1}^{n} \phi(l,i) d_l^\dagger$, respectively. So we have

$$c_{i\uparrow F}^\dagger c_{i\uparrow F} c_{i\downarrow F}^\dagger c_{i\downarrow F} = (c_{i\uparrow L}^\dagger + c_{i\uparrow R}^\dagger)(c_{i\uparrow L} + c_{i\uparrow R})(c_{i\downarrow L}^\dagger + c_{i\downarrow R}^\dagger)(c_{i\downarrow L} + c_{i\downarrow R}). \tag{4}$$

Expanding the right hand side of the above equation, we have $2^4 = 16$ terms. Only 14 terms involve both left block operators and right block operators. The other two terms with subscripts LLLL or $RRRR$ are already defined in the left block or right block Hamiltonian. The total number of interaction terms is thus $14N$, $14 \times 60 = 840$ for C_{60}.

The above decomposition of the Hubbard interaction is generic and easy to implement. However, it may not be the best way for systems having specific symmetry. In the next section we use the above interaction decomposition in the DMRG calculation for C_{60}.

4 C_{60} DMRG Results for Small U

In our K-space calculation for the Hubbard model (1), we first transform it into the diagonal single particle representation. These single particle states are molecular orbitals, and their energies are shown in Fig. 2(**b**). We need to pay attention to two energy gaps here: the energy gap ($\epsilon_0 \sim 0.7566$) between the HOMO and LUMO, and the energy gap ($\epsilon_1 \sim 0.2434$) between LUMO and LUMO+1 in Fig. 2(**b**). For the K-space DMRG application to C_{60} molecular orbitals, we do not split the site state into spin up and spin down electron states, and we do not make use of the icosahedra symmetry, either.

We calculate the ground-state energies E_n for $n = 0$ to 6 with n electrons added to a neutral C_{60}. We consider U from 0 to 4. Starting from the $U = 0$ free fermion case, we choose the unique environment blocks given by the exact ground state. For example for the $n = 0$ case, in any environment blocks we choose the orbital to be occupied if it is at or below the HOMO, and choose the orbital being empty if it is at or above the LUMO. For the $U = 0$ case, these environment blocks are exact and we obtain exact ground-state energy in each step of finite-size sweepings. Then we increase U gradually, and do several finite-size DMRG sweepings at each U. In the finite size DMRG calculation, we keep only 50 optimal states to obtain a large amount of data and test many different approaches. There are many different possible orderings for the orbitals in the 1D chain like in Fig. 2(**c**). We order the orbitals by interlacing the levels above and below the Fermi surface. The truncation error increases with U. The biggest truncation error during the finite-size sweeping is about 1×10^{-6}, 1×10^{-4}, and 1×10^{-3} for $U = 0.1$, 1, and 4, respectively.

We show the ground-state energy by drawing $(E_n - E_0 - n\epsilon_0)/U$ vs. U in Fig. 3(**a**). The hopping energy $n\epsilon_0$ at $U = 0$ has been subtracted. The chemical potential is $\mu_n = E_n - E_{n-1}$ and we show $(\mu_n - \epsilon_0)/U$ vs. U in Fig. 3(**b**). In this figure we subtract the chemical potential ϵ_0 at $U = 0$. We consider these two figures as contributions from finite interaction U. At the energy scale of the chemical potential, the lines are quite flat and the error bars are small. In Fig. 3(**b**) the lines are at 0.5 and 0.55, and the deviations from 0.5 and 0.55 are only about ten percent. In this sense, single particle states are a good approximation at the energy scale of the chemical potential. The $n = 1$ to $n = 6$ electrons are filled in the upper Hubbard band, and the chemical potential is the old one, ϵ_0 plus the average interaction $0.5U$.

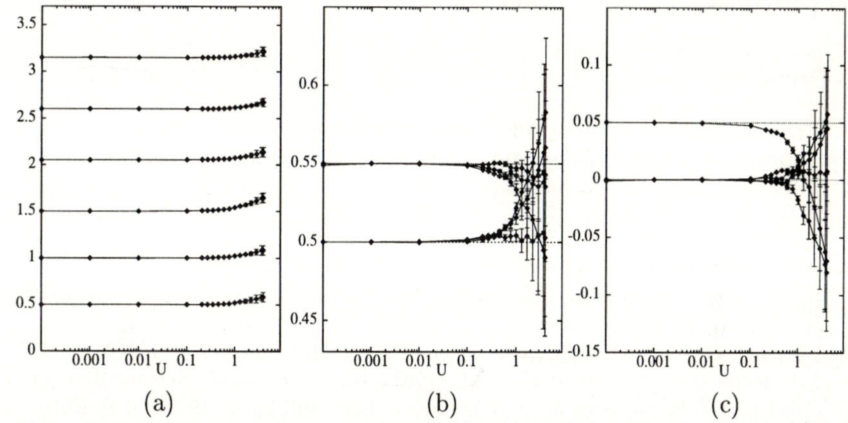

Fig. 3. Ground-state energy of C_{60} with $n = 1$ to 6 added electrons. (**a**) $(E_n - E_0 - n\epsilon_0)/U$ vs. U, $n = 1$ to 6 are from the bottom to the top. (**b**) $(\mu_n - \epsilon_0)/U$ vs. U, $n = 1$ to 6 are from the bottom to the top. (**c**) $(\mu_n - \mu_{n-1})/U$ vs. U., $n = 2, 3, 5, 6$ are close to zero axis, and $n = 4$ is close to 0.05 when U is small. The error bars are indicated by assuming that the numerical error of $E_n - E_0$ is ten times the truncation error in DMRG [27].

When we consider a lower-energy scale, namely the effective Hubbard repulsion U_{eff} for the three-fold degenerate LUMO orbital [10], the single particle state picture cannot explain the energy spectrum. In Fig. 3(**c**), $U_{eff} = \mu_4 - \mu_3 = U/20$, since when the fourth electron is added to C_{60}, the three fold degenerate LUMO orbital will be doubly occupied. Each of the chemical potentials μ_4, μ_5, and μ_6 will have an additional part $3U/60$ because the newly added electron sees the three opposite spin electrons already on the LUMO orbital. In Fig. 3(c), when the interaction U is as big as the energy gap between LUMO and LUMO+1, $\epsilon_1 = 0.2434$, U_{eff} starts to deviate from $U/20$. When the interaction U is as big as the energy gap $\epsilon_0 = 0.7566$ between HOMO and LUMO, the effective Hubbard repulsion reduces to almost half of $U/20$. The error bar is small when U is smaller than 1. One may thus identify the perturbatively good region as $U < \epsilon_0$.

When U is larger than 1 and approaching 4, the error bar becomes bigger. Larger computer resources are required to clarify whether for realistic situation ($U \sim 10$) the single particle state picture and approximations [10] for C_{60} are good and to check the probability of on-ball pairing [13]. These are open questions for further studies.

Shaojin Qin would like to thank Dr. Shoudan Liang for helpful discussions. He also thanks the ICTP for the hospitality during his visit. The major part of the computations was carried out on IBM SP2/16 at SISSA. The work was partially supported by NSFC and CAS.

References

1. S.D. Liang and H.B. Pang, Phys. Rev. B **49**, 9214 (1994)
2. S. Qin, S.D. Liang, Z.B. Su and L. Yu, Phys. Rev. B **52**, 5475 (1995)
3. T. Xiang, Phys. Rev. B **53**, 10445 (1996)
4. S. Qin and L. Yu, Phys. Rev. B **54**, 1447 (1996)
5. K. Penc, K. Hallberg, F. Mila and H. Shiba, Phys. Rev. Lett. **77**, 1390 (1996)
6. S. Moukouri and L.G. Caron, Phys. Rev. B **52**, 15723 (1995)
7. H.W. Kroto, J.R. Heath, S.C. O'Brien and R.F. Curl, R.E. Smalley, Nature **31**, 162 (1985); W. Krätschmer, L.D. Lamb, K. Fostiropoulos and D.R. Huffman, Nature **347**, 354 (1990); H. Ajie *et al.* Phys. Chem. **94**, 8630 (1990)
8. A.F. Hebard *et al.*, Nature **350**, 600 (1991); K. Holczer *et al.*, Science **252**, 1154 (1991); M.J. Rosseinsky *et al.*, Phys. Rev. Lett. **66**, 2830 (1991); S.P. Kelty *et al.*, Nature **352**, 223 (1991)
9. L. Degiorgi, Adv. Phys. **47**, 207 (1998)
10. O. Gunnarsson, Rev. Mod. Phys. **69**, 575 (1997)
11. S. Chakravarty and S. Kivelson, Europhys. Lett. **16**, 751 (1991)
12. R.W. Lof, M.A. van Veenendaal, B. Koopmans, H.T. Jonkman and G.A. Sawatzky, Phys. Rev. Lett. **68**, 3924 (1992)
13. S. Chakravarty, M.P. Gelfand and S. Kivelson, Science **254**, 970 (1991)
14. W.E. Goff and P. Phillips, Phys. Rev. B **46**, 603 (1992)
15. G.N. Murthy and A. Auerbach, Phys. Rev. B **46**, 331 (1992)
16. R. Friedberg, T.D. Lee and H.C. Ren, Phys. Rev. B **46**, 14150 (1992)
17. P.E. Lammert and D.S. Rokhsar, Phys. Rev. B **48**, 4103 (1993)
18. S.L. Sondhi, M.P. Gelfand, H.Q. Lin, and D.K. Campbell, Phys. Rev. B **51**, 5943 (1995)
19. J.P. Lu, Phys. Rev. B **49**, 5687 (1994)
20. M.B.J. Meinders, J. Vandenbrink, J. Lorenzana and G.A. Sawatzky, Phys. Rev. B **52**, 2484 (1995)
21. L. Bergomi and Th. Jolicoeur, Acad. Sci. Paris **328**, 283 (1994)
22. M. Okada, Y. Yuang, T. Yamabe and K. Tanaka, J. Phys. Chem. Sol. **58**, 1901 (1997)
23. V.Y. Krivnov, I.L. Shamovsky, E.E. Tornau and A. Rosengren, Phys. Rev. B **54**, 12144 (1994)
24. B. Srinivasan, S. Ramasesha and H.R. Krishnamurthy, Phys. Rev. B **54**, 1692 (1996)
25. J. Gonzalez and J.V. Alvarez, Phys. Rev. B **53**, 11729 (1996)
26. S.D. Liang, Phys. Rev. Lett. **75**, 3493 (1995)
27. O. Legeza and G. Fath, Phys. Rev. B **53**, 14349 (1996)
28. C.L. Wang, W.Z. Wang and Z.B. Su, Phys. Rev. B **49**, 17468 (1994); J. Phys. Cond. Matt. **5**, 5851 (1993)

3 Classical Statistical Physics

3.1 Non-Hermitian Problems and Some Other Aspects

Ingo Peschel and Matthias Kaulke

Fachbereich Physik, Freie Universität Berlin
Arnimallee 14, D-14195 Berlin, Germany

The success of the DMRG method in the area of one-dimensional quantum systems suggest its application also to other problems. At the same time, a better understanding of the density matrix and its properties is desirable. In the following, three topics are discussed which belong into this context.

1 Non-Hermitian Problems

Problems involving a non-hermitian operator whose ground state is of particular interest can be found in equilibrium and in non-equilibrium physics. In the first case, transfer matrices may have this property and a number of situations have been treated by DMRG: a model for the quantum Hall effect [1], the quantum transfer matrices in Chaps. 4(I), 6(I) and the q-symmetric Heisenberg chain related to the conformal series of critical models [2]. In the second case, on which we want to concentrate here, Fokker-Planck and master equations in the absence of detailed balance belong to this category. The ground state then is the stationary state corresponding to eigenvalue zero. The application of the DMRG to such problems is only beginning [3,4].

An example is the model shown in Fig. 1. Classical hard-core particles move stochastically on a chain and can enter and leave the system at the ends. Denoting (un)occupied sites by $\sigma = +1(-1)$, the master equation can

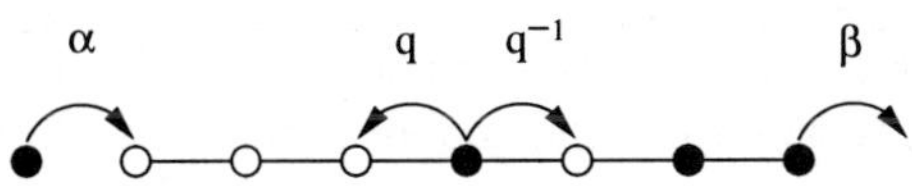

Fig. 1. Hopping model, elementary processes and rates. Particles (empty sites) are shown as black (white) circles

be formulated in a spin one-half language and the time-evolution operator for the probability $p(\boldsymbol{\sigma},t)$ to find a configuration $\boldsymbol{\sigma} = \{\sigma_1,\ldots,\sigma_L\}$ becomes $H = H_\mathrm{q} + H_\mathrm{b}$, where

$$H_\mathrm{q} = -\frac{1}{4}\sum_{n=1}^{L-1}\left[q^{-1}\sigma_n^-\sigma_{n+1}^+ + q\sigma_n^+\sigma_{n+1}^- + \left(q+q^{-1}\right)\left(\sigma_n^z\sigma_{n+1}^z - 1\right)\right.$$
$$\left. + \left(q-q^{-1}\right)\left(\sigma_n^z - \sigma_{n+1}^z\right)\right] \qquad (1)$$

describes the hopping on the chain and

$$H_\mathrm{b} = -\frac{\alpha}{2}\left(\sigma_1^+ + \sigma_1^z - 1\right) - \frac{\beta}{2}\left(\sigma_L^- - \sigma_L^z - 1\right), \qquad (2)$$

which is non-hermitian, corresponds to the processes at the ends. This model, which has been studied intensely in the last years [5], shows three different phases with different current and density profiles. It belongs to a whole class of one-dimensional reaction-diffusion models which have been formulated in quantum language [6].

In the argument for using the density matrix to obtain an optimal approximation (see Chap. 2(I)), the operator H does not appear. Therefore the method is applicable also in the non-hermitian case, where the eigenvectors $\langle\Phi_\mathrm{l}|$ and $|\Phi_\mathrm{r}\rangle$ are in general different. One then is lead to the density matrix

$$\varrho_\mathrm{r} = |\Phi_\mathrm{r}\rangle\langle\Phi_\mathrm{r}| \qquad (3)$$

and the analogous quantity ϱ_l. In this way, calculations for the q-symmetric chain [2], for the hopping model above [7] and for coagulation models [4] have been done. Due to the absence of a variational principle, the ground-state eigenvalue E_0 does not vary monotonically e.g. with m. This is shown in Fig. 2 for the Hamiltonian (1) with antiferromagnetic sign. Nevertheless,

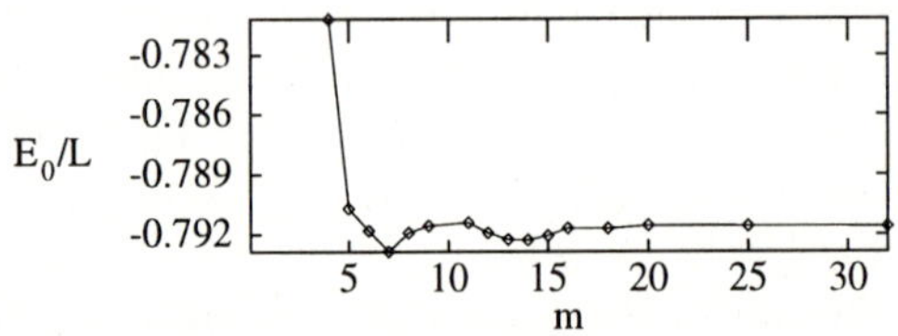

Fig. 2. Ground-state energy per site of the q-symmetric Heisenberg chain for $q = \exp(i\pi/4)$ vs. the number of kept states m

very accurate results could be obtained for E_0 in [2]. The situation is less favourable in the reaction-diffusion models, where the finite-size algorithm had to be applied to improve the accuracy. This is discussed in detail in [4].

A mixture of the density matrices ϱ_r and ϱ_l was used in [1] and very recently in [4] where also different choices were compared. In transfer-matrix calculations, and also in a treatment of the above hopping model with parallel dynamics [3], the non-symmetric quantity

$$\varrho_\mathrm{rl} = |\Phi_\mathrm{r}\rangle\langle\Phi_\mathrm{l}| \qquad (4)$$

was used. This can be motivated by its appearance in expectation values

$$\langle A \rangle = \langle \Phi_\text{l} | A | \Phi_\text{r} \rangle = \text{Tr}\,(\varrho_\text{rl} A) \tag{5}$$

and similarly, with the appropriate reduced density matrix, for quantities involving only half of the system. A truncation in this case corresponds to an approximation of such traces. One has to expect complex eigenvalues, however, and the truncation procedure is not unique. In calculations on an *XY* chain with imaginary boundary fields this could be seen explicitely and, for this case, the choice (4) turned out to be inferior to (3) [7]. In the other examples cited above, the situation was more favourable. In the transfer-matrix calculations, only spurious complex eigenvalues were observed, see Chap. 6(I), and in [3] density profiles were obtained which are in good agreement with previous results. For stochastic systems also a simplification occurs, because the left ground state is connected with the probability conservation and therefore known explicitely [6]. This can be used in the calculation of expectation values.

2 Special Ground States

Normally one uses the DMRG method to study systems with presumably complicated ground states. However, the procedure may also be used to investigate the opposite situation, where the state in question is relatively simple. This will be shown here with examples from Sect. 1.

Imagine that $|\Phi\rangle$ is the superposition of a small number of products

$$|\Phi\rangle = \sum_{\alpha=1}^{p} |\varphi_\alpha\rangle |\psi_\alpha\rangle \tag{6}$$

where $\{|\varphi_\alpha\rangle\}$ and $\{|\psi_\alpha\rangle\}$ are sets of linearly independent states in the two parts of the system. Then the density matrices ϱ_1, ϱ_2 which follow from $|\Phi\rangle$ have at most p non-zero eigenvalues and therefore reflect directly its structure.

Such a situation occurs for the simple ferromagnetic Heisenberg model and also for its q-symmetric generalization given by (1). In this case (6) arises from a Clebsch-Gordan decomposition of the ground state and p increases with the chain length [2].

Here we want to consider another example, namely finite-dimensional matrix-product states (MPS). For such states, the weight of a configuration is given by a product of matrices M, one for each site and spin direction, in an auxiliary space. For an open chain $|\Phi\rangle$ then reads

$$|\Phi\rangle = \sum_{\boldsymbol{\sigma}} [\boldsymbol{u} M(\sigma_1) M(\sigma_2) \ldots M(\sigma_L) \boldsymbol{v}]\, |\boldsymbol{\sigma}\rangle \tag{7}$$

where $\boldsymbol{u}$ and $\boldsymbol{v}$ are boundary vectors, so that the expression in the bracket becomes a number. Dividing the system into two parts then amounts to opening the matrix product and one finds

$$|\varphi_\alpha\rangle = \sum_{\boldsymbol{\sigma}_1} \left[\boldsymbol{u} M(\sigma_1) M(\sigma_2) \ldots M(\sigma_{L/2})\right]_\alpha |\boldsymbol{\sigma}_1\rangle \quad ; \quad \alpha = 1, 2, \ldots, p \tag{8}$$

where $\boldsymbol{\sigma}_1 = \{\sigma_1, \sigma_2, \ldots, \sigma_{L/2}\}$. A similar expression holds for $|\psi_\alpha\rangle$.

Such MPS, which generalize simple product states, appear as exact ground states for a number of equilibrium [8] and non-equilibrium [5,9] systems. Also the DMRG ground state (after truncation) has this structure [10] (see also Chap. 3(I)). The conventional MPS arise from an interplay between different terms in the Hamiltonian and thus appear only for certain values of the parameters. Usually they are found via an ansatz but they can also be observed in a standard DMRG calculation.

The result of such a calculation for the hopping model of Sect. 1 is shown in Fig. 3. Plotted is the density-matrix spectrum for $\beta = 2\alpha$ as a function

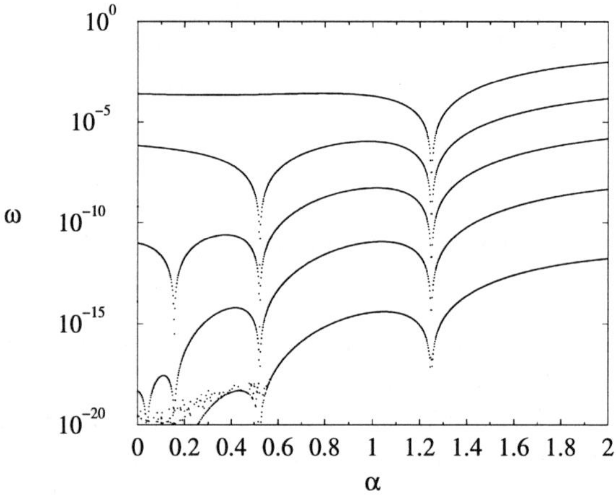

Fig. 3. Density-matrix eigenvalues of the hopping model with 10 sites, calculated with 8 states for $q = 0.25$

of α. The largest eigenvalue is close to one and thus not visible on the logarithmic scale. First, one notices that always only a few eigenvalues are larger than 10^{-15}, so that the system is on the whole rather well-behaved [3]. But at certain values of α the spectrum simplifies even more. For $\alpha_1 = 1.25$ all eigenvalues except one vanish, for $\alpha_2 \simeq 0.5224$ all except two and for $\alpha_3 \simeq 0.1554$ all except three. These are the points where the corresponding finite-dimensional MPS (which are known from the general solution [5,11]) appear. As L is varied, the curves move, but the singular points remain. This is an important test, since spurious structures can sometimes appear for small systems. At the points α_n one can also compare with analytical results and verify e.g. the L-dependence of the non-zero eigenvalues. One can even go further and scan the whole α-β-plane. The result is shown in Fig. 4, where the third eigenvalue is plotted for a short chain. The two ditches in the con-

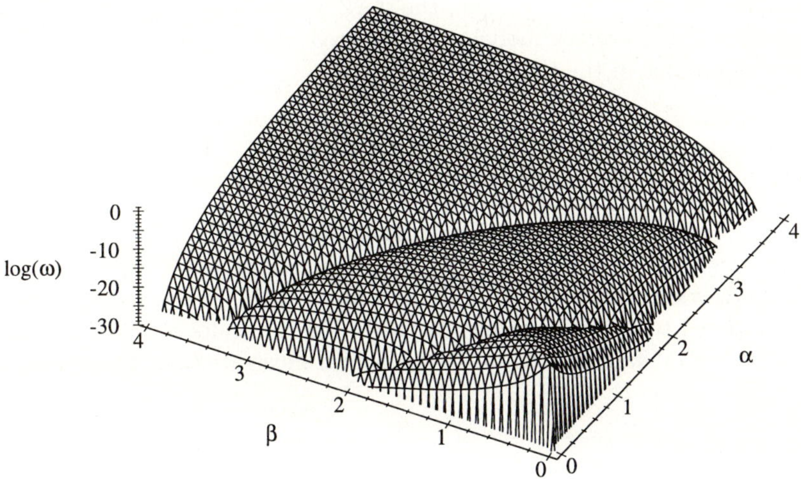

Fig. 4. Third density-matrix eigenvalue of the hopping model with 4 sites, $q = 0.25$

tour plot correspond to the two lines, along which the ground state is a one- and two-dimensional MPS, respectively.

These features can be found in all models which possess matrix-product states. For example, the four-dimensional MPS [9] occuring for the reaction-diffusion model studied in [12] shows up in the same way. In tracking down such states one does not have to hit exactly the right points since the spectra deform already in the vicinity. A practical condition is, however, that not too many parameters should be involved.

3 Exponential Spectra

Usually, the density-matrix spectra are more complicated than those in the previous section and involve many levels. When plotted, however, a relatively simple pattern becomes visible: the eigenvalues decrease rapidly and follow roughly an exponential law [13,2]. This behaviour is essential for the success of the DMRG method, since it guarantees that a small number of states already leads to a good approximation. One would therefore like to understand these spectra and their origin in more detail. This is possible for non-critical integrable models [14].

The argument uses the relation between quantum chains and two-dimensional classical models (cf. Chap. 5(I)) by which the ϱ_α can be expressed as partition functions of a long strip with a perpendicular cut. Such partition functions, in turn, can be obtained via corner transfer matrices (CTM) à la Baxter [15], for which the spectrum is known in the thermodynamic limit and has exponential form. For example, the XXZ Heisenberg chain with $\Delta > 1$ is related to the six-vertex model in its ordered phase, and from the CTM-results one obtains in the thermodynamic limit

$$\omega = C \exp\left(-\varepsilon \sum_{j\geq 0} 2jn_j\right) \tag{9}$$

where $n_j = 0, 1$ and $\varepsilon = \operatorname{arcosh}\Delta$. Thus all eigenvalues have the form $\exp(-2\varepsilon k)$ with integer k.

This behaviour can be seen already in finite chains if the system size is large compared to the correlation length. In Fig. 5 results of DMRG calculations for several values of Δ are shown, which illustrate this. The values of

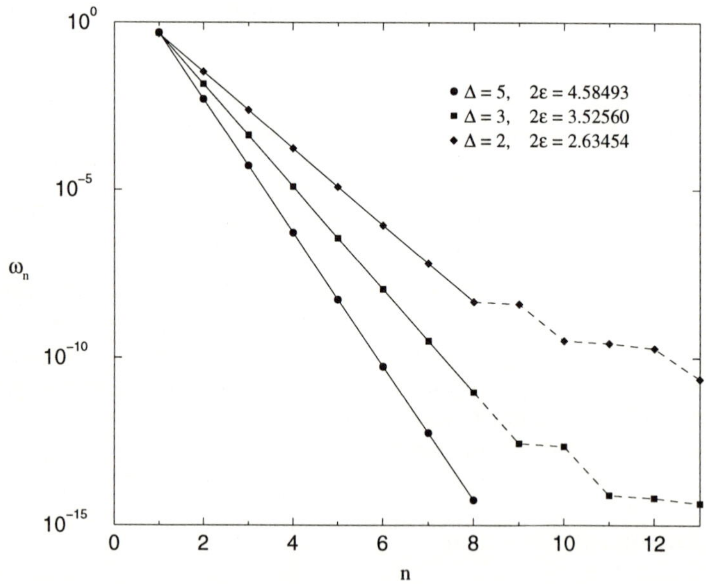

Fig. 5. Density-matrix spectra of Heisenberg chains with 98 sites, calculated with 64 states for three values of Δ. The degeneracies have already been removed

ε obtained numerically agree with the theoretical ones up to several decimal places. Also the degeneracies predicted by (9) show up perfectly. Deviations from the exponential law only occur at the lower end of the spectrum and can be ascribed to the finite size and the limited accuracy. The situation is very similar for the transverse Ising chain, which also has been tested, and will generally be found for integrable models. In all these cases, the spectrum is ultimately a consequence of the Yang-Baxter equations.

For non-integrable, non-critical systems the experience shows that the spectra are less regular but still have the same general features. This kind of universality has been investigated recently for several models [16,17]. The spectra observed for critical systems have yet to be understood in the context of conformal invariance.

References

1. J. Kondev and J.B. Marston, Nucl. Phys. B **497**, 639 (1997)
2. M. Kaulke and I. Peschel, Eur. Phys. J. B **5**, 727 (1998)
3. Y. Hieida, J. Phys. Soc. Japan **67**, 369 (1998)
4. E. Carlon, M. Henkel, U. Schollwöck, cond-mat/9902041 (1999)
5. see, for example B. Derrida, M.R. Evans, V. Hakim and V. Pasquier, J. Phys. A: Math. Gen. **26**, 1493 (1993); F. Eßler and V. Rittenberg, J. Phys. A: Math. Gen. **29**, 3375 (1996)
6. F.C. Alcaraz, M. Droz, M. Henkel and V. Rittenberg, Ann. Phys. **230**, 250 (1994)
7. M. Kaulke, PhD Thesis, Freie Universität, Berlin (1999)
8. see, for example I. Affleck, T. Kennedy, E.H. Lieb and H. Tasaki, Comm. Math. Phys. **115**, 477 (1988); A. Klümper, A. Schadschneider and J. Zittartz, J. Phys. A: Math. Gen. **24**, L955 (1991); Europhys. Lett. **24**, 293 (1993); A.K. Kolezhuk and H.-J. Mikeska, Phys. Rev. Lett. **80**, 2709 (1998)
9. H. Hinrichsen, S. Sandow and I. Peschel, J. Phys. A: Math. Gen. **29**, 2643 (1996)
10. S. Östlund and S. Rommer, Phys. Rev. Lett. **75**, 3537 (1995)
11. K. Mallick and S. Sandow, J. Phys. A: Math. Gen. **30**, 4513 (1997)
12. H. Hinrichsen, K. Krebs and I. Peschel, Z. Phys. B **100**, 105 (1996)
13. S.R. White, Phys. Rev. Lett. **69**, 2863 (1992); Phys. Rev. B **48**, 10345 (1993)
14. I. Peschel, M. Kaulke and Ö. Legeza, Ann. Physik (Leipzig) **8**, 153 (1999)
15. R.J. Baxter, Exactly Solved Models in Statistical Mechanics, Academic Press, London (1982)
16. K. Okunishi, Y. Hieida and Y. Akutsu, cond-mat/9810239 (1998)
17. C. Ritter, Private communication

3.2 Walls, Wetting and Surface Criticality

Enrico Carlon

Laboratoire de Physique des Matériaux
Université Henri Poincaré, Nancy I
F-54506 Vandœuvre-lès-Nancy Cedex, France

After the discovery of the density matrix renormalization group by White [1] a lot of work has been done to extend the method in various directions. We review here some recent applications of the DMRG to classical two-dimensional systems in confined geometry, where we used Nishino's adaptation [2] of White's algorithm for the renormalization of classical transfer matrices (see Chap. 5(I)).

The DMRG is particularly suited to study systems in a confined geometry since it naturally deals with lattices of size $L \times \infty$, i.e. of finite extension along one direction only, which is the situation most commonly considered in theoretical studies. It is also well-known that DMRG works best with open boundary conditions, which makes the technique well-suited to study the effects of surfaces. For two-dimensional systems the DMRG algorithm is superior to the classical Monte Carlo (MC) method in accuracy, speed and in the possibility of treating much larger systems.

When a system is placed in a confined geometry, say a fluid in a small tube or in a slab, the interaction with the walls of the containing vessel may strongly influence its thermodynamical properties. A lot of studies have been performed on such systems in the past years [3]. From a theoretical viewpoint typical approaches are mean-field theories, which are known to describe the critical properties of a system correctly above an upper critical dimension, or MC simulations. The latter are performed when one is interested in the effect of thermal fluctuations, which are not included in a mean-field treatment and are the strongest at low spatial dimensions. In this respect the two-dimensional case is of particular interest and the DMRG method represents a good alternative to MC. In simple cases, as in the Ising model, DMRG calculations are extremely fast since it turns out that even with few states kept ($m \approx 30 - 40$) one gets already extremely accurate results.

In this brief review we present some aspects of the applications of the DMRG to two examples, namely the Ising and the Potts model. Without entering too deeply into the details of the physics we show certain aspects of the calculations and try to stress the advantages offered by DMRG with respect to the more traditional MC method.

1 Influence of Wetting on Capillary Condensation

As a first example we consider an Ising model on a $L \times \infty$ strip with surface fields equal to h_1 at the two boundaries. These fields acting only on the edge spins indicate the preference of the walls for the phase with either positive or negative magnetization. In the bulk limit $L \to \infty$ the two phases coexist (having the same free energy) for $h = 0$ and $0 \leq T < T_c$, where h is a bulk magnetic field and T_c the critical temperature; the two-phase coexistence region is shown as thick dashed line in Fig.1(a). The presence of surface fields induces a wetting transition at a temperature T_w, which depends on h_1, and is known exactly for the two-dimensional Ising model [4]. Wetting occurs when a thick layer of spins of magnetization equal to that of the surface fields is adsorbed at the wall. By approaching the wet regime $T_w \leq T < T_c$ and $h = 0$ either from the temperature ($h = 0$ and $T \to T_w^-$) or from the field ($h \to 0$ and $T_w \leq T < T_c$ fixed) direction, the thickness of the wetting layer l diverges [5] according to certain critical exponents (obviously l can diverge in a semi-infinite system only). The aim of this study is that of investigating the influence of wetting on the coexistence between phases.

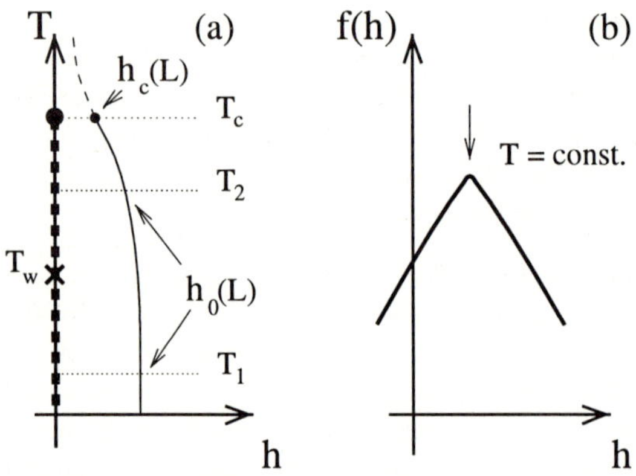

Fig. 1. (**a**) Phase diagram of the Ising model in the bulk limit $L \to \infty$ (thick dashed line) and capillary condensation line for a finite L (thin solid line). (**b**) Identification of the capillary condensation point $h_0(L)$ from the free energy maximum at fixed T and L.

In the strip geometry phase coexistence is shifted to a finite value of the bulk field $h_0(L)$, which has sign opposite to that of the surface fields. The locus of points $h_0(L)$, shown as a thin line in Fig. 1(a), is known as capillary condensation line; we stress that this line does not correspond to true thermodynamic singularities, since there cannot be phase transitions for finite L, but to a rounded-off or *pseudo*-transition. The calculation of $h_0(L)$ from

DMRG was done as follows: from the largest eigenvalue $\lambda_{\max}$ of the transfer matrix one finds the total free energy which is given by $f(h) = -k_B T \ln \lambda_{\max}$, where k_B is the Boltzmann constant and T the temperature. We fixed the temperature and varied the bulk field h; the maximum of $f(h)$ corresponds to $h_0(L)$, i.e. the point where the two phases with positive $(\partial f(h)/\partial h < 0)$ and negative $(\partial f(h)/\partial h > 0)$ magnetizations meet (see Fig. 1(b)).

According to theoretical predictions [6,7] one expects for $L \to \infty$:

$$h_0(L) = \frac{A}{L} + \frac{B}{L^\gamma} + \ldots, \tag{1}$$

where the dominant $1/L$ term gives the so-called Kelvin equation (see [6–8] and references therein) and $\gamma > 1$ is a correction-to-scaling exponent. The interest in this scaling behavior stems from the fact that one expects [6,7] different values for γ above and below the wetting temperature T_w. The theoretical predictions are $\gamma = 2$ for $T < T_w$ and $\gamma = 5/3$ for $T \geq T_w$. It is clear that to determine γ one needs good numerical accuracy in the calculation of $h_0(L)$; a recent MC study of this problem did not provide the sufficient accuracy to resolve this issue [8].

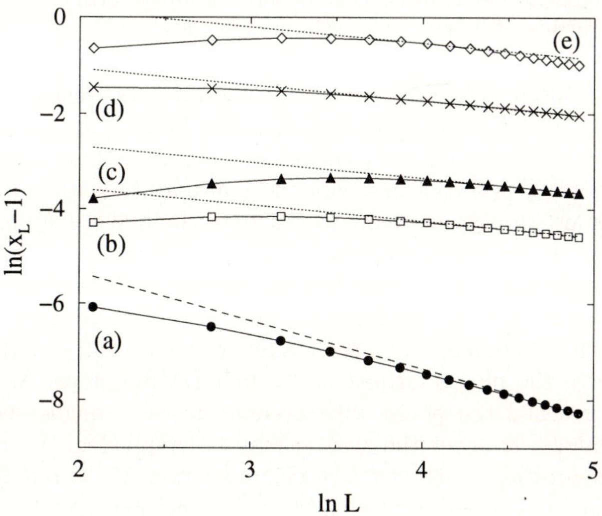

Fig. 2. Plot of $\ln(x_L - 1)$ vs. $\ln L$ below (a) and above (b-e) the wetting temperature. Error bars are smaller than symbol sizes. The dashed line has slope equal to -1, while the dotted lines have slope $-1/3$.

Using the finite-system DMRG algorithm we calculated $h_0(L)$ with an estimated accuracy of about $5-6$ decimals for $m \approx 40$ states kept and for

$L \leq 150$. From a set of data for $h_0(L)$ for $L = 8, 16, 24, \ldots 144$ we formed the logarithmic derivatives $x_L = \log[h_0(L+8)/h_0(L)]/\log(1+8/L)$ and verified the dominant $1/L$ behavior as predicted from (1) from the finite-size analysis of x_L [9]. For the study of the correction-to-scaling term we considered:

$$\ln(x_L - 1) = (1-\gamma)\ln L + \ln\left|\frac{A}{B}\right| + \ldots \qquad (2)$$

a relation valid in the limit $L \to \infty$, which can be easily obtained from (1) and from the definition of x_L.

Figure 2 shows plots of $\ln(x_L - 1)$ vs. $\ln L$ for various values of surface fields and temperatures; from the asymptotic slope of the data one can determine the correction-to-scaling exponent γ. Below the wetting temperature (set (a) of Fig. 2) there is good agreement with the theory: the dashed line has slope $1 - \gamma = -1$, which corresponds to a correction-to-scaling exponent $\gamma = 2$. However above the wetting temperature the numerical DMRG data do not agree with the expected value $\gamma = 5/3$; we rather find a good fit with $\gamma = 4/3$. An explanation of this discrepancy is discussed in detail in [9]; for the pourposes of this short review we stress the fact that DMRG calculations are capable of distinguishing between the scaling behavior of $h_0(L)$ in the wet and non-wet regimes even if the correction-to-scaling term is rather small (note the magnitude of $\ln(x_L - 1)$ in the plots of Fig. 2).

One point that is worth mentioning concerns metastable states. As is well-known MC encounters difficulties in locating first order transitions points due to metastability, which leads to hysteresis loops. Sometimes problems with metastability can be found also in DMRG calculations; these problems can however be avoided, provided one uses the appropriate density matrix. As an example let us suppose we want to calculate the free energy of an Ising strip with surface and bulk fields such that $h_1 < 0$ and $h > 0$ and for L large enough such that the system has on average positive magnetization. For strips of small widths, the starting point of a DMRG calculation, the surface fields will "dominate" with the tendency of giving an overall negative magnetization to the blocks formed in the first DMRG steps. As a certain value of L is reached the phase with average positive magnetization gets a lower free energy, however the system may be trapped in the previously generated configurations with negative magnetization. If the temperature is not too low the application of the finite-system method fixes the problem. However at low temperatures this is not always sufficient; in this case we found that it is necessary to use the following density matrix:

$$\rho = \frac{1}{2}\hat{tr}\left\{|\psi_0\rangle\langle\psi_0| + |\psi_1\rangle\langle\psi_1|\right\} \qquad (3)$$

where $\hat{tr}$ denotes the partial trace on part of the system (either the left or the right part) and $|\psi_0\rangle$, $|\psi_1\rangle$ are the eigenstates of the transfer matrix corresponding to the largest and to the second-largest eigenvalues, respectively.

In this mixed density matrix the part containing $|\psi_1\rangle$ carries the information of the metastable state throughout all the steps of the DMRG calculation.

2 Magnetization Profiles and Surface Critical Properties of Potts Models

Besides the example discussed in the previous section a number of other DMRG calculations have been performed on Ising models, to study the effects of a non-homogeneous field and the critical point shift in systems of finite size [10], and on the competing effects of surface and bulk fields in the vicinity of the critical point (see Chap. 3.3(II)). As next example we consider the Q-state Potts model defined by the Hamiltonian:

$$H = -J \sum_{\langle ij \rangle} \delta_{s_i s_j} \tag{4}$$

where spins at each site assume Q possible values $s_i = 0, 1, \dots Q-1$. The model generalizes the Ising model (which corresponds to $Q = 2$) and has a second-order transition for $Q \leq 4$ and a first order one for $Q > 4$.

Although the latter case has also been studied recently by DMRG [11] as well, we restrict ourselves here to the second-order regime $Q \leq 4$ [12], since one of our aims is to test the performance of DMRG for a critical system, when the bulk correlation length becomes infinite and the gap vanishes in the thermodynamic limit. In particular we are interested in the calculation of quantities as the magnetization profile:

$$m(l) = \frac{Q\langle\delta(s_l)\rangle - 1}{Q - 1} \tag{5}$$

where $1 \leq l \leq L$ labels the distance between the two walls and $\langle\delta(s_l)\rangle$ is the density of spins equal to zero at position l. The shape of $m(l)$ at the critical temperature T_c has been derived from conformal invariance, but never before from a miscroscopic calculation, except for the Ising case ($Q = 2$). Obviously $m(l)$ is non-zero only if one imposes symmetry breaking boundary conditions (BC). The most interesting case is that of fixed/free BC [1] for which conformal invariance predicts the following profile [13]:

$$m(l) = \mathcal{A} \left\{ \frac{L}{\pi} \sin\left(\frac{\pi l}{L}\right) \right\}^{-x_m} \cos\left(\frac{\pi l}{2L}\right)^{x_m^s} . \tag{6}$$

Here x_m and x_m^s are the bulk and surface magnetic exponents. Notice that the quantity $L^{x_m} m(l)$, the scaled magnetization, depends on l only through the scaled variable l/L. This agrees with the general scaling theory of Fisher and de Gennes [14]. The advantage of fixed/free BC is that from the decay of the magnetization in the middle of the strip $m(L/2) \sim L^{-x_m}$ and at the

[1] We fix in this case the spins at $l = 0$ to $s = 0$

free surface $m_s \sim L^{-x_m^s}$ one can find simultaneously bulk and surface critical exponents.

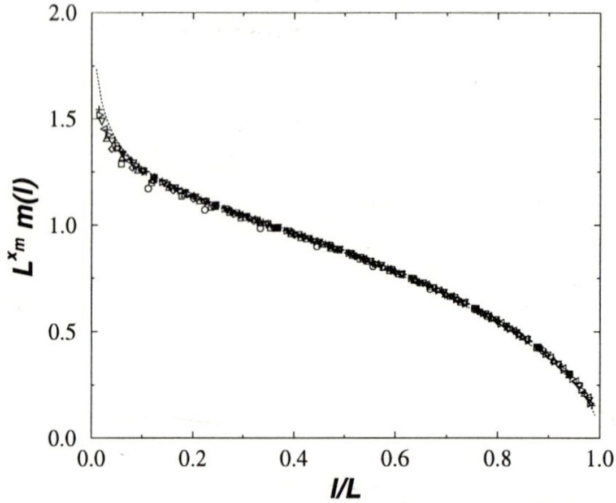

Fig. 3. Plot of the scaled magnetization $L^{x_m}m(l)$ as function of l/L for the three state Potts model and $L = 8$, 16, 24 ...64. The DMRG data collapse into a single curve which follows accurately the prediction of conformal invariance given by (6) and shown as dashed line.

We calculated the magnetization profiles for systems of sizes $L = 8$, 16, 24 ...64 at the bulk critical temperature T_c. To obtain the best accuracy we used the finite-system method to generate strips of all the wanted sizes L_0, L_1, ...in a single run in the following way: After the iterations of the finite-system method for a strip of width L_0 have been completed, we used the two blocks of size $L_0/2$ as starting point for generating larger left and right blocks of sizes $L_0/2 + 1$, $L_0/2 + 2$ ...until the system of size L_1 has been generated. At this point the finite-system sweeps are started again and the procedure is repeated. The advantage of this method is clear: although it requires many iterations, one can use White's acceleration method for the finite system algorithm [15], which reduces substantially the required CPU time. For the calculation of the whole profile we did not store spin operators in the blocks; in each step of the finite-system method while sweeping through a system of fixed size the two spins of the configuration $B_l \bullet \bullet B_r$ "move" through the whole system from the two boundaries, allowing the calculation of the profiles at different DMRG steps.

Fig. 3 shows a plot of the scaled magnetizations $L^{x_m}m(l)$ for the $Q = 3$ state Potts model and for different values of L. All the data referring to different sizes collapse into a single curve and follow the shape predicted from conformal invariance given by (6) and shown as a dashed line in the plot. For the scaling plot and in (6) we used the exponents predicted by conformal invariance [13]: $x_m = 2/15$ and $x_m^s = 2/3$.

Table 1. BST extrapolants for the calculation of the surface magnetic exponent for $Q = 3$.

0.554998	0.673224	0.668052	0.667014	0.666865	0.666938	0.666666
0.584425	0.670860	0.667623	0.666936	0.666799	0.666574	
0.600929	0.669696	0.667388	0.666880	0.666762		
0.611585	0.669012	0.667238	0.666839			
0.619076	0.668565	0.667136				
0.624650	0.668252					
0.628972						

Such exponents can also be derived from finite-size scaling; as an illustration we consider the calculation of the surface magnetic exponent for the $Q = 3$ case. From the calculated values of the surface magnetization $m_s(L)$ for systems with different L we formed the approximants (or logarithmic derivatives) x_L as done in the previous section for $h_0(L)$. For $L \to \infty$, x_L converges to the surface exponent x_m^s. The extrapolation is done using the BST method [13] which consists in obtaining from a the original sequence of, say, N data, succesive sequences of $N-1$, $N-2$ ... data which should get closer at each step to the asymptotic value. For the surface exponent of the $Q = 3$ model this is illustrated in Table 1. Each new iteration is reported in a new column: a clear increase of convergence is visible in the first five columns, which does not improve from the sixth column on; this leads to the estimate $x_m^s = 0.6667(1)$ in excellent agreement with the expected value 2/3.

Table 2. Comparison between exact (†) or conjectured (‡) values bulk and surface exponents for the $Q \leq 4$ Potts model and the numerical results obtained from finite size scaling extrapolation of DMRG data (from [12]).

Q	x_m	x_m (DMRG)	x_m^s	x_m^s (DMRG)	x_ϵ	x_ϵ (DMRG)
2	$1/8^\dagger$	0.125000(2)	$1/2^\dagger$	0.50000(5)	$1^\dagger$	1.00000(1)
3	$2/15^\ddagger$	0.1334(1)	$2/3^\ddagger$	0.6667(1)	$4/5^\ddagger$	0.800(1)
4	$1/8^\ddagger$	0.120(5)	$1^\ddagger$	1.02(2)	$1/2^\ddagger$	0.49(2)

For the DMRG calculation we kept $m \approx 60-80$ states and we estimated a numerical accuracy of $8-9$ digits. Obviously the DMRG allows the study of larger systems than $L = 64$, however the numerical accuracy will not be so high. For an accurate calculation of critical exponents using powerful extrapolation techniques as the BST method one needs an accuracy of at least

6 − 7 digits, which is difficult to obtain for larger systems, in particular at criticality. It is known that in the thermodynamic limit the DMRG is unable to describe gapless spectra [16]. However, as in the present case, one can extract critical exponents from an accurate scaling analysis of finite systems, which are always gapped even at T_c [17]. Table 2 collects the values of the critical exponents calculated from BST extrapolation of the DMRG data, including the thermal exponent x_ϵ which can be obtained from the energy-density profiles. In general one obtains excellent agreement with the iexactly known ($Q = 2$) or conjectured values ($Q > 2$) for the critical exponents. The accuracy for the case $Q = 4$ is not very high due to the well-known problem of logarithmic corrections (for more details see discussion in [12]).

In conclusion we have presented two examples of application of the DMRG to classical two-dimensional systems in a strip geometry, where we focused on the effect of open boundary conditions and of surface and bulk fields on the system. The high accuracy both at and off criticality and the possibility of treating large systems make of the DMRG a very good method for investigating classical systems in two dimensions, in particular for studies of surface effects or of properties of interfaces.

References

1. S.R. White, Phys. Rev. Lett. **69**, 2863 (1992)
2. T. Nishino, J. Phys. Soc. Jpn. **64**, 3598 (1995)
3. R. Evans, J. Phys. Cond. Matt, **2**, 8989 (1990)
4. D. Abraham, Phys. Rev. Lett. **44**, 1165 (1980)
5. For a general introduction on wetting and related topics see: S. Dietrich in Phase Transition and Critical Phenomena, Vol. 12, Eds. C. Domb and J.L. Lebowitz, Academic Press (1988)
6. E.V. Albano, K. Binder, D.W. Heermann and W. Paul, J. Chem. Phys. **91**, 3700 (1989)
7. A.O. Parry and R. Evans, J. Phys. A: Math. Gen. **25**, 275 (1992)
8. E.V. Albano, K. Binder and W. Paul, J. Phys. A: Math. Gen. **30**, 3285 (1997)
9. E. Carlon, A. Drzewiński and J. Rogiers, Phys. Rev. B **58**, 5070 (1998)
10. E. Carlon and A. Drzewiński, Phys. Rev. Lett. **79**, 1591 (1997); Phys. Rev. E **57**, 2626 (1998)
11. F. Iglói and E. Carlon, cond-mat/9805083
12. E. Carlon and F. Iglói, Phys. Rev. B **57**, 7877 (1998)
13. See: P. Christe and M. Henkel, Introduction to Conformal Invariance and Its Applications to Critical Phenomena, Springer (1993)
14. M. Fisher and P.G. de Gennes, C.R. Acad. Sci. Paris B **287**, 207 (1978)
15. S.R. White, Phys. Rev. Lett. **77**, 3633 (1996)
16. S. Östlund and S. Rommer, Phys. Rev. Lett. **75**, 3537 (1995)
17. Bulk exponents for the case $Q = 2$ and $Q = 3$ have also been calculated by the corner transfer matrix method in: T. Nishino and K. Okunishi, J. Phys. Soc. Jpn. **66**, 3040 (1997)

3.3 Critical Two-Dimensional Ising Films With Fields

Andrzej Drzewiński

Institute of Low Temperature and Structure Research, Polish Academy of Sciences
P.O.Box 1410, PL-50 950 Wrocław 2, Poland

The statistical physics of fluids confined in narrow pores or capillaries is a subject that has attracted much research effort [1]. Unfortunately, because of the high ratio of the surface to the volume, its influence on a system can be extremely strong and theories for the bulk critical phenomena can fail here – not only quantitatively but also qualitatively. Therefore, a method was needed, that would be able to allow numerical calculations for larger systems than with existing approaches. Such a method should be applicable to systems with various fields and boundary conditions, giving results with high and controlled accuracy.

The DMRG method for the study of ground-state properties of quantum chains has proved to be the desired idea [2,3]. It was extended by Nishino to $2D$ classical systems as a transfer-matrix renormalization [4], see Chap. 5(I), which gave an excellent numerical tool for studying systems in a confined geometry [5–7].

The Ising model can be studied as a model of a fluid or a ferromagnet. The numerical diagonalizations for the standard transfer-matrix method in an Ising system are restricted to strips of rather small widths ($L \leq 20$), whereas the DMRG method has enabled us to consider strips with a width up to $L = 200$ without extraordinary efforts. More importantly, only a relatively small number of states, which are kept ($m \leq 40$ here) ensures a very high accuracy for the DMRG results (Chap. 3.2(II)).

In this brief review we present some applications of the DMRG method to studies of $2D$ strips with competing bulk and surface fields near criticality [8,9]. Physically, our system describes for example a single idealized pore with attractive walls containing a nearly critical fluid in contact with a bulk reservoir.

1 Depletion Effect

Description of the phenomenon

Critical adsorption occurs when a simple fluid, ferromagnet, or another system in a single-phase region approaches its bulk critical point in the presence of an attracting external wall or another distinct physical interface. Away from criticality the order parameter (OP) differs from its bulk value (fixed by the properties of the reservoir far from the wall) only over microscopic distances and the amount adsorbed is finite. But close to criticality, where the correlation length is $\xi \sim\mid \tau \mid^{-\nu}$ $(\tau \equiv (T - T_c)/T_c)$, the influence of the wall extends to macroscopic distances and the amount adsorbed can diverge. Near criticality, ξ is the controlling length scale and as $\tau \to 0$ the OP profile should be described in terms of a universal scaling function as proposed by Fisher and de Gennes [10]

$$m(l) = \tau^{\beta} \mathcal{M}_{si}\Big(\frac{l}{\xi}, y\Big), \tag{1}$$

where the subscript '*si*' denotes the semi-infinite system. The adsorption (coverage) Γ has in the semi-infinite system the form

$$\Gamma \equiv \int_0^{\infty} m(l) dr = \tau^{\beta} \xi \mathcal{G}_{si}(y), \tag{2}$$

where l is the distance from the wall, located at $l = 0$. For the magnetic case $m(l)$ is the magnetization profile, β is the critical exponent describing the vanishing of the bulk magnetization and Γ is the (excess) magnetization. For a pure fluid $m(l) = \rho(l) - \rho_c$, where $\rho(l)$ is the density profile and ρ_c is the bulk density. The above formulae refer to $\tau > 0$ and the situation where the bulk magnetic field is zero. $\mathcal{M}_{si}$ and $\mathcal{G}_{si}$ are universal scaling functions and $y = \tau^{-\nu} H_1^{\nu/\Delta_1}$ is the scaling variable describing the coupling to a surface magnetic field H_1. The quantity $y \sim \xi/l_1$ is the ratio of the two length scales ξ and $l_1 \approx H_1^{-\nu/\Delta_1}$, where l_1 denotes an approximate distance from the surface up to which the OP responds linearly to H_1. Δ_1 is the surface gap exponent [11].

Within the general context of surface critical phenomena [11] critical adsorption corresponds to the $H_1 \neq 0$ fixed point so that $y \to \infty$ as $\tau \to 0$. In this limit $\mathcal{G}_{si}(y)$ is asymptotically constant and, from (2), Γ takes on the asymptotic form (for $\tau \to 0$ and for any non-zero value of H_1), which presents a monotonous behaviour

$$\Gamma \sim \tau^{\beta - \nu}. \tag{3}$$

Recent experiments by Thommes *et al.* [12] designed to test the above prediction for SF_6 on a colloidal graphite adsorbent found that as T was going down along a near critical isochore Γ first increased (as predicted) but then decreased very rapidly, taking on negative values for $T - T_c < 1K$. Such a deviation from the monotonous behaviour is called *critical depletion*.

Maciołek *et al.* [13] proposed recently that this phenomenon is associated with a fluid in a single idealized pore and results from a competition between positive adsorption (the walls or surface fields favour the dense liquid phase) and negative adsorption associated with a bulk-like field, which favours the dilute gas phase. Such a bulk-like field is present when the bulk reservoir is not exactly at the critical density, because then the chemical potential for the pore becomes uncritical. Near the bulk critical point even a very small difference between the actual and the critical values of the chemical potential may lead to pronounced differences in the density (due to the large compressibility $\kappa_T \sim \tau^{-\gamma}$).

Therefore, for a system between two walls (the model of a single idealized pore), we can approximate the value of adsorption by

$$\Gamma \sim 2\tau^{\beta}\xi\mathcal{G}_{si} + \mathcal{A}H\tau^{-\gamma}L. \qquad (4)$$

The critical exponents for the $2D$ Ising model are $\nu = 1$, $\Delta = 15/8$, $\gamma = 7/4$, and $\Delta_1 = 1/2$.

When the surface and the bulk fields favour different phases, say $H_1 > 0$ and $H < 0$, the bulk and the surface contributions to Γ compete. As $\xi \sim \tau^{-\nu}$ and $\gamma > \nu - \beta$, for not too weak H the bulk term exceeds the surface term for sufficiently small τ, so the Fisher-de Gennes (FdeG) form is no longer valid and critical depletion can result. This explanation, proposed first in [13], was supported by the critical scaling analysis and a mean-field (MF) calculation for a lattice gas confined between two identical planar walls.

Our first goal is to test whether this prediction is still valid, when the critical fluctuations are taken into account. Although the critical scaling analysis takes the fluctuations into account, it is restricted to the asymptotic regime and, what is more, is not able to give some details (*e.g.* a form of scaling functions). Then we investigate whether in the case of a critical film the fluctuations may lead to a new behaviour of the OP profile.

DMRG results

We present the results of DMRG calculations for a $2D$ Ising film defined on a $L \times M$ square lattice with $M \to \infty$. We assume nearest-neighbour interactions of strength J and a Hamiltonian of the form:

$$\mathcal{H} = -J\Big[\sum_{<i,j>}\sigma_i\sigma_j - H\sum_i\sigma_i - H_1\sum_i^{(1)}\sigma_i - H_L\sum_i^{(L)}\sigma_i\Big], \qquad (5)$$

where the first sum runs over all nearest-neighbour pairs of sites while the last two sums run, respectively over the first and the L-th column. H is the bulk magnetic field and H_1 and H_L are the surface fields corresponding to a direct, short range interaction between the walls and the spins in the film. H, H_1 and H_L are all dimensionless fields, measured in units of J. We assume positive surface fields which lead to preferential adsorption of the positively magnetized bulk phase on the walls ($H_1 = H_L = 0.8$).

We have used the finite-system DMRG algorithm [3] and found that for a strip with $L = 96$ a number of $m = 40$ kept states is sufficient to guarantee a very high accuracy of the magnetization profiles [6].

The field H was taken negative, ranging from very small values $H = -10^{-7}$ up to $H = -0.8$. The results of these calculations for the adsorption $\Gamma \equiv \sum_{l=1}^{L} m_l$ as a function of τ are presented in Fig. 1.

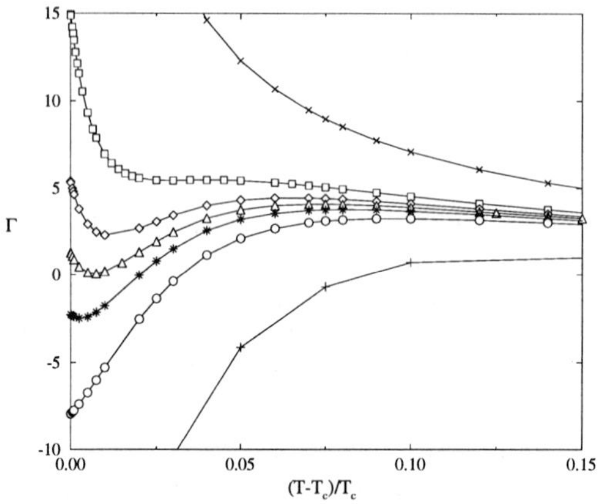

Fig. 1. Adsorption Γ as a function of the reduced deviation from the bulk temperature τ calculated by the DMRG method. The fields H from top to bottom are -0.001; -0.0022; -0.0024; -0.0025; -0.0026; -0.0028 and -0.004.

For the weakest bulk fields $H \geq -10^{-3}$ the behaviour of $\Gamma(\tau)$ is similar to the case of vanishing bulk field [13] - see the typical curve $H = -10^{-3}$ in Fig. 1. Γ is positive for all values of τ. It first increases monotonically as T decreases and in the temperature range corresponding to τ between 0.2 and 0.8, $\log \Gamma$ vs $\log \tau$ is approximately linear with a slope equal to $-7/8$, which agrees with the FdeG result $\Gamma \sim \tau^{\beta-\nu}$. When T goes down further the adsorption continues to increase but the curve deviates from the FdeG power law behaviour.

For the strongest bulk fields $H \leq -4 \times 10^{-3}$ the adsorption is negative for all $\tau < 1$ - see the typical curve $H = -4 \times 10^{-3}$ in Fig. 1. Γ decreases monotonically when we approach T_c and it saturates for $\tau < 10^{-3}$. This is the behaviour typical for the 'strong field' regime (we estimated that the strong field region corresponds to $|H| > 10^{-2}$).

In the case of intermediate bulk fields, for the fields between -10^{-3} and -4×10^{-3} a very interesting behaviour of the adsorption as a function of temperature can be seen in Fig. 1. Apart from the expected *depletion* of the adsorption (a maximum of $\Gamma(\tau)$) there is also a qualitatively new behaviour: a minimum not found in the MF approximation of [13]. These dramatic changes of the shape of the function $\Gamma(\tau)$ take place in a very narrow range of values of H. It is worth stressing that $\Gamma(\tau)$ approaches the FdeG form for enough high τ. The locations of the maximum and the minimum change with H. The maximum shifts slightly towards higher values of τ, whereas the minimum shifts towards smaller values of τ and becomes deeper as $\mid H \mid$ is increased.

At the same time the value at which Γ saturates, which for $H = -0.0022$ and $H = -0.0024$ is greater than the value of Γ at the maximum, quickly decreases with growing $|H|$, so that the minimum disappears for $H = -0.003$.

This new behaviour occurs when the bulk correlation length is comparable to the system size $\xi \sim L$ and can be explained as the effect of the very strong fluctuations. More details are given in [8].

2 Magnetization Profiles for Weak Surface Fields

Similar relevant qualitative differences in physical quantities were discovered recently in the semi-infinite system [14]. Contrary to the MF predictions, in the critical region the OP profiles near the weakly adsorbing walls were nonmonotonic functions of the distance from the wall. Close to the surface, m(l) increases and only for $l > l_1$ a decaying OP profile occurs.

It was shown, that the range within which this growth occurs at the bulk criticality is determined by the surface field H_1 alone, the characteristic length scale being l_1 and for $l \ll l_1$ the magnetization is described by

$$m(l, H_1, T_c) \sim H_1 l^{1-\eta_\perp} \ln(H_1 l), \tag{6}$$

where for the $2D$ Ising model $1 - \eta_\perp = 3/8$, where $\eta_\perp$ is the anomalous dimension governing the decay of the correlations in the direction parallel to the surface [15]. Therefore for a given value of H_1 the magnetization $m(l)$ grows as $\sim l^{3/8} \ln l$ for $l \leq l_1$. For $l \sim l_1$ the profile has a maximum, whereas much farther away from the surface, the typical decay of the OP profile is found, $m(l) \approx l^{-\beta/\nu}$ [15].

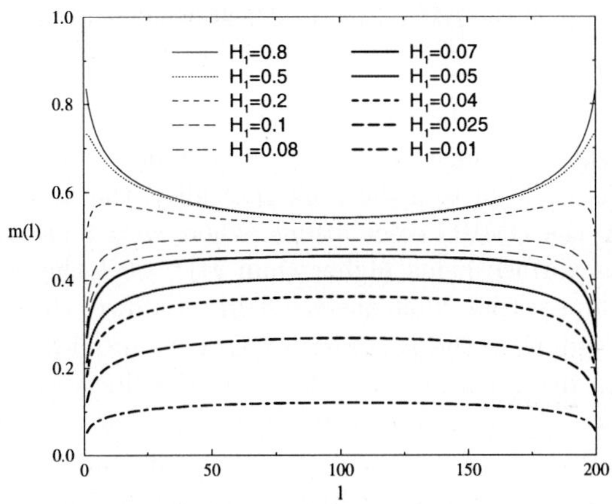

Fig. 2. Selection of the magnetization profiles calculated by the DMRG method for different surface fields at T_c. The bulk field is $H = 0$.

Motivated by the above results one may ask how weak surface magnetic fields (macroscopic l_1) influence the behaviour of a confined system [9]. We expect that in confined systems the shape of the OP profile for weak H_1 can be even more surprising than in the semi-infinite system (especially in the critical region) when the OP near one wall is modified due to the presence of the other wall and when the finite-size effects become important. In order to answer this question we study here the behaviour of Ising films of the width $L = 200$ at the bulk critical temperature $T = T_c$ for various surface fields $H_1 = H_L$ ranging from $H_1 = 10^{-7}$ to 10. For simplicity the bulk field H is equal to zero. In this case exact results are available [16]. A selection of profiles for $L = 200$ is shown in Fig. 2.

Profiles obtained from exact and approximate methods show remarkable agreement. In order to discuss it in detail we present the errors defined as $(m_l^{DMRG} - m_l^{exTM})/m_l^{exTM}$ in Fig. 3.

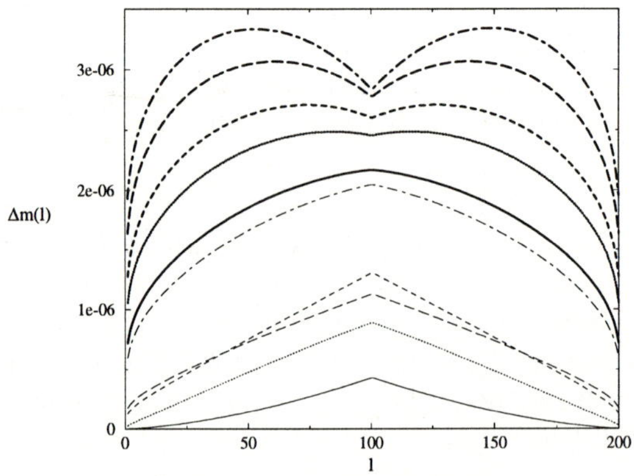

Fig. 3. Reduced errors of the DMRG profiles. The meaning of lines is as in Fig. 2.

Generally speaking the larger are fluctuations of spins, the larger are errors of magnetization. That is why, when we decrease a value of the surface field, which means we approach the bulk criticality, the accuracy decreases. When we perform the DMRG calculations, where $m = 32$ states are kept in each iteration, for surface fields higher than $H_1 > 0.1$ the error is smaller than 10^{-6}. One can see that when surface fields go down, the error increases.

Also it turns out that the accuracy of DMRG profiles depends on the distance from the surface in a way related to the fluctuations. For strong fields ($l_1 << 1$) the system is most strongly ordered close to the walls, that is why the error is the smallest there and has a maximum in the center of the strip. When the surface fields decrease, the region of highest order shifts to the middle of the film. The result is that now the profile has two maxima disposed symmetrically with respect to the center of the strip. When the surface fields decrease further, both maxima move towards the walls.

As far as the shapes of profiles in Fig. 2 are concerned we can distinguish three different regimes of H_1.

- For the strongest surface fields ($L/l_1 \gg 1$), H_1 between 10 and approximately 0.5, the profiles take the shape one would expect, i.e they monotonously decrease towards the center of the film.
- In the intermediate regime the length l_1 becomes macroscopic and, in agreement with the behaviour found for the semi-infinite systems, the maximum order is shifted away from the walls so that two symmetric maxima appear at $l \sim l_1$, each one in the vicinity of each wall. As H_1 is lowered further the maxima become flat and extended and move towards the center of the film. Eventually they merge into one for $H_1/J \approx 0.07$, where $l_1 \approx L$.
- For very weak surface fields ($L/l_1 \ll 1$) the profiles increase monotonously towards the center of the film but now the value of the magnetization both at the surface and in the middle of the film rapidly goes to zero with H_1.

In order to understand better an origin of the growth of the magnetization near the walls, we have to consider the influence of the surface magnetization (generated by H_1) on the next spins. First, it is proportional to correlations in planes parallel to the wall. In the surface they are suppressed (because of $H_1 \neq 0$ the surface is not critical), but in the vicinity of the wall they grow linearly with a distance l [14]. The maximum of the magnetization is observed, since for $l < l_1$ the increase of parallel correlations is not compensated by the decay of the perpendicular correlations.

More details will be presented elsewhere [9].

As a supplement, we consider the reduced errors for the free energy per spin Δf. When the surface field goes up the error increases: linearly $\approx H_1$ for small H_1 ($H_1 < 0.08$) and logarithmically $\approx \ln(H_1)$ for large H_1 ($H_1 > 0.2$). Since approximately the error of the partition function is $\Delta Z \approx \Delta f$ and as well ΔZ is proportional to the errors during truncation (proportional to the eigenvalues of truncated states), the behaviour of Δf gives also a certain information about the spectrum of the density matrix.

In conclusion we have presented two examples of the application of the DMRG to $2D$ Ising films in a strip geometry. We considered the effect of fluctuations near and at criticality, where the MF approach is not valid. That is why the DMRG results not only verify some predictions, but present also some qualitatively new phenomena.

We are very grateful to A. Maciołek for making available her exact results for $2D$ Ising strips with surface fields. We are indebted to I. Peschel, E. Carlon and A. Maciołek for critical reading of the manuscript.

References

1. R. Evans, J. Phys.: Cond. Matter. **2**, 8989 (1990) and references therein
2. S.R. White, Phys. Rev. Lett. **69**, 2863 (1992)

3. S.R. White, Phys. Rev. B **48**, 10345 (1993)
4. T. Nishino, J. Phys. Soc. Jpn. **64**, 3598 (1995)
5. E. Carlon and A. Drzewiński, Phys. Rev. Lett. **79**, 1591 (1997); Phys. Rev. E **57**, 2626 (1998)
6. E. Carlon, A. Drzewiński and J. Rogiers, Phys. Rev. B **58**, 5070 (1998)
7. E. Carlon and F. Iglói, Phys. Rev. B **57**, 7877 (1998)
8. A. Drzewiński, A. Ciach and A. Maciołek, Eur. Phys. J. B **5**, 825 (1998)
9. A. Maciołek, A. Ciach and A. Drzewiński, to appear in Phys. Rev. E
10. M.E. Fisher and P.G. de Gennes, C.R.Acad. Sci. Ser. B **287**, 207 (1978)
11. K. Binder, in Phase Transitions and Critical Phenomena, Vol. 8, Eds. C. Domb and J. L. Lebowitz, Academic Press (1983)
12. M. Thommes, G.H. Findenegg and H. Lewandowski and Ber. Bunsenges. Phys. Chem. **98**, 477 (1994)
13. A. Maciołek, A. Ciach and R. Evans, J. Chem. Phys. **108**, 9765 (1998)
14. U. Ritschel and P. Czerner, Phys. Rev. Lett. **77**, 3645 (1996)
15. H. Diehl, Phase Transitions and Critical Phenomena, Vol. 10, Eds. C. Domb and J.L. Lebowitz, Academic Press (1986)
16. A. Maciołek and J. Stecki, Phys. Rev. B **54**, 1128 (1996)

4 Thermodynamic Properties

4.1 One-Dimensional Kondo Lattices

Naokazu Shibata

Institute of Applied Physics, University of Tsukuba
Tsukuba 305-8573, Japan

The characteristic features of strongly correlated systems are clearly shown in their dynamic properties. The dynamic properties of interacting systems are temperature dependent and this dependence causes unique features. Among typical strongly correlated systems, the Kondo insulators have attracted much attention. In these systems an excitation gap develops at low temperatures and unusual dynamic properties appear. In the following we study the dynamic properties of the half-filled Kondo lattice (KL), which is a theoretical model of the Kondo insulators, using DMRG techniques.

1 Hamiltonian

The KL model consists of two different kinds of electrons, the conduction electrons that propagate in the lattice as itinerant electrons and the localized f-electron spins whose charge degrees of freedoms are completely suppressed. The antiferromagnetic exchange interaction couples the conduction spins and the localized spins, and it is the source of various interesting physical phenomena [1].

The Hamiltonian is given by

$$\mathcal{H} = -t\sum_i\sum_s(c^\dagger_{is}c_{i+1s} + \text{h.c.}) + J\sum_i\sum_{s,s'}\boldsymbol{S}_i\cdot\frac{1}{2}\boldsymbol{\sigma}_{ss'}c^\dagger_{is}c_{is'} \qquad (1)$$

where $\boldsymbol{\sigma} = (\sigma_x, \sigma_y, \sigma_z)$ are the Pauli matrices and $\boldsymbol{S}_i = \sum_{s,s'}\frac{1}{2}\boldsymbol{\sigma}_{ss'}f^\dagger_{is}f_{is'}$ is the f-electron spin at site i. The model has hopping $-t$ between the nearest neighbor pairs. At half-filling the number of conduction electrons is same as the number of localized f-electron spins.

2 Zero-Temperature Spin and Charge Gaps

For one dimension, it has been shown that the ground state is always insulating at half-filling [1–4]. This insulating state has a spin gap, whose size is smaller than that of the charge gap [5–7]. This is clear in the limit of strong exchange coupling, $J \to \infty$. In this limit the localized f-spins together with a conduction electron form a local singlet on every site. The spin excitations need a minimum energy of J, which is the energy difference between the local spin-singlet state and the local spin-triplet state. On the other hand, the creation of a charge excitations requires a minimum energy of $3J/2$ which corresponds to the energy cost for breaking two local singlets by transferring a conduction electron to a neighboring site.

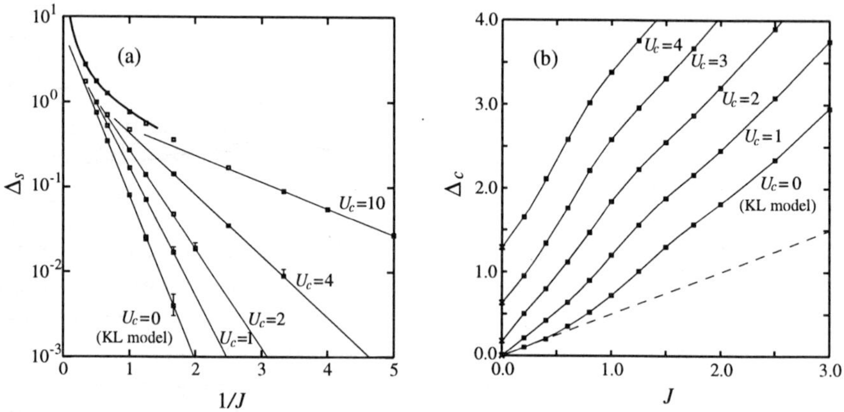

Fig. 1. **(a)** Spin gap, **(b)** Charge gap of the half-filled one-dimensional Kondo lattice model calculated by the zero-temperature DMRG [8]. U_c is the on-site Coulomb interaction between the conduction electrons, and $U_c = 0$ corresponds to the original KL model. The thick curve in **(a)** represents the result of the perturbation theory in terms of t/J for $U_c = 10t$. Results on the vertical axis in **(b)** are obtained from the exact solution by Lieb-Wu [9]. The typical truncation error in the DMRG calculation is 10^{-6} for $J = 1$. Error bars are estimated from L^{-1} and L^{-2} scalings. The gap energies, J and U_c are all in units of t.

These excitation gaps monotonically decrease with decreasing exchange coupling J. The spin gap becomes exponentially small [1–4] as shown in Fig. 1(a):

$$\Delta_s \propto \exp\left(-\frac{1}{\alpha\rho J}\right), \tag{2}$$

where α is the enhancement factor which is 1.4(1) [7] and ρ is the density of states of the conduction band at the Fermi level. This J-dependence is similar to that of the Kondo temperature of the single impurity Kondo model, $T_K \sim D \exp(-\frac{1}{\rho J})$, and it suggests that the spin gap is due to the singlet formation between the conduction electrons and the localized spins.

In contrast to the spin gap, the charge gap is linear in J in the limit of $J \to 0$ [6,7]. This is shown in Fig. 1(b). The linear dependence is explained as follows. In the weak-coupling regime the charge gap is much bigger than the spin gap [6,7]. It implies that the correlation length for the spin degrees of freedom is much longer than the charge correlation length. Thus it is justified to assume that the spin-spin correlation length is infinitely long for the discussion of the charge gap. Under this assumption, the charge gap is

$$\Delta_c = \frac{J}{2}. \tag{3}$$

3 Susceptibilities at Finite Temperatures

The charge gap determined at zero temperature is much bigger than the spin gap. The big charge gap originates from staggering internal magnetic fields induced by long correlation length of the f-spins. At finite temperatures, however, the f-spin correlations are subject to thermal fluctuations. When temperature becomes comparable to the spin gap, the spin correlation length gets smaller and the whole electronic properties including the charge excitation spectrum are reconstructed. Here we calculate spin and charge susceptibilities and see the interplay between the spin and charge excitations at finite temperatures.

In order to calculate thermodynamic quantities, we use the finite-temperature density-matrix renormalization group (finite-T DMRG) method [10–12]. In this method we iteratively increase the Trotter number of the quantum transfer matrix, and obtain its eigenvector for the largest eigenvalue with desired accuracy. The thermodynamic quantities are directly calculated from the largest eigenvalue and corresponding eigenvector, and the extrapolation in the system size is not needed. Since this method is free from statistical errors and the negative sign problem, it has advantages compared with the quantum Monte Carlo method.

The spin and charge susceptibilities are calculated from the derivatives of the free energy with respect to the external magnetic field or the chemical potential. The results are shown in Fig. 2 [13].

At low temperatures, spin susceptibility χ_s drops sharply with decreasing temperature for finite exchange coupling. This drastic decrease is due to the appearance of the low-energy scale for the spin sector [14,13]. The energy scale at low temperatures is estimated by fitting the susceptibility to an exponential form, and it is shown that the smaller one of the spin gap and the quasiparticle gap determines this energy scale. For small exchange coupling, the spin gap is smaller than the quasiparticle gap, and the low-temperature energy scale of χ_s is given by Δ_s.

The charge susceptibilities χ_c are shown in Fig. 2(b). At $J/t = 0$, χ_c in the limit of $T = 0$ is determined by the density of states of the conduction electrons. The slight increase in χ_c around $T \sim 0.6t$ is a characteristic feature of the one-dimensional system where the density of states diverges at the band edges. A finite value of J produces a sharp drop in χ_c at low temperatures. Similarly to χ_s, this drop is due to the appearance of the low-energy scale Δ_c

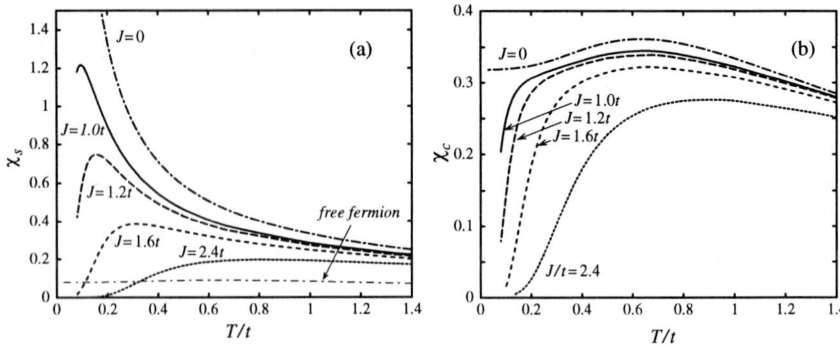

Fig. 2. **(a)** Spin susceptibility, **(b)** Charge susceptibility of the half-filled one-dimensional Kondo lattice model. The calculations were performed by the infinite-system algorithm keeping 40 states per block. The truncation errors in the calculations are typically 10^{-3} and at the lowest temperature 10^{-2} for the Trotter number $M = 50$.

for the charge sector. By fitting χ_c with an exponential form, the activation energy is obtained, which is equal to the quasiparticle gap $\Delta_{qp} = \Delta_c/2$ [13].

Although the quasiparticle gap determines the exponential temperature dependence at low temperatures, it is not the unique characteristic energy for χ_c. This is clearly shown for $J/t = 1.0$. A sharp decrease in χ_c is seen at $T \sim 0.1t$ which is much smaller than the quasiparticle gap $0.36t$ but rather close to the value of the spin gap $0.08t$. This fact suggests that the charge excitation spectrum is reconstructed when temperature is raised up to the spin gap. We will discuss this aspect in more detail in connection with the temperature dependence of various excitation spectra.

4 Calculation of Dynamic Quantities

The dynamic quantities are calculated from the correlation functions in the imaginary time direction [15,16], see also Chap.7(I). Since this calculation requires good accuracy for the eigenvectors of the quantum transfer matrix $\langle \Psi^L |$ and $| \Psi^R \rangle$, it is necessary to use the finite-system algorithm of the DMRG.

The Green's function in the β direction is calculated from the left and the right eigenvectors of the transfer matrix whose eigenvalue is the largest (See Fig. 3):

$$\begin{aligned} G(\tau_j) &\equiv -\mathrm{Tr}\{\mathrm{e}^{-\beta H} c_{i\sigma}(\tau_j)\, c_{i\sigma}^\dagger(0)\}/Z \\ &= -\langle \Psi^L | c_{i\sigma}(\tau_j)\, c_{i\sigma}^\dagger(0) | \Psi^R \rangle . \end{aligned} \tag{4}$$

Similarly the local dynamic correlation function $\chi_{AB}(\tau_j)$ is obtained as

$$\begin{aligned} \chi_{AB}(\tau_j) &\equiv \mathrm{Tr}\{\mathrm{e}^{-\beta H} A_i(\tau_j)\, B_i(0)\}/Z \\ &= \langle \Psi^L | A_i(\tau_j)\, B_i(0) | \Psi^R \rangle . \end{aligned} \tag{5}$$

By Fourier transformation, the Green's function and the dynamic correlation function as functions of the imaginary frequencies are calculated as

$$G(i\omega_n) = \frac{\beta}{M} \sum_j \mathrm{e}^{i\omega_n \tau_j} G(\tau_j), \tag{6}$$

$$\chi_{AB}(i\omega_n) = \frac{\beta}{M} \sum_j \mathrm{e}^{i\omega_n \tau_j} \chi_{AB}(\tau_j), \tag{7}$$

where ω_n is the Matsubara frequency that is $\pi(2n+1)/\beta$ for fermionic operators and $2\pi n/\beta$ for bosonic operators.

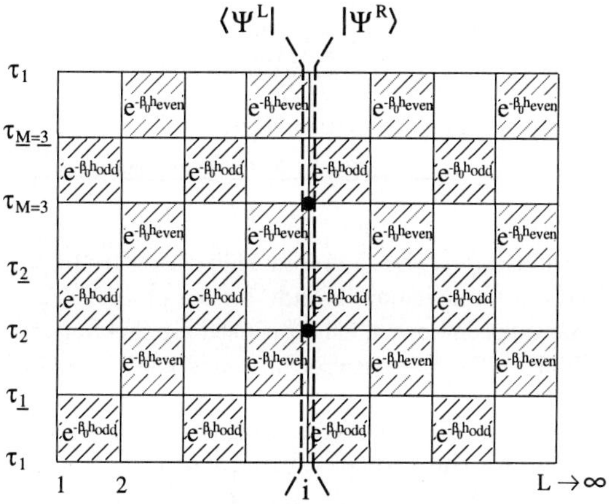

Fig. 3. Schematic diagram of the calculation of the imaginary time correlation function $\langle \Psi^L | c_{i\sigma}(\tau_3)\ c^\dagger_{i\sigma}(\tau_2) | \Psi^R \rangle$. The Trotter number of this example is $M = 3$. The quantum transfer matrix runs in the horizontal direction.

The real frequency Green's function and the dynamic susceptibility are obtained by analytic continuation to the real frequency axis. The standard methods for the analytic continuation are the Padé approximations and the maximum entropy method. The former method is based on the fitting of $G(i\omega_n)$ or $\chi_{AB}(i\omega_n)$ by rational functions of the frequency $i\omega_n$ which are analytically continued to the real axis by $i\omega_n \to \omega + i\delta$. The maximum entropy method is based on the spectral representations

$$G(\tau) = \int_{-\infty}^{\infty} \rho(\omega) \frac{\mathrm{e}^{-\tau\omega}}{1+\mathrm{e}^{-\beta\omega}} d\omega, \tag{8}$$

$$\chi_{AB}(\tau) = \int_{-\infty}^{\infty} \frac{1}{\pi} \mathrm{Im}\chi_{AB}(\omega) \frac{\mathrm{e}^{-\tau\omega}}{1-\mathrm{e}^{-\beta\omega}} d\omega, \tag{9}$$

with $\rho(\omega) = -\frac{1}{\pi}\mathrm{Im}G(\omega + i\delta)$ being the density of states. Starting from a flat spectrum, this method finally finds the optimal $\rho(\omega)$ and $\chi_{AB}(\omega)$ that reproduce $G(\tau)$ and $\chi_{AB}(\tau)$ best.

The dynamical structure factor $S_{AB}(\omega)$ is obtained from the imaginary part of $\chi_{AB}(\omega)$ through the fluctuation dissipation theorem,

$$\mathrm{Im}\chi_{AB}(\omega) = \pi(1 - \mathrm{e}^{-\beta\omega}) S_{AB}(\omega). \tag{10}$$

5 Dynamic Properties

The temperature dependence of the dynamic correlation functions obtained by the finite-T DMRG with the maximum entropy method are shown in Figs. 4 and 5 [15,16]. Fig. 4(a) shows the temperature dependence of the

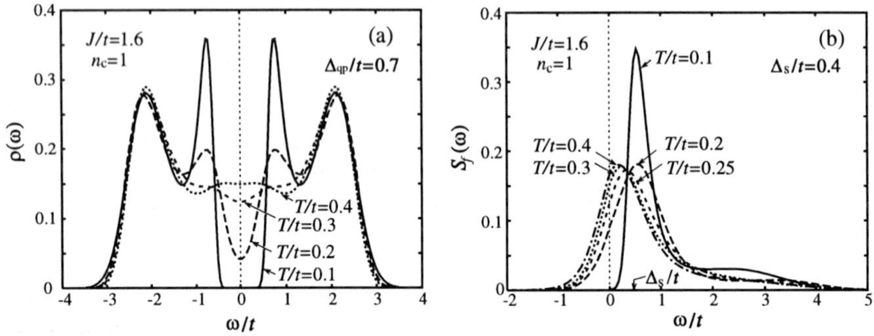

Fig. 4. (a) Quasiparticle density of states, $\rho(\omega)$, **(b)** Dynamic spin structure factor of the f-spins, $S_f(\omega)$, of the half-filled one-dimensional Kondo lattice model: $J/t = 1.6$, $\Delta_{qp}/t = 0.7$, $\Delta_s/t = 0.4$. The maximum entropy method is used for the analytic continuations. The Trotter number is $M = 60$, the number of states kept $m = 50$.

quasiparticle density of states, $\rho(\omega)$. At the lowest temperature, $T = 0.1t$, the existence of the quasiparticle gap, $\Delta_{qp} = \Delta_c/2 = 0.7t$, is clearly seen. There are sharp peaks at $\omega = \pm\Delta_{qp}$. These peaks suggest the formation of heavy-quasiparticle bands at the gap edges. As the temperature is increased, the sharp peaks at the threshold get broadened, and at temperatures $T \sim \Delta_s = 0.4t$, they disappear completely. This result shows that although the sharp peaks of the quasiparticle density of states are located at the frequency $\omega = \pm\Delta_{qp}$, they disappear around the temperature $T \sim \Delta_s$ which is much lower than $T \sim \Delta_{qp}$.

A significant change in the excitation spectrum at temperatures $T \sim \Delta_s$ is also shown in the f-spin dynamic structure factor $S_f(\omega)$, Fig. 4(b). At the lowest temperature a clear spin gap is seen with a sharp peak at the gap edge. As the temperature is increased, the peak structure at $\omega = \Delta_s$ becomes broad and the spectral intensity increases around the zero frequency $|\omega| < \Delta_s$. At the temperatures $T \sim \Delta_s$, the peak position of the spectrum shifts to zero frequency, and the peak height becomes almost temperature independent.

The spectral intensity $S_f(\omega)$ at zero frequency is related to the NMR relaxation rate $1/T_1$. The present results show that $1/T_1$ is nearly temperature independent at high temperatures and drastically decreases with decreasing the temperature below the characteristic temperature of the order Δ_s.

The dynamic spin structure factor for the conduction electrons $S_c(\omega)$, Fig. 5(a), also shows a clear change at temperatures $T \sim \Delta_s$. At low tem-

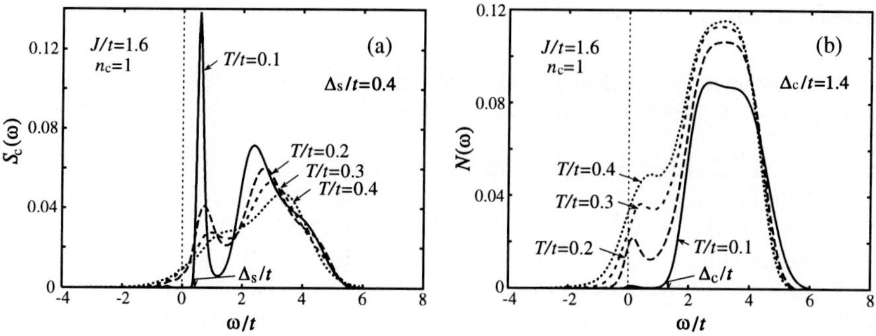

Fig. 5. **(a)** Dynamic spin structure factor of the conduction electrons, $S_c(\omega)$, **(b)** Dynamic charge structure factor of the conduction electrons, $N_c(\omega)$, of the half-filled one-dimensional Kondo lattice model: $J/t = 1.6$, $\Delta_{qp}/t = 0.7$, $\Delta_s/t = 0.4$.

peratures, $S_c(\omega)$ has two peaks. The peak in the low-frequency side is located at the energy of Δ_s similarly to $S_f(\omega)$. This peak corresponds to the spin excitations of the spin-singlet bound states of conduction electrons with f-spins to the spin-triplet states. The high-frequency peak, which corresponds to the spin excitations of quasiparticles, is located slightly above the charge gap. With increasing temperature both peaks lose their intensity, and above $T \sim \Delta_s$ the low-frequency peak disappears. The spectrum at the high-frequency side becomes similar to that of the dynamic charge structure factor $N(\omega)$.

The dynamic charge structure factor $N(\omega)$ is shown in Fig. 5(b). At the lowest temperature, $N(\omega)$ is very small below the frequency $\omega = \Delta_c$, and the spectral intensities sharply rise at $\omega = \Delta_c$. With increasing temperature, a peak grows at $\omega = 0$, but at $T \sim \Delta_s$, the peak structure is smeared and the gap structure in $N(\omega)$ becomes unclear.

The dynamic quantities studied by the finite-T DMRG have revealed the many-body nature of the gap formation in the Kondo insulators. Different kinds of excitation gaps are clearly visible in the dynamic quantities at low temperatures, and three low-energy scales corresponding to the spin gap, the quasiparticle gap, and the charge gap appear.These different excitation gaps are a characteristic feature of the Kondo insulators compared with the ordinary band insulators. Among the various energy scales the lowest one, the spin gap, plays a special role. At temperatures higher than the spin gap, the excitation spectra in the charge sector are also modified strongly. Thus the whole excitation spectrum is reconstructed above the temperature corresponding to the lowest energy scale.

The author would like to acknowledge the collaborations with Kazuo Ueda, Tetsuya Mutou, and Manfred Sigrist. This work was financially supported by the Inoue Foundation for Science.

References

1. H. Tsunetsugu, M. Sigrist and K. Ueda, Rev. Mod. Phys. **69**, 809 (1997)
2. H. Tsunetsugu, Y. Hatsugai, K. Ueda and M. Sigrist, Phys. Rev. B **46**, 3175 (1992)
3. A.M. Tsvelik, Phys. Rev. Lett. **72**, 1048 (1994)
4. S. Fujimoto and N. Kawakami, J. Phys. Soc. Jpn. **66**, 2157 (1997)
5. C.C. Yu, and S.R. White, Phys. Rev. Lett. **71**, 3866 (1993)
6. T. Nishino and K. Ueda, Phys. Rev. B **47**, 12451 (1993)
7. N. Shibata, T. Nishino, K. Ueda, and C. Ishii, Phys. Rev. B **53**, R8828 (1996)
8. S.R. White, Phys. Rev. Lett. **69**, 2863 (1992); Phys. Rev. B **48**, 10345 (1993)
9. E.H. Lieb and F.Y. Wu, Phys. Rev. Lett. **20**, 1445 (1968)
10. R.J. Bursill, T. Xiang and G.A. Gehring, J. Phys.: Cond. Matter. **8**, L583 (1996)
11. X. Wang and T. Xiang, Phys. Rev. B **56**, 5061 (1997)
12. N. Shibata, J. Phys. Soc. Jpn, **66**, 2221 (1997)
13. N. Shibata, B. Ammon, M. Troyer, M. Sigrist and K. Ueda, J. Phys. Soc. Jpn **67**, 1086 (1998)
14. R.M. Fye, and D.J. Scalapino, Phys. Rev. Lett. **65**, 3177 (1990); Phys. Rev. B **44**, 7486 (1991)
15. T. Mutou, N. Shibata and K. Ueda, Phys. Rev. Lett. **81** 4939 (1998)
16. N. Shibata and K. Ueda, J. Phys.: Cond. Matt. **11**, R1 (1999)

4.2 Impurities in Spin Chains

Stefan Rommer** and Sebastian Eggert

Institute for Theoretical Physics, Chalmers University of Technology
SE - 412 96 Göteborg, Sweden

The problem of impurities in quantum systems has been extensively studied in recent years. The Kondo problem [1] as well as impurities in quantum wires [2] and spin chains [3] are examples of systems that have attracted great interest. In the following we will be concerned with impurities in the spin-1/2 Heisenberg chain. We will discuss how the transfer-matrix DMRG method can be used to calculate thermodynamic impurity properties and we will show results for two impurity configurations: two weak links in a periodic chain and an external impurity spin coupled to one spin in a periodic chain. The numerical results presented agree well with predictions from field theory calculations.

Other numerical methods have been used to study similar systems. The finite-size scaling of energy gaps of systems with different impurity configurations has been studied using exact diagonalization [3–5] and zero-temperature DMRG [5,6] techniques. Local properties at finite temperature have been investigated by quantum Monte Carlo techniques [7–9]. Also a finite-temperature DMRG method, first introduced for uniform systems by Moukouri and Caron [10] (see Chap. 4.4(II)), has been applied to impurity systems by Zhang *et al.* [8]. In this method the low-lying energy spectrum is estimated and thermodynamic properties are calculated directly from the resulting energy eigenvalues.

The transfer-matrix DMRG was first introduced by Nishino to calculate the partition function of a two-dimensional classical system [11] (see Chap. 5(I)). Bursill, Xiang and Gehring have then shown that the same idea can be used to calculate thermodynamic properties of the quantum spin-1/2 XY-chain [12]. The method has later been improved by Wang and Xiang [13] (see Chap. 6(I)) as well as Shibata [14] (see Chap. 4.1(II)) and been applied to the anisotropic spin-1/2 Heisenberg chain.

The transfer-matrix DMRG method directly determines the partition function in the thermodynamic limit and, as we will see, we can directly retrieve impurity contributions to extensive thermodynamic properties like the free energy (i.e. that contribution that is independent of the system size). Methods like quantum Monte Carlo and the finite-temperature DMRG by Moukouri and Caron on the other hand require an extensive finite-size analysis to reach the thermodynamic limit and to extract impurity contributions.

** Present address: Department of Physics and Astronomy, University of California, Irvine, CA 92697-4575

Much effort has also been devoted to analytical studies of quantum impurity problems. In quantum field theory studies it is found that at zero-temperature impurities in spin chains and several other one-dimensional quantum systems effectively behave as boundary conditions and that the low-temperature behavior can be described by a crossover between fixed points [1–3,6]. In this work we are able to confirm this crossover behavior for the thermodynamic properties in the spin-1/2 chain.

In Sect. 1 we discuss how the transfer-matrix DMRG method can be used to calculate impurity properties. We also present the impurity configurations considered. In Sect. 2 we give a brief review of predictions from field theory analysis of the impurity systems. In Sect. 3 we show numerical results for two different impurity models in the spin-1/2 chain and are able to confirm the predictions from field theory calculations. Section 4 concludes with a discussion of the results and an analysis of accuracy.

1 The Method

The method of the transfer-matrix DMRG is treated in Chaps. 5(I), 6(I) in this book and is also explained in detail in [12–14]. Before discussing impurity systems we will therefore only give a very brief review of parts of the method that are of relevance for the impurity problems considered in this text.

Let us consider the Heisenberg spin-1/2 chain

$$H = \sum_{x=1}^{N} J_x \, \mathbf{S}_x \cdot \mathbf{S}_{x+1} + B_x S_x^z \,, \tag{1}$$

where J_x is the exchange coupling between sites x and $x+1$, and B_x is an external magnetic field in the z-direction at site x. Periodic boundary conditions, $\mathbf{S}_{N+1} \equiv \mathbf{S}_1$, are assumed. The partition function is defined by $Z = \mathrm{tr}\left(e^{-\beta H}\right)$, where $\beta = 1/k_B T$. Using the Trotter-Suzuki decomposition [15], it can approximately be written as a trace over a product of transfer matrices

$$Z_M = \mathrm{tr} \prod_{x=1}^{N/2} T_M(2x-1), \tag{2}$$

where M is the Trotter number and $T_M(2x-1)$ is the $(2^{2M} \times 2^{2M})$-dimensional quantum transfer matrix from lattice site $2x-1$ to site $2x+1$. For a uniform system, the transfer matrix is independent of lattice site, $T_M(2x-1) \equiv T_M$, and the partition function is then simply given by $Z_M = \mathrm{tr}\; T_M^{N/2}$. In the thermodynamic limit of the uniform system we have $\lim_{N\to\infty} Z_M = \lambda^{N/2}$, where λ is the largest eigenvalue of T_M. The transfer-matrix DMRG technique can now be applied to find the largest eigenvalue λ.

Since the reduced density matrix, ρ, used in the transfer-matrix DMRG in general is non-symmetric [13,14] it is not obvious that the eigenvalues are real, but because ρ represents a density matrix we expect it to be positive definite. However, because of numerical inaccuracies complex eigenvalues tend to appear in the course of iterations. This is usually connected to level crossings in the eigenvalue spectrum of the density matrix as M is increased. In addition, we also observe that just before the complex eigenvalues appear, multiplet symmetries in the eigenvalue spectrum of ρ may split up (often due to "level repulsion" from lower-lying states which were previously neglected). To overcome this problem it is important to keep all original symmetries (i.e. eigenvalue multiplets of ρ) even as M increases. For that purpose we found it necessary to increase the number of states m kept per block just before a multiplet tends to split up, which avoids the numerical error that leads to the symmetry breaking. In case of a level crossing of two multiplets that do not split up, complex eigenvalues may still appear and it is then possible to numerically transform the complex eigenstate pair into a real pair spanning the same space.

Let us now assume that the system has a single impurity. We will study two systems; a single spin-1/2 weakly coupled to the ends of an open spin-1/2 chain and an external spin-1/2 coupled to one spin in a periodic chain,

$$H_1 = J \sum_{x=1}^{N-1} \mathbf{S}_x \cdot \mathbf{S}_{x+1} + J' \, (\mathbf{S}_N + \mathbf{S}_1) \cdot \mathbf{S}_0 \tag{3}$$

$$H_2 = J \sum_{x=1}^{N} \mathbf{S}_x \cdot \mathbf{S}_{x+1} + J' \, \mathbf{S}_1 \cdot \mathbf{S}_f \, , \tag{4}$$

where $\mathbf{S}_f$ is the spin operator of an external spin-1/2. The models are depicted in Fig. 1.

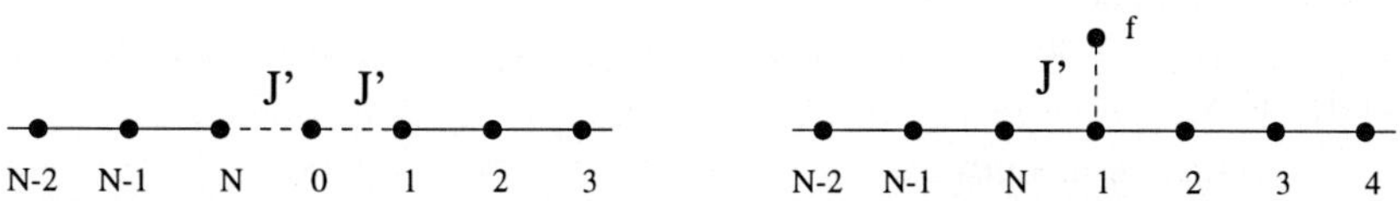

Fig. 1. The impurity configurations considered in (3) and (4). The solid lines represent the couplings J of the pure chain.

For systems with such a local impurity, which is contained within two neighboring links, only one of the $T_M(x)$ in (2) will differ from all other T_M and we get

$$Z_M = \mathrm{tr} \left(T_M^{N/2-1} T_{\mathrm{imp}} \right) , \tag{5}$$

where T_{imp} is the transfer matrix of the two links containing the impurity and T_M is the transfer matrix describing the bulk. In the thermodynamic limit the partition function will still be dominated by the largest eigenvalue of the "pure" transfer matrix. From (5) we have

$$\lim_{N\to\infty} Z_M = \lambda^{N/2-1}\langle\psi^L|T_{\text{imp}}|\psi^R\rangle, \tag{6}$$

where $\langle\psi^L|$ and $|\psi^R\rangle$ are the left and right eigenvectors respectively of the *pure* transfer matrix T_M corresponding to the largest eigenvalue λ. The generalization of (5) and (6) to impurity configurations ranging over more than two links is straightforward.

Let us define $\lambda_{\text{imp}} \equiv \langle\psi^L|T_{\text{imp}}|\psi^R\rangle$. The total free energy of the system given by $F_{\text{total}} = -\ln(Z)/\beta$ is then

$$F_{\text{total}} = -\frac{1}{\beta}\ln\left(\lambda^{N/2-1}\lambda_{\text{imp}}\right) . \tag{7}$$

By the impurity correction F_{imp} to the free energy we mean that part of the total free energy F_{total} that does not scale with system size N,

$$F_{\text{total}} = NF_{\text{pure}} + F_{\text{imp}} + \mathcal{O}(1/N), \tag{8}$$

where F_{pure} is the free energy per site for an infinite system without impurities. By comparing (8) with (7) we can retrieve the pure and impure parts

$$F_{\text{pure}} = -\frac{1}{2\beta}\ln\lambda, \quad F_{\text{imp}} = -\frac{1}{\beta}\ln\frac{\lambda_{\text{imp}}}{\lambda}. \tag{9}$$

The impurity susceptibility can be found from the change of F_{imp} in a small magnetic field by

$$\chi_{\text{imp}} = -\frac{\partial^2}{\partial B^2}F_{\text{imp}}. \tag{10}$$

Due to the difficulty to accurately calculate a second derivative numerically we will instead use an approximation to estimate χ_{imp}.

Local properties such as the magnetization of the impurity spin can be determined by

$$\langle S^z_{\text{imp}}\rangle = \frac{\langle\psi^L|T^{sz}_{\text{imp}}|\psi^R\rangle}{\lambda_{\text{imp}}}, \tag{11}$$

where T^{sz}_{imp} is defined similar to T_{imp} but with the additional operator S^z included in addition to the Boltzmann weights [13]. The magnetization of spins close to the impurity is readily obtained by

$$\langle S^z\rangle = \frac{\langle\psi^L|T^{sz}_M(T_M)^x T_{\text{imp}}|\psi^R\rangle}{\lambda^{x+1}\,\lambda_{\text{imp}}}, \tag{12}$$

where $2x$ is the number of sites between the impurity and the spin of interest. Note that since a transfer matrix involves a total of three lattice sites, T^{sz}_M can be constructed to measure the spin at any of these sites (or the mean value).

In our calculations for the impurity systems we have used the same reduced density matrix as for the *pure* system. From a computational point of view, this method is very convenient; by storing all target states and projection operators from the DMRG run for the *pure* system, all local impurities can be studied by simply using the same projection operators and target states. This makes subsequent DMRG runs for different impurity parameters very fast. This choice of density matrix is not the only one possible [16] but it is a very convenient choice and we found it to give good results.

2 Field Theory Predictions

The effective low-energy spectrum of the spin-1/2 chain is well described by a free boson Hamiltonian density, where we can introduce the impurities in (3) and (4) in a straightforward way as perturbations. The field theoretical expressions for these perturbations can then be analyzed in terms of their leading scaling dimensions. Local perturbations can be classified as irrelevant, marginal or relevant depending on their leading scaling dimension. Relevant and in some cases also marginal perturbations drive the system to a different fixed point. Hence, we can predict a systematic renormalization flow towards or away from the corresponding fixed point, respectively.

Such an analysis has been made in [3] and the renormalization flows have been confirmed by determining the finite-size corrections to the low-energy spectrum [3]. In particular, an impurity spin with a weak antiferromagnetic coupling $0 < J' \ll J$ to the ends of an open chain, as in (3), has been found to be a logarithmically relevant perturbation. The open chain ($J' = 0$) is therefore an unstable fixed point in the antiferromagnetic direction [3]. The periodic chain ($J' = J$) on the other hand is a stable fixed point. Hence we expect a renormalization flow between the two fixed points as the temperature is lowered, and the temperature dependence of the impurity susceptibility as well as local properties will be described by a crossover function. Below a certain crossover temperature T_K this crossover function describes the behavior of the stable fixed point (the periodic chain) while above T_K the system may exhibit a completely different behavior. The crossover temperature T_K itself is determined by the initial coupling strength J'. In particular we find $T_K \to 0$ as $J' \to 0$ while $T_K \to \infty$ as $J' \to J$. In other words, close to the unstable fixed point the crossover temperature is very small, indicating that we have to go to extremely low temperatures before we can expect to observe the behavior of the stable fixed point.

A different scenario holds for the impurity model with one external spin $\mathbf{S}_f$ coupled to a single spin in a periodic chain as in (4). In this case, the periodic chain ($J' = 0$) is the unstable fixed point, while the open chain with a decoupled singlet ($J' \to \infty$) is the stable fixed point [3,6].

For more details on the field theory treatment of impurities in the spin-1/2 chain see [3].

3 Results

Let us first consider the two-weak-link case as defined by the Hamiltonian in (3). While our method is in principle capable of extracting the impurity susceptibility in (10) directly, we were able to obtain results at much lower temperatures by considering the linear response of the impurity spin to a uniform field B

$$\chi_{\mathrm{imp}} \approx \left.\frac{d\langle S_0^z\rangle}{dB}\right|_{B=0} - \chi_{\mathrm{pure}} \,, \tag{13}$$

which gives a good indication of the true impurity susceptibility. Here χ_{pure} denotes the susceptibility per site far away from the impurity at $B = 0$.

It is known that the impurity susceptibility is logarithmically divergent below T_K while it shows a Curie-law behavior above [1]. Interestingly, this cross-over shows a universal data collapse, because changing the coupling strength (i.e. T_K) is equivalent to rescaling the temperature, i.e. there is only one independent variable T/T_K

$$\chi_{\mathrm{imp}}(T) = f(T/T_K)/T_K, \tag{14}$$

where $f(x)$ is a universal function (ignoring higher order operators). This data collapse is clearly seen in Fig. 2 with an appropriate choice of T_K as a function of J' (inset), showing the predicted logarithmic scaling at low T.

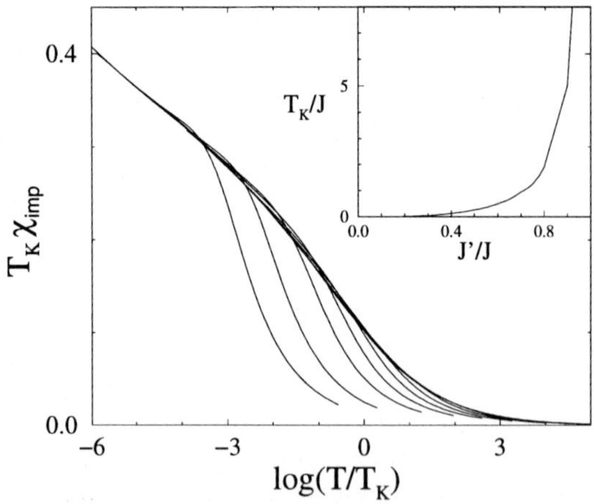

Fig. 2. The scaled impurity susceptibility $T_K\chi_{\mathrm{imp}}$ from left to right $J' = 0.1J$, ..., $0.95J$. The T_K values are determined from the inset. The non-universal deviations of some of the curves at highest T/T_K are due to the fact that in those cases T_K was so large that regions above the cutoff below which scale invariance is valid were included.

We now consider the local susceptibility (Knight shift) over a large range around the impurity as a function of site index x

$$\chi_{\mathrm{local}}(x) = \left.\frac{d\langle S_x^z\rangle}{dB}\right|_{B=0} , \tag{15}$$

where B is a uniform magnetic field on the entire chain. It is known from the field theory analysis that the local susceptibility can to a good approximation be written as a sum of an uniform part χ_{pure} independent of x and a staggered part $(-1)^x\chi_{\mathrm{alt}}(x)$, [7]

$$\chi_{\mathrm{local}}(x) \approx \chi_{\mathrm{pure}} + (-1)^x\chi_{\mathrm{alt}}(x) . \tag{16}$$

Field theory analysis further predicts that open boundary conditions induce a staggering part which first *increases* with the distance from the boundary and then decreases with larger x [7]

$$\chi_{\mathrm{alt}}(x) \propto \frac{x\sqrt{T}}{\sqrt{\sinh 4xT}} , \tag{17}$$

where T is measured in units of J and x is measured from the boundary. This surprising effect has been confirmed by NMR experiments [17].

For our particular case with an impurity spin symmetrically coupled to the ends of an open chain field theory predicts also another staggered component which decreases with distance from the boundary [18]

$$\chi_{\mathrm{alt}}(x) \propto \log[\tanh(xT)] . \tag{18}$$

Clearly, for a small coupling J' and at larger temperatures we expect the open chain behavior (17), but as the temperature is lowered (or the coupling is increased) the alternating part in (18) will start to dominate, which has the opposite sign [18]. This competition between the two contributions is clearly seen in Fig. 3. Moreover, below T_K the total staggered contribution always fits extremely well to a superposition

$$\chi_{\mathrm{alt}}(x) = c_1 \log[\tanh(xT)] + c_2\frac{x\sqrt{T}}{\sqrt{\sinh 4xT}} . \tag{19}$$

The coefficients c_1 and c_2 are determined by fitting (19) to the transfer-matrix DMRG results for all couplings $J' \geq 0.2J$. The result is plotted in Fig. 3.

Let us now consider the model with one external spin (4). This model has been of great interest in a number of studies [3–6,8,9]. In [3] it was first shown that the stable fixed point corresponds to open boundary conditions with a decoupled singlet which has also been confirmed numerically [5]. The local response of the impurity spin to a local magnetic field B_f acting only at the impurity site,

$$\chi_f = \left.\frac{d\langle S_f^z\rangle}{dB_f}\right|_{B_f=0} , \tag{20}$$

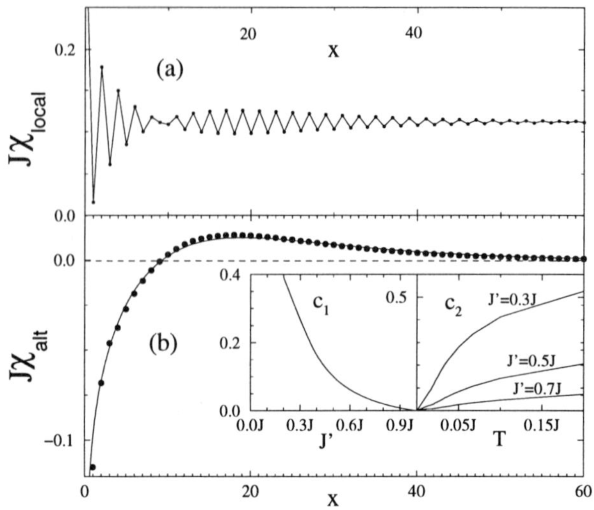

Fig. 3. (**a**) Local susceptibility as a function of the distance from the impurity, x, for $T = 0.05J$ and $J' = 0.7J$. (**b**) Fit of the alternating amplitude to (19) (solid line) with the appropriate coefficients (inset).

is found to be a constant as $T \to 0$ with a linear term: $\chi_f(T) \stackrel{T\to 0}{\longrightarrow}$ const. + $\mathcal{O}(T)$ [6,16]. Figure 4 shows χ_f for several coupling strengths J'. Our data fits well to the predicted form. Moreover, we find a scaling behavior which holds for all coupling strengths

$$\chi_f(T) = g(T/T_K)/T_K. \tag{21}$$

A similar scaling relation was observed before [9]. Our results are consistent with previous numerical studies [8], but distinctly larger in the low-temperature region. We attribute this to the method used in [8], which becomes unreliable when the temperature falls below the finite-size gap of the system. At large J' our findings are completely consistent with the expectation that the impurity spin is locked into a singlet at the stable fixed point.

4 Discussion

To accurately calculate impurity properties we have to determine not only the largest eigenvalue of the transfer matrix, T_M, but also the corresponding eigenvectors to high accuracy. To estimate the error of the impurity properties is difficult. Errors come both from the finite Trotter number, M, and from the finite number of states, m, in the DMRG. The scaling of pure properties with M and m usually turn out to be simpler than the scaling of impurity properties, which show a less clear form of the errors.

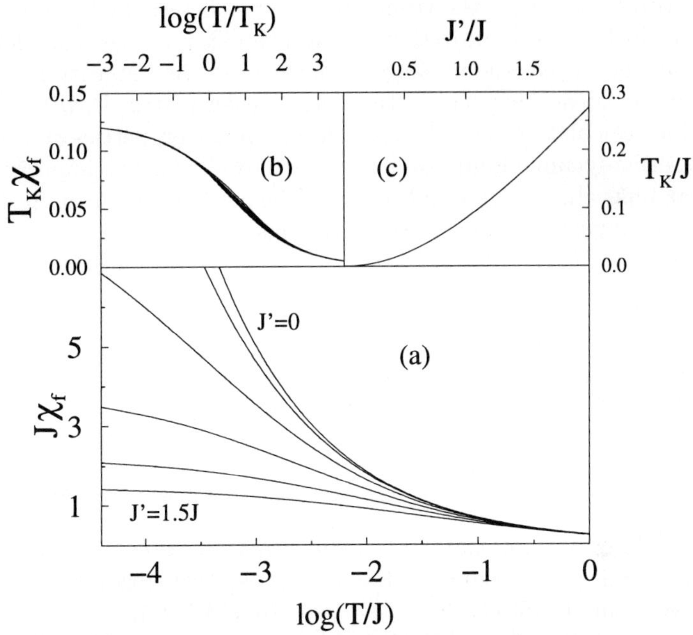

Fig. 4. Figure (**a**) shows the local response of the impurity spin χ_f for different coupling strengths $J'/J = 0, 0.3, 0.6, 0.9, 1.2, 1.5$. Figure (**b**) shows the data collapse according to (21). In Fig. (**c**) the corresponding crossover temperature T_K is shown, which is expected to be $T_K \propto J'^2$ for $J' \ll J$ and $T_K \propto J'$ for $J' \gg J$.

We have used the value $\beta/M = 0.05$ for all calculations presented in this article. With this value, the error due to the finite M should be small, which is also confirmed by test-runs.

We have tested the error due to the finite Trotter number M by doing separate DMRG runs for different values of β/M. For the pure case at moderate temperatures we find that the eigenvalue λ scale as $1/M^2$, as is expected, while at the lowest temperatures, the error due to the finite m is larger making it difficult to see the expected $1/M^2$ scaling. For the impure case, the convergence of λ_{imp} with M is more complicated, but the overall scaling is still roughly as $1/M^2$.

We have also tested the convergence with the number of basis states m. For both the pure and the impure cases we find a rapid convergence with increasing m. We have used a maximum of $m = 65$ for the calculations on the two weak link impurity and $m = 38$ in the external spin case. We found however no noticeable difference between $m = 38$ and $m = 65$ down to $T = 0.02$ in a test run for the external spin.

The truncation error, i.e. $1 - \sum_{i=1}^{m} w_i$, where w_i are the largest eigenvalues of the density matrix, is less than about 10^{-5} for $m = 65$ at the lowest temperatures. Note that the truncation error is determined during the pure sweep in which also the projection operators and target states are determined and is thus not a realistic measure of the error of the impurity properties.

In summary we have shown that the transfer-matrix DMRG is a useful method for calculating finite-temperature *impurity* properties of a spin chain in the thermodynamic limit. We have shown results for two impurity models: two weak links and one external spin, but the method can be applied to other impurity configurations and electron systems.

References

1. For a summary see I. Affleck, Acta Physica Polonica B **26**, 1869 (1995); preprint cond-mat/9512099
2. C.L. Kane and M.P.A. Fisher, Phys. Rev. B **46**, 7268 (1992)
3. S. Eggert and I. Affleck, Phys. Rev. B **46**, 10866 (1992)
4. J. Igarashi, T. Tonegawa, M. Kaburagi and P. Fulde, Phys. Rev. B **51**, 5814 (1995)
5. W. Zhang, J. Igarashi and P. Fulde, Phys. Rev. B **56**, 654 (1997)
6. A. Furusaki and T. Hikihara, Phys. Rev. B **58**, 5529 (1998)
7. S. Eggert and I. Affleck, Phys. Rev. Lett. **75**, 934 (1995)
8. W. Zhang, J. Igarashi and P. Fulde, J. Phys. Soc. Japan **66**, 1912 (1997)
9. W. Zhang, J. Igarashi and P. Fulde, Phys. Rev. B **54**, 15171 (1996)
10. S. Moukouri and L.G. Caron, Phys. Rev. Lett., **77**, 4640 (1996)
11. T. Nishino, J. Phys. Soc. Japan **64**, 3598 (1995)
12. R.J. Bursill, T. Xiang and G.A. Gehring, J. Phys C **8**, L583 (1996)
13. X. Wang and T. Xiang, Phys. Rev. B **56**, 5061 (1996)
14. N. Shibata, J. Phys. Soc. Jpn **66**, 2221 (1997)
15. H.F. Trotter, Proc. Am. Math. Soc. **10**, 545 (1959); M. Suzuki, Prog. Theor. Phys. **56**, 1454 (1976)
16. S. Rommer and S. Eggert, Phys. Rev. B **59**, 6301 (1999)
17. M. Takigawa, N. Motoyama, H. Eisaki and S. Uchida, Phys. Rev. B **55**, 14129 (1997); Phys. Rev. Lett. **76**, 4612 (1996)
18. S. Eggert and S. Rommer, Phys. Rev. Lett. **81**, 1690 (1998)

4.3 Thermodynamics of Ferrimagnets

Ulrich Schollwöck

Sektion Physik, Ludwig-Maximilians-Universität München
Theresienstr. 37, D-80333 München, Germany

Ferrimagnetic spin chains are antiferromagnetic quantum spin chains with different spin lengths S, s on sublattices A and B, such that the unit cell comprises two spins. For a chain of N unit cells, the Hamiltonian of a (S, s) ferrimagnet reads

$$\mathcal{H} = J \sum_{j=1}^{N} (\boldsymbol{S}_j \cdot \boldsymbol{s}_j + \boldsymbol{s}_j \cdot \boldsymbol{S}_{j+1}) - g\mu_{\mathrm{B}} H \sum_{j=1}^{N} (S_j^z + s_j^z) \quad (J > 0). \tag{1}$$

All numerical results are for the generic example $(1, \frac{1}{2})$. This ferrimagnet should show quantum effects most clearly. We will set $g\mu_B = J = k_B = 1$.

Ferrimagnetic spin chains are characterized at $T = 0$ by an intricate interplay of classical ferromagnetic (FM) order and antiferromagnetic (AFM) quantum fluctuations. At $T > 0$, this has dramatic, observable consequences for thermodynamic quantities. For ferrimagnets, these can be investigated in the framework of spin wave theory despite the low dimensionality. On the other hand, quantitative results need numerical techniques. Spin wave theory indicates that the most interesting physical behavior should occur at very low temperatures, which quantum Monte Carlo cannot access and where exact diagonalisation is hampered by the presence of very long-ranged fluctuations. The transfer-matrix DMRG ([1–3], Chap. 6(I)) will be shown to overcome both limitations, making it the best tool for studying these chains.

Let us first develop a naive classical picture at $T = 0$. We expect Néel order with macroscopic spin $N(S - s)$. If we consider quantum fluctuations, the Lieb-Mattis theorem [4] states that the ground state will still be a spin of length $N(S - s)$, implying macroscopic magnetization. This FM order is essentially classical and provides a starting point for spin wave analysis. On the other hand, the FM order is formed by pairs of big and small spins, which can be broken by quantum fluctuations. In fact, as the unit cell has two spins, we find two excitation branches. One decreases, the other increases the magnetisation. The former are FM excitations with a gapless dispersion relation: Goldstone modes are present due to the broken symmetry. The latter are short-ranged gapped AFM quantum fluctuations.

Since the mid-eighties, various ferrimagnetic chains have been synthesized and thermodynamic measurements been carried out [5]. The first samples were $(\frac{5}{2}, \frac{1}{2})$ ferrimagnets with Mn carrying spin-$\frac{5}{2}$ and Cu carrying spin-$\frac{1}{2}$. The big spin makes them almost classical and less interesting than e.g. the

$(1, \frac{1}{2})$ ferrimagnet [6] NiCu(pba)$(D_2O)_3 \cdot D_2O$ (pba = 1,3 propylenebis (oxamato)), where Ni and Cu carry big and small spin. Experimental comparisons here will be for this compound.

1 Zero-Temperature Spin Wave Theory

Spin wave theory works if magnetic order is present and merely renormalized by quantum fluctuations, so it should work here [7]. Starting from the classical ground state for $H = 0$ with total magnetisation $N(S-s)$, one expands (1) in powers of the bosonic Holstein-Primakoff operators. Up to quadratic order, one can diagonalise (1) using Fourier and Bogoliubov transformations and finds two excitation branches and a quantum correction to the ground-state energy. The two branches are given by

$$\omega_k^{\mp} = \sqrt{(S-s)^2 + 4Ss\sin^2(k/2)} \mp (S-s), \tag{2}$$

with a gapless quadratic FM branch to magnetisation sector $N(S-s) - 1$ with

$$\omega_k^{-} = \frac{Ssk^2}{2(S-s)} \tag{3}$$

for $k \to 0$ and a gapped (gap $\Delta = 2(S-s)$) AFM branch to magnetisation sector $N(S-s) + 1$

$$\omega_k^{+} = 2(S-s) + \frac{Ssk^2}{2(S-s)} \tag{4}$$

for $k \to 0$. The quartic term, if treated in a mean-field approximation, yields further quantum corrections to ground-state energy and excitations. While the FM branch is hardly modified, the gap of the AFM branch is greatly increased (Fig. 1).

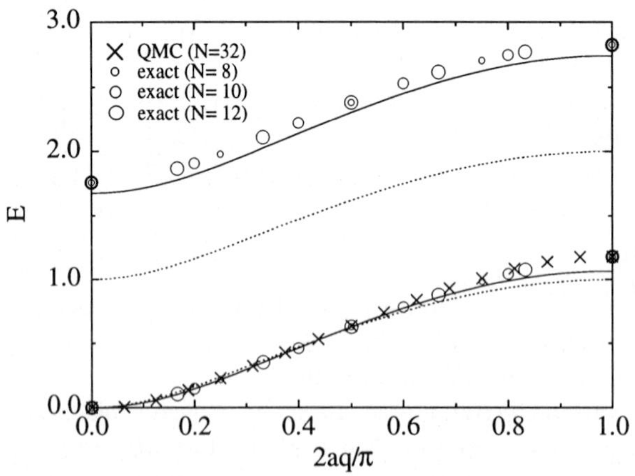

Fig. 1. AFM (top) and FM (bottom) dispersion relations. Dashed line: quadratic spin wave theory; solid line: quartic spin wave theory; compared to QMC and exact diagonalization of finite chains.

Numerical studies show that quartic spin wave theory is quite accurate: $\Delta_{SWT} = 1.676$ vs. $\Delta_{num} = 1.759$ [7]. The AFM quantum fluctuations are very short ranged, the decay length being of the order of the unit cell size.

2 Thermodynamics in Zero Field

Qualitative analysis

The presence of low-energy FM excitations and rather high-energy AFM excitations leads one to expect that low-T properties are governed by FM excitations, medium-T properties by AFM excitations, while at high T the chains are essentially paramagnetic. The specific heat should thus go as $C \propto \sqrt{T}$ for $T \to 0$, while showing a Schottky peak for $T \sim \Delta$ and decaying as $C \propto T^{-2}$ for $T \to \infty$. For the magnetic susceptibility, it is preferable to consider χT, a constant in a Curie paramagnet. Antiferromagnetic interactions suppress susceptibility at low T, so χT increases monotonically with temperature. Similarly, it decreases monotonically with T for ferromagnets. The presence of ferromagnetic excitations lets us expect $\chi T \propto T^{-1}$ for $T \to 0$ [8], then a crossover from decrease to increase in χT and finally, from Curie's law, $\chi T \to (S(S+1) + s(s+1))/3 = 11/12$ for $T \to \infty$.

From the spin wave theory dispersion relations, one can build both internal energy and entropy. This gives only the leading order of C, and is useless for χ, as there is no constraint on the number of bosons. In ferromagnets, Takahashi [9] has introduced a zero-magnetisation constraint, $\langle S^z + s^z \rangle = 0$, which introduces an effective chemical potential and limits the number of bosons. One finds as specific heat per unit cell [10]

$$C = \frac{3}{4}\left(\frac{S-s}{Ss}\right)^{\frac{1}{2}} \frac{\zeta(\frac{3}{2})}{\sqrt{2\pi}} \widetilde{t}^{\frac{1}{2}} - \frac{1}{Ss}\widetilde{t} + O(\widetilde{t}^{\frac{3}{2}}) + O(\widetilde{t}^{2}) \tag{5}$$

and as susceptibility per unit cell

$$\chi = \frac{Ss(S-s)^2}{3}\widetilde{t}^{-2} - (Ss)^{\frac{1}{2}}(S-s)^{\frac{3}{2}} \frac{\zeta(\frac{1}{2})}{\sqrt{2\pi}} \widetilde{t}^{-\frac{3}{2}} + O(\widetilde{t}^{-1}) \tag{6}$$

where $\widetilde{t} = T/\gamma$. These are exactly the results for a ferromagnet with a quantum renormalization factor $\gamma \approx 0.7 < 1$, implying that in the ferrimagnet all features are shifted to lower temperatures. This captures however only low-T features. At sufficiently high temperatures, this constraint is not sensitive to non-physical breaking of unit cell Néel order. To catch this feature, the right approach is a zero *staggered* magnetisation constraint,

$$\langle : S^z - s^z : \rangle = (S+s)N - (S+s)\sum_k \sum_{\sigma=\pm} \frac{\widetilde{n}_k^\sigma}{\omega_k} = 0. \tag{7}$$

This leads to numerically solvable equations, reproducing low-T results, but now also yielding meaningful results at higher T [10]. Both C and χ show the predicted behaviour, but already at rather low temperatures results are only qualitative, calling for a numerical treatment.

Transfer-matrix DMRG

To study the thermodynamics of ferrimagnets, essentially four numerical methods compete. Exact diagonalization allows the exact treatment of thermodynamics, is however limited to very small systems, of the order of 10 spins. This method fails to yield good results at low T where increasingly long-range FM excitations dominate. Quantum Monte Carlo (QMC) suffers less from finite-size effects, but has intrinsic problems of accessing low temperatures. In the present problem, it cannot go far below $T = 0.2$, before the FM behavior sets in. The quantum transfer matrix method [11] is able to capture the FM excitations properly, as it works in the thermodynamic limit. However, it is limited to Trotter numbers $M \approx 10$, and as one should have $\beta/M \ll 1$, low temperatures are again inaccessible. The transfer-matrix DMRG ([1,2], Chaps. 5(I), 6(I)) combines working in the thermodynamic limit, which is essential for this problem, with being able to access much higher Trotter numbers, essentially up to $M \approx 1000$.

Our implementation follows the lines of Wang, Xiang [3] and Shibata [12], Chap. 4.1(II); one effectively treats a three-spin piece of the full chain. Numerically, it is most effective to use a small spin–big spin–small spin decomposition. We found that the use of asymmetric density matrices gave much more reliable results. Internal energy and magnetisation were evaluated directly, to reduce the number of numerical differentiations. The algorithm, by the way, lends itself easily for parallelisation.

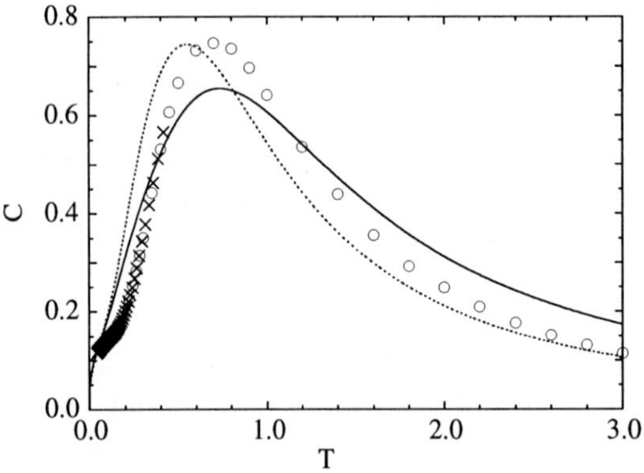

Fig. 2. Specific heat: dashed and solid lines show two different quartic spin wave approximations, circles QMC results, crosses DMRG results

Let me comment on error control. The transfer-matrix DMRG error is controlled by the Trotter decomposition (the analogue of real space length in conventional DMRG) $M = \beta/\beta_0$, where β_0 is the starting temperature, and m, the number of states kept to describe the chain. At high T, the error mainly influenced by the Trotter decomposition, as the effective Trotter number $M = \beta/\beta_0$ is still very small, while few decimations have been carried out. At low T, the effective Trotter number is very big, as is the number of decimation steps, such that the error is mostly due to the number m of states kept.

Here, we use Trotter numbers up to $M \sim 1000$, with starting temperatures $\beta_0 = 0.05, 0.1, 0.2$, which is rather coarse, but use up to $m = 140$ states, as we are mainly interested in low-T properties. All results presented have been extrapolated to $\beta_0 \to 0$, while the extrapolation in m was done by increasing it such that results did not change appreciably any more. The truncation error was 10^{-7} or better. At $H = 0$, we obtained full convergence down to $T \sim 0.06J$, while we got no useful results below $T \sim 0.04J$. Comparing to numerical completely satisfying results for the Heisenberg ferromagnet, which is also critical with a divergent density of states close to the ground state, we conclude that these difficulties are probably due to the additional fluctuations in the ferrimagnetic unit cell. For $H \neq 0$, we could, even at quite small fields, obtain convergence down to $T \sim 0.01J$.

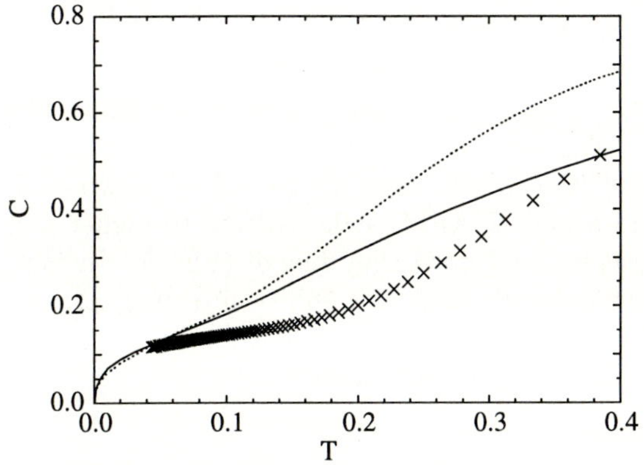

Fig. 3. Low-temperature specific heat, which cannot be captured by QMC. Symbols as in Fig. 2

Numerical results (Figs. 2, 3) indeed show the predicted deviation from a linear AFM behavior in the specific heat at very low T, below the temperatures accessible to other methods. The numerical result can be convincingly linked to spin wave calculations at very low T. The same holds true for the susceptibility (Fig. 4).

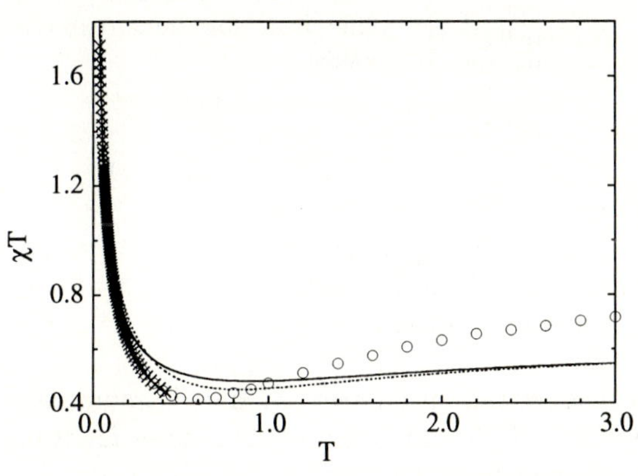

Fig. 4. Susceptibility times temperature vs. temperature. Symbols as in Fig. 2

3 Thermodynamics in Finite Field

The magnetic field term commutes with the field-free Hamiltonian; its only effect is therefore a shifting of dispersion relations. The FM dispersion relation is increased in energy and acquires a gap H, while the AFM dispersion relation shows a gap reduced by H. AFM excitations become gapless for $H = \Delta$, defining a *lower critical field* $H_{c1} = 1.76$. On the other hand, a spin wave analysis starting from a fully polarized ground state (all spins aligned in a sufficiently strong field) yields (for spins $(1, \frac{1}{2})$) a two-branched dispersion relation

$$\omega_k^{\mp} = H - \frac{3}{2} \pm \frac{1}{2}\sqrt{5 + 4\cos k}, \tag{8}$$

which shows an instability (negative energy) below an *upper critical field* $H_{c2} = 3$. The presence of two gapped excitation branches lets us expect a generic double peak in C for $0 < H < H_{c1}$, opposed to a Luttinger like behavior $C \propto T$ for $H_{c1} < H < H_{c2}$, due to criticality.

Below the lower critical field, the specific heat shows indeed a two-peak structure, low-T FM peak, and a high-T AFM peak, shifting to higher and lower temperatures with increasing field until they merge (Fig. 5). In fact, comparing the low-T part to the specific heat of a spin-$\frac{1}{2}$ ferromagnet, one sees that the quantum renormalization of the ferromagnetic result is very weak, much weaker than predicted by spin wave theory.

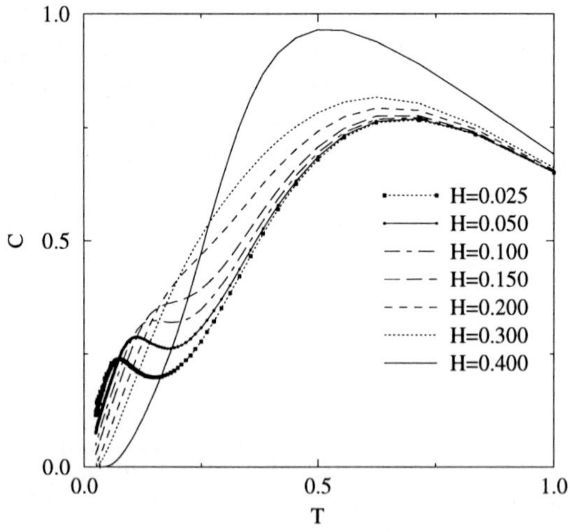

Fig. 5. Low-temperature specific heat for small external fields

Let us briefly discuss also magnetisation curves [13]. Indeed, numerical magnetisation curves confirm $H_{c1} = 1.76$ and $H_{c2} = 3$. For fixed H, there is interesting structure in $M(T)$, which we have not investigated yet. For

$T \to 0$, $M(H)$ shows a square-root like behavior at both critical fields. The same behavior is observed in integer quantum spin chains [14].

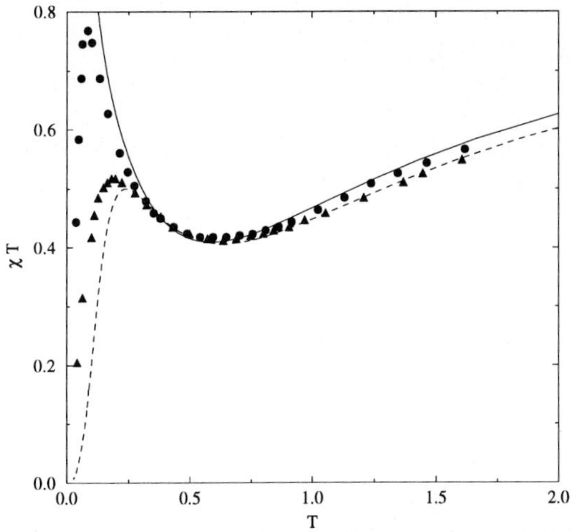

Fig. 6. Susceptibility times temperature vs. temperature. Circles and triangles: experiment for $H = 0.01T$ and $H = 7T$ [5]; solid and dashed lines: theoretical fits for $H/J = 0$ and $H/J = 0.06$, normalized to $J = 121\text{K}$ and $J = 135\text{K}$ respectively.

Turning lastly to comparison to experiment [6], we can compare theoretical and experimental results (Fig. 6) down to much lower T than exact diagonalization [6]; fitting more of the experimental features makes DMRG interesting for experimentalists. We tried to fit down to $T/J \approx 0.2$ (at lower T, the onset of 3D ordering makes comparisons impossible). Our results indicate that further study is necessary: while we can fit the zero-field curve using the parameters extracted in the experiment, the 7T curve is best reproduced by assuming $H/J \approx 0.06$ (vs. $H/J \approx 0.08$), $J \approx 135\text{K}$ (vs. $J = 121\text{K}$) and the g-values rather close to 2. Work along these lines in progress. One could also imagine that "differential" measurements of C at various fields, introducing low-T structure, would allow to eliminate problematic lattice contributions and allow for rather precise fits.

4 Conclusion

Concluding, I think it has been demonstrated without doubt that the transfer-matrix DMRG has provided a complete and satisfying quantitative picture of ferrimagnetic spin chains, which should be relevant to further experimental studies.

Finally, I would like to thank my collaborators Klaus Maisinger, Takahiro Fukui, Shoji Yamamoto, Sven Brehmer and Hans-Jürgen Mikeska. It is our joint work that has been presented here.

References

1. T. Nishino, J. Phys. Soc. Jpn. **64**, L3598 (1995)
2. R.J. Bursill, T. Xiang and G.A. Gehring, J. Phys. Cond. Matt. **8**, L583 (1996)
3. X. Wang and T. Xiang, Phys. Rev. B **56**, 5061 (1997)
4. E. Lieb and D. Mattis, J. Math. Phys. **3**, 749 (1962)
5. M. Verdaguer et al., Phys. Rev. B **29**, 5144 (1984), Y. Pei et al., Inorg. Chem. **26**, 138 (1988), Y. Pei et al., Inorg. Chem. **27**, 47 (1988), O. Kahn, Y. Pei and Y. Journaux, in: Inorganic Materials, ed. D.W. Bruce and D. O'Hare, Wiley (1995)
6. M. Hagiwara, K. Minami, Y. Narumi, K. Tatani and K. Kindo, J. Phys. Soc. Jpn. **67**, 2209 (1998)
7. A.K. Kolezhuk, H.-J. Mikeska and S. Yamamoto, Phys. Rev. B **55**, R3336 (1997); S. Brehmer, H.-J. Mikeska and S. Yamamoto, J. Phys. Cond. Matt. **9**, 3921 (1997); S.K. Pati, S. Ramasesha and D. Sen, Phys. Rev. B **55**, 8894 (1997); J. Phys. Cond. Matt. **9**, 8707 (1997); S. Yamamoto, S. Brehmer and H.-J. Mikeska, Phys. Rev. B **57**, 13610 (1998)
8. M. Takahashi, Prog. Theor. Phys. **46**, 401 (1971), Prog. Theor. Phys. **51**, 1348 (1974); M. Gaudin, Phys. Rev. Lett. **26**, 1301 (1971); S. Eggert, I. Affleck and M. Takahashi, Phys. Rev. Lett. **73**, 332 (1994)
9. M. Takahashi, Prog. Theor. Phys. Supp. **87**, 233 (1986); Phys. Rev. Lett. **58**, 168 (1987); M. Takahashi, Phys. Rev. B **40**, 2494 (1989)
10. S. Yamamoto and T. Fukui, Phys. Rev. B **57**, R14008 (1998); S. Yamamoto, T. Fukui, K. Maisinger and U. Schollwöck, J. Phys.: Cond. Matt. **10**, 11033 (1998)
11. H. Betsuyaku, Phys. Rev. Lett. **53**, 629 (1984); Prog. Theor. Phys. **73**, 319 (1985)
12. N. Shibata, J. Phys. Soc. Jpn. **66**, 2221 (1997)
13. K. Maisinger, U. Schollwöck, S. Brehmer, H.-J. Mikeska and S. Yamamoto, Phys. Rev. B **58**, R5908 (1998)
14. I. Affleck, Phys. Rev. B **41**, 6697 (1990); A. Tsvelik, Phys. Rev. B **42**, 10499 (1990); M. Takahashi and Sakai, J. Phys. Soc. Jpn. **60**, 760 (1991); *ibid.*, 3615 (1991); Phys. Rev. B **43**, 13383 (1991)

4.4 Thermodynamics of Metallic Kondo Lattices

Samuel Moukouri and Laurent G. Caron

Centre de Recherche en Physique du Solide et Département de Physique, Université de Sherbrooke, Sherbrooke, Québec, Canada J1K 2R1

The exceptional accuracy provided by the density matrix renormalization group (DMRG) [1] in the study of the ground-state properties of many one-dimensional quantum and two-dimensional classical models with a relatively modest computer effort lead to a dramatic burst in popularity of this technique in last few years. It is now expected that the extension of the DMRG to finite temperatures [2,3] or frequencies [4,5] may also lead to results superior to that of exact diagonalization or quantum Monte Carlo (QMC) methods.

For finite temperatures, we applied in an earlier study [2] an approach which consists in targeting a few superblock states. We computed the magnetic susceptibility of quantum spin chains with $S = \frac{1}{2}$ and $\frac{3}{2}$. Our result was compared to the Bethe-ansatz calculation and showed excellent agreement. Zhang et al. [6] applied the same method in the study of a magnetic impurity embedded in a quantum spin chain. Their results agree quite well with the QMC at high temperatures. At low temperatures where QMC is plagued by important statistical uncertainties or the sign problem, they found that the DMRG was still robust.

Another route to the study of systems at finite temperatures was taken by Bursill et al. and Wang and Xiang [3]. They used the ideas primarily proposed by Nishino [7] who extended the DMRG to two dimensional classical systems. Their approach, the transfer matrix renormalization group, is treated in Chap. 6(I).

In this study we propose a more general interpretation of the process of targeting which is at the core of the DMRG. This interpretation shows a richer nature of the internal structure of the DMRG. It focuses on the target space (instead of target state) and its evolution when many states are targeted. Tests on the Hubbard model (HM) are used to illustrate the idea that targeting a few superblock states is enough for low-tempearture investigations. The main part of this contribution is devoted to the study of the low-temperature thermodynamics of the one-dimensional metallic Kondo lattice model (KLM). We show that despite the reduced dimension, the susceptibility displays a behavior reminiscent to that found in heavy-fermion compounds.

1 Nature of the Target Space

In the standard DMRG, one considers a state Ψ (called the target state) which is usually the ground state of the superblock $G \equiv B^L \bullet \bullet B^R$,

$$\Psi = \sum_{k,l} \Psi_{k,l} \left|k\right\rangle \otimes \left|l\right\rangle \tag{1}$$

where $\left|k\right\rangle$, $k = 1, ..., K$, $\left|l\right\rangle$, $l = 1, ..., L$ are the basis of the subspaces $E \equiv B^L \bullet$and $F \equiv \bullet B^R$ respectively. The system generated by $\left|k\right\rangle \otimes \left|l\right\rangle$ gives the canonical basis of the tensor product $G \equiv E \otimes F$. One searches an approximation

$$\widetilde{\Psi} = \sum_{\alpha} \widetilde{\Psi}_\alpha \left|u_\alpha\right\rangle \otimes \left|v_\alpha\right\rangle \tag{2}$$

to Ψ when $\alpha = 1, ..., m < K$ states are kept in the left block. The best approximate wave function is obtained through the singular-value decomposition of the matrix $[\Psi_{k,l}]$. The m states are the states associated with the m largest singular values. In practical calculations, one uses the reduced density matrix

$$\rho = \Psi^t \Psi \tag{3}$$

where ${}^t\Psi$ is the transpose of Ψ. The eigenvalues of ρ are the squares of the singular values of Ψ.

It was expected that this method might be easily extended to non zero temperatures [1]. One simply has to target as many states as needed. The resulting density matrix is

$$\rho = \sum_k \omega_k \rho_k \tag{4}$$

where ρ_k and ω_k are respectively the reduced density matrix of the kth state and its weigth. Unfortunately, the number of states needed for a thermodynamic analysis is impossible to obtain by the direct application of this prescription. But it was already noticed that the truncated Hamiltonian obtained by targeting the ground state leads to a good description of not only this particular state but also to many low-lying states. That is why quite accurate values of the spin and charge gaps can be obtained whitout targeting the first excited state of the charge or spin sectors. This fact lead us to believe that a good accuracy for the thermodynamic quantities at low temperatures may be obtained by targeting a few states of the superblock [2]. Despite the excellent accuracy obtained in [2], the method was somewhat empirical. In the following, we wish to present a systematic procedure for a thermodynamic analysis. Since the standard DMRG describes only the targeted states, a different mathematical presentation is necessary in order to understand why by targeting a small number of states, we can nevertheless obtain accurate results.

A given state Ψ of G can always be decomposed in the following form (for more details, see the algebra book of Bourbaki [8])

$$\Psi = \sum_{i=1}^{s} \phi_i^L \otimes \phi_i^R, \quad \phi_i^L \in E, \quad \phi_i^R \in F. \tag{5}$$

where s is an integer little or equal to the dimension of the superblock. There exists a linear mapping u_Ψ wich connects the dual space of F , which is usually written as F^*, to E. For each $\phi^*$of F^*, u_Ψ is defined as follows

$$u_\Psi(\phi^*) = \sum_{i=1}^{s} \langle \phi_i^R, \phi^* \rangle \phi_i^L, \qquad \langle \phi_i^R, \phi^* \rangle = \phi^*(\phi_i^R) \tag{6}$$

where the $\phi^*(\phi_i^R)$ are complex numbers. The matrix elements of u_Ψ in the basis $\phi_{l'}^*$ of F^* associated to $|l\rangle$ which is defined by

$$\phi_{l'}^*(|l\rangle) = \delta_{l,l'} \tag{7}$$

are

$$[u_\Psi]_{k,l} = \sum_{i=1}^{s} \phi_{ik}^L \phi_{il}^R \equiv \Psi_{k,l} \tag{8}$$

ϕ_{ik}^L, ϕ_{il}^R are respectively the coordinates of ϕ_i^L, ϕ_i^R in the basis $|k\rangle$ and $|l\rangle$. The image subspace $Im[u_\Psi]$ associated to Ψ and its rank r are given by the singular-value decomposition of Ψ. The rank r is equal to the number of non-zero singular values. It may be obtained more easily by using the reduced density matrix $\rho_\Psi = u_\Psi \circ {}^t u_\Psi$, where ${}^t u_\Psi$ is the transpose of u_Ψ. This connects our presentation with the original derivation. This presentation has the advantage to focus not on the target state but on its associated subspace. This allows a global analysis of the method when many states are targeted. The understanding of the behavior of r will determine the possibility of a thermodynamic study. This is analyzed in the next section.

2 Test Case: The One-Dimensional Hubbard Model

In the usual notation, the HM may be written as follows

$$H = -t \sum_i (c_{i\sigma}^+ c_{i+1\sigma} + hc) + U \sum_i n_{i\uparrow} n_{i\downarrow} \tag{9}$$

Let us consider for instance a superblock of 8 sites in the case of the HM at quarter filling and $U = 8$. The dimension of the $S_z = 0$ subspace is 784. In Fig. 1, we show r as a function of the number M of targeted states; we see a rapid growth of r before it saturates. It reaches its maximum value well before all the states are targeted. Any further increase of M after this maximum is reached has only the effect of changing the magnitude of the

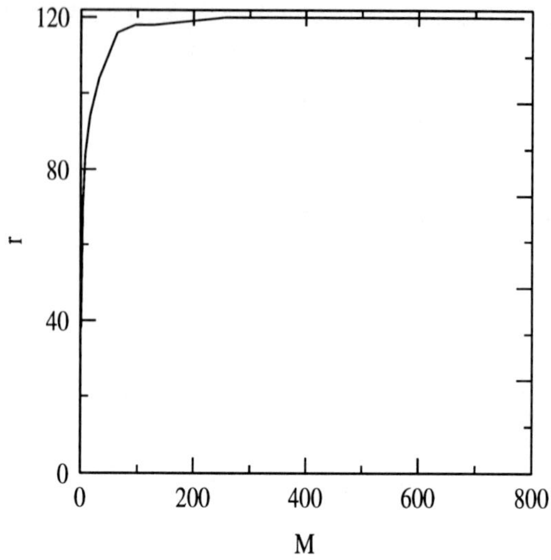

Fig. 1. The rank r as function of the number of targeted states M in the $S_z = 0$ subspace of a Hubbard chain of $N = 8$ sites for $U = 8$

eigenvalues. But since their corresponding eigenvalues are smaller, they are less well described. But in pratice, we never reach the saturation value which increases with N. Hence, we are always in the region where r grows sharply. This means that when a few M lowest states of a given charge-spin sector are targeted, many of the excited states are also taken into account.

It is now clear how a thermodynamical analysis can be made in a systematic way. One starts by fixing the temperature range. The desired physical quantity is calculated for initial values of m and M. Then the dependence of this quantity is analyzed by varying m and M. We attributed the same weigth to each target state, $\omega_k = 1/M$. The importance of a target space in the reduced density matrix is thus given by the number of target states fixed for this space. In practice, the computations rapidly become cumbersome when M is increased. At low temperatures, the convergence is reached for relatively small values of M. This is seen in Fig. 2 where we computed the magnetic susceptibility of chains of $N = 8$ to 20. Our assumptions are checked for $N = 8$ and 20 chains (Fig. 2 (a) and (b)). For $N = 8$ sites, keeping $m = 64$ states, we show in Fig. 2(a) that χ converges to it exact values after only $M = 4$ states are targeted to each spin sector. For $N = 20$ in Fig. 2(b), we fixed $M = 4$ for all $S_z \neq 0$ spin sectors; for $S_z = 0$, M_0 is varied from 4 to 32. The extrapolated curve in Fig. 2(c) obtained from $N = 8$ to 20 is in a very good agreement with the exact result [9] within the range of validity of the method which is $0.1 \leq T \leq 0.3$. These bounds are estimated as in [2].

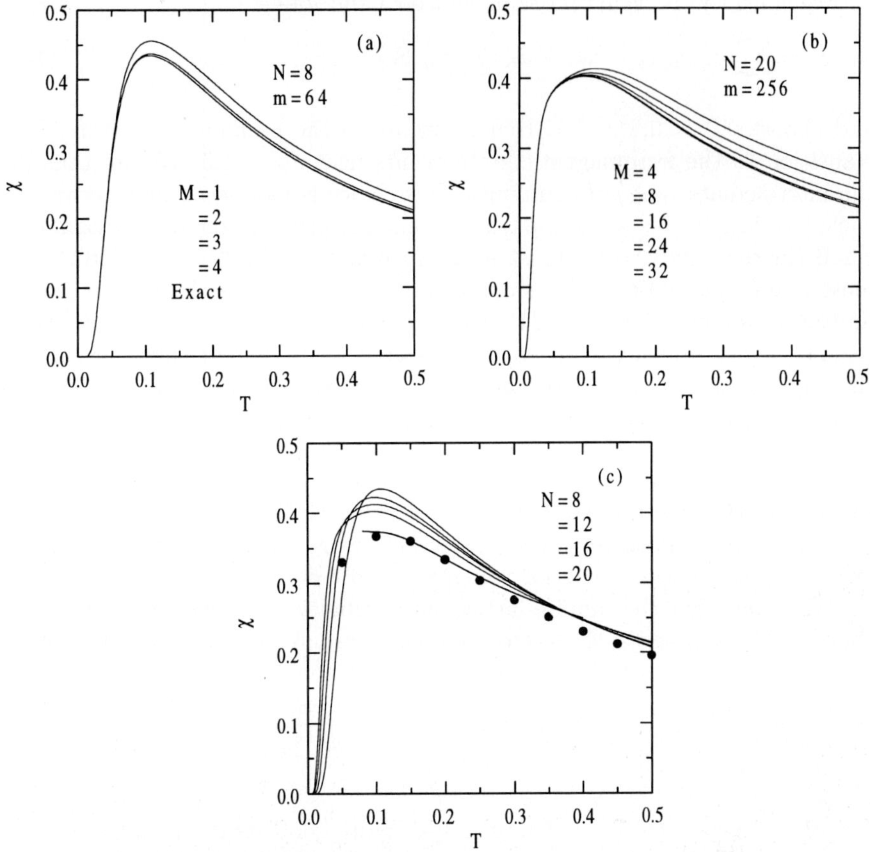

Fig. 2. Magnetic susceptibility χ of Hubbard chains. Evolution of χ as a function of M (**a**) for 8 sites, (**b**) for 20 sites (for $N = 20$, $M = 4$ in each spin sector, except $S_z = 0$ for which M_0 states are targeted). (**c**) comparison of the converged (as a function of M) and N-extrapolated χ with the exact result (dots).

3 Metallic Kondo Lattice

At low temperatures, the heavy-fermion compounds are known to behave like Fermi liquids with enormous values of the specific heat coefficient γ and of the magnetic susceptibility χ [10]. The Kondo lattice model (KLM) has been proposed for the description of these systems. The one-dimensional version of this model is expected to retain some features of this striking behavior. Recently, a converging view concerning the nature of its ground state has emerged through different studies [11], Chap. 4.1(II). But since the introduction of the model, the expected heavy-fermion character has only been reported in non-rigorous studies. A previous QMC study [12] of the metallic phase was unable to reach low temperatures because of the sign problem.

We consider the usual KLM at quarter band filling,

$$H = \sum_{i\sigma}(c^+_{i\sigma}c_{i+1\sigma} + hc) + J_K \sum_i \mathbf{S}_{ic}.\mathbf{S}_{if} \tag{10}$$

and choose $J_K = 0.8$ and 1. Hence, we are in the paramagnetic state; the transition to the ferromagnetic state occurs near $J_K = 1.5$. We are thus in the intermediate range of couplings. This choice is motivated by a technical consideration: in this range the truncation error $p(m, M)$ remains reasonably small (in this study, the largest $p(m, M)$ was 1.1×10^{-4} for m=256). The most interesting limit of very small coupling is hard to reach because of the limitation imposed by $p(m, M)$. We recall that even in a $T = 0$ study, this regime is very difficult to reach. In this limit, the different energy levels of the superblock are very close and many eigenvalues of the reduced density matrix are nearly degenerate. Hence, $p(m, M)$ decreases only slowly when m is increased. Nevertheless, we believe that the low J_K regime will only show quantitative modifications. For the computation of χ, it was necessary to keep all the spin sectors; for the the longest chain $N = 10$ of our study, we thus have 8 spin sectors. Two states were targeted in each spin subspace, except the subspace $S_Z = 0$ for which we used 8 states.

We found that the infinite-lattice susceptibility is bracketed by those of chains with even and odd electron numbers (Fig. 3(a)). The position of the maximum for $J_K = 1$ is $T^* = 0.013$, it marks the onset of the strong-coupling regime between the conduction electrons and the localized spins. Since χ converges rapidly as function of N, a natural procedure for the estimation of the thermodynamic limit value is to take the average of the values of the two longest chains studied $N = 10$ and $N = 8$. This procedure, althought not free of finite-size errors was found to give quite good results in the study of spin chains [2]. The lowest temperature we were able to reach is $T = 0.005$. This is approximatively the value of the finite-size spin gap . These values of χ are nearly two order of magnitude greater than that of the free Fermi gas $\chi_0 \approx 0.1$. Larger values of the ratio χ/χ_0 may be obtained by going to smaller couplings. This is seen in Fig. 3(b) where χ for $J_K = 1$ and 0.8 are compared. The deviation from a Curie-Weiss behavior first occurs for $J_K = 1$. This behavior of χ as a function of J_K is consistent with the expected physics; it reflects the normal tendency of the Kondo coupling: for stronger couplings the individual spins are more screened, as a consequence the magnetic susceptibility decreases. The onset of the strong-coupling regime occurs at extremely low temperatures. The maximum of χ is an order of magnitude smaller than the value of the maximum of χ found in a conventional strongly correlated electron system such as the Hubbard model studied in the previous section.

The product $T\chi$ which is usually defined as the square of the effective local moment is shown in Fig. 3(c). At high temperature (not shown here), χ has a Curie-like behavior, the effective local moment is a constant. As the temperature is lowered, $T\chi$ slowly increases above T^* in the two cases. This indicates that the coupling of the conduction spins with the localized ones

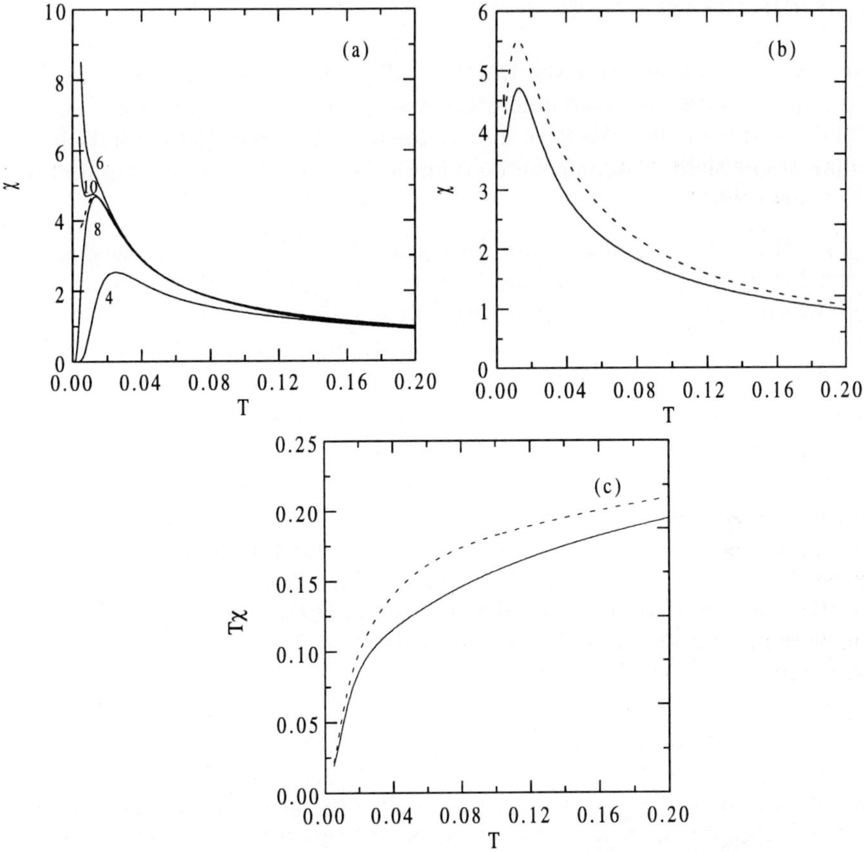

Fig. 3. (**a**) The temperature dependence of the magnetic susceptibility χ of the KLM for $N = 4$, 6, 8 and 10 for $J_K = 1$. (**b**) χ for $J_K = 0.8$ (dashed line) and $J_K = 1$ (full line) . (**c**) The corresponding effective moment $T\chi$.

has set in. Below T^*, a significant drop of $T\chi$ is observed: the conduction electron spins are now strongly coupled to the f spins.

Finally, it has been suggested in previous studies [11] that the KLM may belong to the universality class of Luttinger liquids (LL). An indirect proof of a possible LL character is to compare the low-temperature values of χ to that obtained at $T = 0$ with the standard DMRG ($m = 256$ and $N = 20$) by using the relation $\chi = 2/N\Delta$, where Δ is the finite-size spin gap. For $J_K = 1$ we found $\chi = 5.7.4$. This value is larger than that we may obtain by extrapolating the curve of Fig. 3(b) to T=0. We cannot however rule out the LL character in the KLM; the large error in our result is due to the finite size and the truncation of the Hilbert space effects: for $N > 8$, the values of Δ are close to $p(m, M)$.

4 Conclusion

We have shown that using the DMRG and our proposed procedure, we can reliably compute the low-temperature susceptibility of the one-dimensional Kondo lattice model. We have clearly found that the model exhibits large values of χ. Finally, the same method can easily be applied to the computation of dynamic quantities.

We thank Dr. G. Jüttner for sharing his data. We also acknowledge the critical reading of A-M. Tremblay. This work was supported by a grant from the Natural Sciences and Engineering Research Council (NSERC) of Canada and the Fonds pour la formation de Chercheurs et d'Aide à la Recherche (FCAR) of the Québec government.

References

1. S. R. White, Phys. Rev. Lett. **69**, 2863 (1992); Phys. Rev. B **48**, 10345 (1993)
2. S. Moukouri and L. G. Caron, Phys. Rev. Lett. **77**, 4640 (1996)
3. R. J. Bursill, T. Xiang and G. A. Gehring, J. Phys. Cond. Matt. **8**, L583 (1996); X. Wang and T. Xiang, Phys. Rev. B **56**, 5061 (1997)
4. K. Hallberg, Phys. Rev. B **52** R9827 (1995)
5. H. Pang, H. Akhlaghpour and M. Jarrell, Phys. Rev. B **53**, 5086 (1996)
6. W. Zhang, J. Igarashi and P. Fulde, J. Phys. Soc. Jap. **66**, 1912 (1997)
7. T. Nishino, J. Phys. Soc. Jpn. **64**, L3598 (1995)
8. N. Bourbaki, Élém. de math., Algèbre I, p. II 111, Hermann, Paris (1970)
9. G. Jüttner, A. Klümper and J. Suzuki, cond-mat 9711310
10. P. A. Lee, T. M. Rice, J. W. Serene, L. J. Sham and J. W. Wilkins, Com. Cond. Mat. Phys. **12**, 99 (1986)
11. H. Tsunetsugu, M. Sigrist and K. Ueda, Rev. Mod. Phys. **69**, 809 (1997); S. Moukouri and L. G. Caron, Phys. Rev. B **52**, R15 723 (1995). S. Caprara and A. Rosengren, Europhys. Lett. **39**, 55 (1997); G. Honner and M. Gulàcsi, Phys. Rev. Lett. **78**, 2180, (1997)
12. M. Troyer and D. Würtz, Phys. Rev. B **47**, 2886 (1993)

5 Phonons and Disorder

5.1 Methods for Electron-Phonon Systems

Eric Jeckelmann[1], Chunli Zhang[2], and Steven R. White[2]

[1] Fachbereich Physik, Philipps-Universität Marburg
D-35032 Marburg, Germany
[2] Department of Physics and Astronomy, University of California
Irvine, CA 92697, USA

During the past decade there have been great strides in the development of numerical techniques for simulating strongly correlated systems. A significant limitation to many of these methods – for example, the density matrix renormalization group [1,2] (DMRG) – is that they require a finite basis. In an electron-phonon lattice model, the number of phonons is not conserved and the Hilbert space is infinite. Of course, the number of phonons can be artificially constrained, but the number of phonons needed for an accurate treatment may be quite large. This often severely constrains the size of the systems or the regime of coupling which may be studied with exact diagonalizations (for recent examples, see [3–7]) and DMRG [8,9].

Here, we describe two methods for treating systems with a large number of degrees of freedom per site. Both methods use the information contained in a density matrix to reduce the computational effort required for the study of such systems. The first method [10] is just a modification of the usual DMRG technique [see Chap. 2(I)] which allows us to deal more efficiently with a large bosonic Hilbert space. The basic idea is to transform each boson site into several artificial interacting two-state sites (pseudo-sites) and then to use DMRG techniques to treat this interacting system. The second method [11] is a procedure for generating a controlled truncation of a large Hilbert space, which allows the use of a very small optimal basis without significant loss of accuracy. This procedure is closely related to DMRG, in that the optimal basis is generated using a density matrix. It is used here in conjunction with exact diagonalization but could be incorporated into DMRG. A similar DMRG algorithm has been developed to solve periodic systems with a large number of degrees of freedom per site [12].

We illustrate both methods on the Holstein model [13], a model of noninteracting electrons on a lattice coupled to phonons represented by one Einstein oscillator on each site. Its Hamiltonian is given by

$$H = \omega \sum_i b_i^\dagger b_i - \gamma \sum_i \left(b_i^\dagger + b_i \right) n_i - t \sum_{i\sigma} \left(c_{i+1\sigma}^\dagger c_{i\sigma} + c_{i\sigma}^\dagger c_{i+1\sigma} \right) , \quad (1)$$

where $c_{i\sigma}^\dagger$ and $c_{i\sigma}$ are electron operators, $n_i = c_{i\uparrow}^\dagger c_{i\uparrow} + c_{i\downarrow}^\dagger c_{i\downarrow}$, and $b_i^\dagger$ and b_i are phonon operators. t is the hopping integral, γ is the electron-phonon coupling constant, and each oscillator has frequency ω. Numerical studies of the Holstein model have often been limited either to small systems or to a particular regime of the parameters γ and ω because of the difficulty in dealing with the large phononic Hilbert space. With both methods described here, one can keep enough phonons to study all regimes of parameters with great accuracy using modest computer resources [10,11,14].

1 Pseudo-Site Method

The DMRG method is known to be a very successful numerical technique for studying spin and fermion lattice models. Although the DMRG algorithm can easily be generalized to treat systems including bosons, previous applications have often been restricted to problems for which one needs to consider a relatively small number of states M per site [8,9,15,16]. In a standard implementation of the DMRG method, each boson forms one lattice site and thus memory and CPU time requirements increase as M^2 and M^3, respectively. Therefore, performing calculations with $M \approx 10 - 100$, as needed in the strong-coupling regime of the Holstein model, requires much more computer resources than computations for bosonic systems with small M or for otherwise similar Heisenberg or Hubbard systems. More sophisticated DMRG techniques must be used to study the Holstein model [10,12].

To understand the basis of the pseudo-site approach [10], it is important to note that, in principle, the computer resources required by DMRG increase linearly with the number of lattice sites (everything else being equal). Thus, DMRG is much better able to handle several few-state sites rather than one many-state site. Experience also indicates that it is possible to improve DMRG performance by substituting several sites with a small Hilbert space for a single site with a larger Hilbert space, for instance in the Hubbard model [17]. Therefore, the key idea of the pseudo-site approach is to transform each boson site with $M = 2^N$ states into N pseudo-sites with 2 states. This approach is motivated by a familiar concept: the representation of a number in binary form. In this case the number is the boson state index going from 0 to $M - 1$. Each binary digit is represented by a pseudo-site, which can be occupied (1) or empty (0). One can think of these pseudo-sites as hard-core bosons. Thus, the level with index 0 is represented by N empty pseudo-sites, while the highest level, $2^N - 1$, is represented by one boson on each of the N pseudo-sites.

Figure 1 illustrates the differences between standard and pseudo-site DMRG approaches. In the standard approach [Fig. 1(a)], a new block (*dashed rectangle*) is built up by adding a boson site (*oval*) with M states to another block (*solid rectangle*) with m states. Initially, the Hilbert space of the new block contains mM states and is truncated to m states according to the

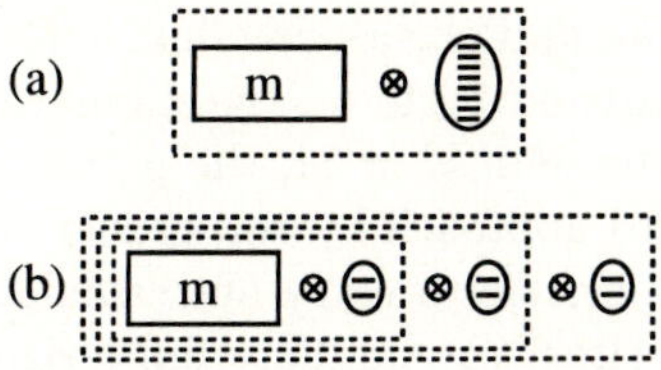

Fig. 1. Symbolic representation of the standard DMRG approach (**a**) and of the pseudo-site approach (**b**) for $M = 8$ ($N = 3$)

DMRG method. In the pseudo-site approach [Fig. 1(b)], a new block is made of the previous block with m states and one pseudo-site with two states. The Hilbert space of this new block contains only $2m$ states and is also truncated to m states according to the DMRG method. It takes N steps to make the final block (*largest dashed rectangle*) including the initial block and all pseudo-sites, which is equivalent to the new block in Fig. 1(a). However, at each step we have to manipulate only a fraction $2/M$ of the bosonic Hilbert space.

To implement this pseudo-site method, we introduce N pseudo-sites $j = 1, ..., N$ with a two-dimensional Hilbert space $\{|r_j\rangle, r_j = 0, 1\}$ and the operators $a_j^\dagger, a_j$ such that $a_j|1\rangle = |0\rangle$, $a_j|0\rangle = 0$ and $a_j^\dagger$ is the hermitian conjugate of a_j. These pseudo-site operators have the same properties as hard-core boson operators: $a_j a_j^\dagger + a_j^\dagger a_j = 1$, and operators on different pseudo-sites commute. The one-to-one mapping between a boson level $|n\rangle$, $n = 0, ..., M-1$, where $b^\dagger b\,|n\rangle = n|n\rangle$, and the N-pseudo-site state $|r_1, r_2, ..., r_N\rangle$ is given by the relation $n = \sum_{j=1}^{N} 2^{j-1}\, r_j$ between an integer number and its binary representation. The next step is to write all boson operators in terms of pseudo-site operators. It is obvious that the boson number operator is given by $N_b = b^\dagger b = \sum_{j=1}^{N} 2^{j-1}\, a_j^\dagger a_j$. Other boson operators take a more complicated form. For instance, to calculate the representation of $b^\dagger$ we first write $b^\dagger = B^\dagger\, \sqrt{N_b + 1}$, where $B^\dagger|n\rangle = |n+1\rangle$. The pseudo-site operator representation of the second term is

$$\sqrt{N_b + 1} = \sum_{n=0}^{M-1} \sqrt{n+1}\; P_1(r_1)\, P_2(r_2) \ldots P_N(r_N) \;, \tag{2}$$

where $P_j(1) = a_j^\dagger a_j$, $P_j(0) = a_j a_j^\dagger$ and the r_j ($j = 1, .., N$) are given by the binary digits of n. For $B^\dagger$ we find

$$B^\dagger = a_1^\dagger + a_2^\dagger\, a_1 + a_3^\dagger\, a_2\, a_1 + \ldots + a_N^\dagger\, a_{N-1}\, a_{N-2} \ldots a_1 \;. \tag{3}$$

The representation of $b^\dagger$ for any number of pseudo-sites N is given by the product of these two operators. Other operators can be obtained in a similar way. Thus, one can substitute $N = \log_2(M)$ pseudo-sites for each boson site in the lattice and rewrite the system Hamiltonian and other operators in terms of the pseudo-site operators. Then, the finite-system DMRG algorithm [see Chap. 2(I)] can be used to calculate the properties of this system of interacting electrons and hard-core bosons.

Performances of the standard and pseudo-site approaches are similar for small M. However, the pseudo-site approach clearly outperforms the standard approach when computations become challenging. For $M = 32$, the pseudo-site approach is already faster than the standard approach by two orders of magnitude. With the pseudo-site method it is possible to carry out calculations on lattices large enough (a few hundred sites) to eliminate finite-size effects while keeping enough states per phonon mode (more than a hundred) to render negligible the errors due to the truncation of the phonon Hilbert space [10]. For the polaron problem (Holstein model with a single electron) it has been shown that the pseudo-site approach is currently the most accurate method, among the various numerical and analytical methods available, to determine the ground-state energy [18].

2 Optimal Phonon Basis

The number of phonon levels M needed for an accurate treatment of a phonon mode can be very large when one uses the bare phonon basis made of the lowest eigenstates of the operators $b_i^\dagger b_i$. However, the number of phonon levels can be strongly reduced by choosing a basis which minimizes the error due to the truncation of the phonon Hilbert space. The key idea of this approach is identical to the key idea of DMRG: in order to eliminate states from a part of a system without loss of accuracy, one should transform to the basis of eigenvectors of the reduced density matrix, and discard states with low probability. The key difference is that here the subsystem is a single site, rather than varying fractions of the entire system. To be specific, consider the Holstein model (1). A site includes both the phonon levels and the electron degrees of freedom. We will consider here only systems with translational invariance, so that all sites are equivalent. Let α label the four possible electronic states of a particular site and let n label the phonon levels of this site. Let j label the combined states of all of the rest of the sites. Then a wavefunction of the system can be written as

$$|\psi\rangle = \sum_{\alpha,n,j} \psi_{\alpha n,j} |\alpha, n\rangle |j\rangle \, . \tag{4}$$

The density matrix ρ for this site for a given electronic state α of the site is

$$\rho^{\alpha}_{n,r} = \sum_{j} \psi_{\alpha n,j} \psi^{*}_{\alpha r,j} \, , \tag{5}$$

where r also labels the phonon levels of this site. Let $w_{\alpha k}$ be the eigenvalues and $\phi_{\alpha k}(n)$ the eigenvectors of ρ, where k labels the different eigenstates for a given electronic state of the site. The $w_{\alpha k}$ are the probabilities of the states $\phi_{\alpha k}$ if the system is in the state (4). If $w_{\alpha k}$ is negligible, then the corresponding eigenvector can be discarded from the basis for the site, without affecting the state (4). If one wishes to keep a limited number of states m for a site, then the best states to keep are the eigenstates of ρ with largest eigenvalues. In electron-phonon systems, these m eigenstates form an optimal phonon basis. We note that we obtain different optimal phonon states for each of the four electron states of the site.

Unfortunately, in order to obtain the optimal phonon states, we need the target state (4), which we do not know – usually we want the optimal states to help get this state. This problem can be circumvented in several ways [11]. Here, we describe one algorithm for obtaining an optimal phonon basis for the ground state and low-lying states of the Holstein model. We consider a bare phonon basis with such a large number M of levels per site that truncation errors would be negligible if we could diagonalize the Hamiltonian in the corresponding Hilbert space. However, to be able to perform an exact diagonalization of the system, each site of the lattice is allowed to have only a small number of optimal phonon levels, $m \sim 3-4$. One site of the system (called the big site) has both the optimal modes and a few extra levels. These extra levels are taken from the set of M bare levels but are explicitly orthogonalized to the current optimal modes. This approach is illustrated in Fig. 2.

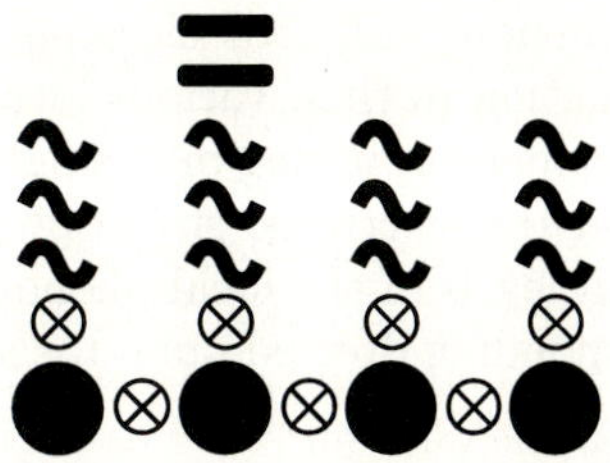

Fig. 2. Algorithm for constructing an optimal basis. Each site (*circles*) has the electronic degrees of freedom and 3 optimal states (*wiggly bars*). The big site (second site from the left) has the optimal levels plus two bare levels (*straight bars*)

The ground state of the Hamiltonian (1) is calculated in this reduced Hilbert space by exact diagonalization using the Davidson algorithm. (In large systems the DMRG method could be used to get the ground state.) Then, the density matrix (5) of the big site is diagonalized. The most probable m eigenstates are the new optimal phonon modes. These optimal phonon modes are used on all the other sites for the next diagonalization. After each diagonalization, the density matrix of the big site is diagonalized and new optimal states are found. At the beginning of the first diagonalization, all of the phonon states are bare. Thus, the optimal modes obtained after the first few diagonalizations are not very accurate because the Hilbert space is severely truncated. Diagonalizations must be repeated until the optimal modes have converged. Each time different extra phonon levels are used for the big site. They allow improvements of the optimal states by mixing in the M bare states little by little. One sweep consists of enough diagonalizations to include all M bare levels as extra levels. A couple of sweeps are needed to reach full convergence of the optimal levels. Each diagonalization uses as the starting wavefunction the converged wavefunction from the last step.

Therefore only two or three Davidson steps are needed for convergence, rather than dozens. In the Holstein model it is possible to transfer optimal phonon levels from small systems to larger ones because of the localized nature of the phonon modes. Therefore, one can first calculate a large optimal phonon basis in a two-site system and then use it instead of the bare phonon basis to start calculations on larger lattices.

To study dynamical properties we can extend the above approach by targeting the ground states as well as some low-lying excited states [14]. In this case, the density matrix of the big site is formed by adding the density matrices (5) of each state. In this way we can include information from several states to select the optimal phonon basis. Not surprisingly, more phonon states must be kept per site to reach a given accuracy when several states are targeted than when only the ground state is targeted. Dynamical properties are computed using the well-known Lanczos algorithm combined with the continued fraction method. This algorithm yields not only the dynamical correlation and response functions of the system, but also the most important eigenstates that contribute to these functions and that we need to build the density matrix (5), see also Chap. 7(I).

A very interesting feature of the optimal basis approach is that it provides a natural way to dress electrons locally with phonons [14]. All electronic operators can be dressed by applying a transformation to their various local components. The matrix elements of an electronic operator acting on a single site are $O_{\alpha n,\beta r} = O_{\alpha,\beta}\,\delta_{n,r}$, where α and β label the electronic states of a particular site, and n and r label the bare phonon levels of the local phonon mode. Using the optimal states $\phi_{\alpha k}(n)$ the corresponding dressed operator is defined as

$$\tilde{O}_{\alpha n,\beta r} = O_{\alpha,\beta} \sum_k \phi_{\alpha k}(n)\, \phi_{\beta k}(r)\,. \tag{6}$$

The operator $\tilde{O}$ not only transforms electronic states, as the bare operator does, but also transforms the local phonon degrees of freedom accordingly. However, the dressing by phonons at a finite distance from the electrons is completely neglected with this method.

The form of the optimal phonon states can sometimes be understood qualitatively. In Fig. 3 we show the most probable optimal phonon wavefunctions for each possible electronic occupation of a site as a function of the oscillator position $q = b^\dagger + b$ in the half-filled Holstein model. In the weak coupling regime ($\gamma/t = 0.5$) optimal states are simply eigenstates of an oscillator with an equilibrium position $q \approx 2\gamma/\omega$ as predicted by a mean-field approximation. In the strong-coupling regime ($\gamma/t = 2$) the most important optimal phonon states for 0 or 2 electrons on the site can be obtained by a unitary Lang-Firsov transformation, in agreement with the strong-coupling theory. However, the optimal phonon states for a singly occupied site is not given by the Lang-Firsov transformation but is approximately the superposition of the

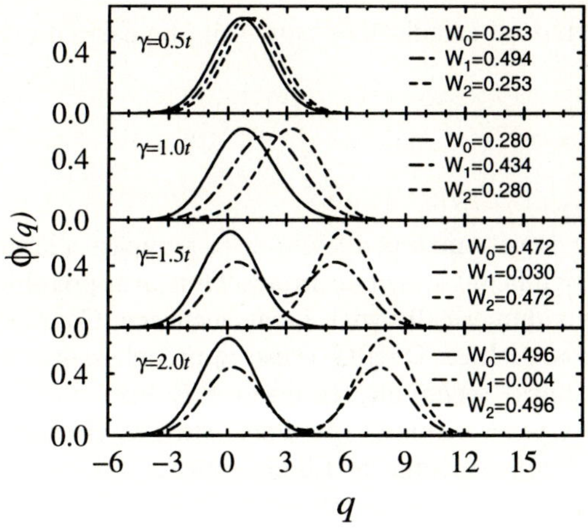

Fig. 3. Optimal phonon wavefunctions as a function of the oscillator position q and their probability W_n for different electron-phonon couplings γ. $n = 0, 1, 2$ is the number of electrons on the site. The system is a six-site half-filled Holstein lattice with $\omega = t$

optimal phonon states for empty or doubly-occupied sites due to retardation effects [11].

The improvement coming from using optimal phonon states instead of bare phonon levels is remarkable. In Fig. 4 we show the ground-state energy of the Holstein model as a function of the number of bare and optimal phonon levels. Using only two optimal modes, the energy is accurate to less than 0.1%, whereas keeping eleven bare modes the error is still greater than 5%. The very rapid convergence of the energy as a function of the number of optimal phonon modes is due to the very tight distribution of their probabilities $w_{\alpha k}$. Typically, the fourth highest eigenvalues for a given electronic state is smaller than 10^{-5} for all couplings. Therefore, two or three optimized phonon states per site can give results as accurate as with a hundred or more bare phonon states per site. For instance, eigenenergies calculated with an optimal basis agree quantitatively with results obtained with a large bare basis using the pseudo-site DMRG [14]. Results for dynamical quantities, such as the optical

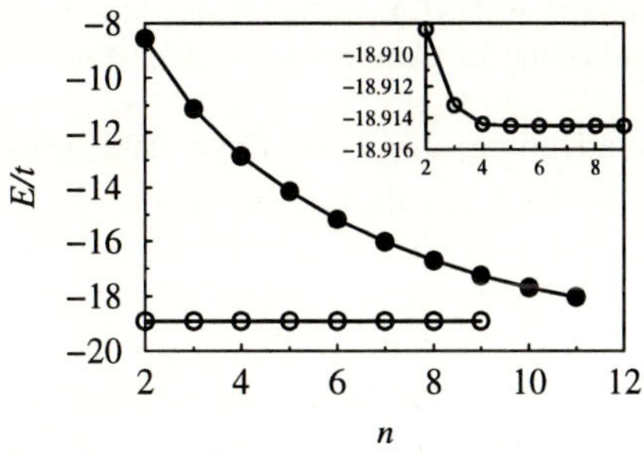

Fig. 4. Ground-state energy of a four-site half-filled Holstein system with $\omega = t$ and $\gamma/t = 1.5$ as a function of the number of bare (*filled circles*) and optimal (*open circles*) phonon states per site. The inset shows an expanded view of the results for an optimal basis

conductivity, also agree qualitatively with results obtained by conventional exact diagonalizations using powerful parallel computers [7], but the optimal phonon approach is so efficient that all calculations can easily be carried out on modest workstations [14].

3 Conclusion

The information contained in the density matrix allows one to truncate a large phononic Hilbert space without significant loss of accuracy. Thus, it is possible to study electron-phonon systems numerically with great accuracy. Coupled to a powerful numerical methods such as DMRG these approaches enable us to treat very large systems. These techniques could greatly improve our capability to perform numerical studies of many problems, such as electron-phonon systems, which involve a large or infinite Hilbert space.

References

1. S.R. White, Phys. Rev. Lett. **69**, 2863 (1992); Phys. Rev. B **48**, 10 345 (1993)
2. S.R. White, Phys. Rev. Lett. **77**, 3633 (1996)
3. G. Wellein and H. Fehske, Phys. Rev. B **58**, 6208 (1998)
4. E.V.L. de Mello and J. Ranninger, Phys. Rev. B **55**, 14 872 (1997)
5. H. Fehske, J. Loos, and G. Wellein, Z. Phys. B **104**, 619 (1997)
6. M. Capone, W. Stephan, and M. Grilli, Phys. Rev. B **56**, 4484 (1997)
7. A. Weisse and H. Fehske, Phys. Rev. B **58**, 13 526 (1998)
8. L.G. Caron and S. Moukouri, Phys. Rev. Lett. **76**, 4050 (1996)
9. L.G. Caron and S. Moukouri, Phys. Rev. B **56**, 8471 (1997)
10. E. Jeckelmann and S.R. White, Phys. Rev. B **57**, 6376 (1998)
11. C. Zhang, E. Jeckelmann, and S.R. White, Phys. Rev. Lett. **80**, 2661 (1998)
12. R.J. Bursill, R.H. McKenzie and C.J. Hamer, Phys. Rev. Lett. **80**, 5607 (1998)
13. T. Holstein, Ann. Phys. (N.Y.) **8**, 325 (1959); **8**, 343 (1959)
14. C. Zhang, E. Jeckelmann and S.R. White, in preparation
15. R.V. Pai, R. Pandit, H.R. Krishnamurthy and S. Ramasesha, Phys. Rev. Lett. **76**, 2937 (1996)
16. T.D. Kühner and H. Monien, Phys. Rev. B **58**, R14741 (1998)
17. R.M. Noack, S.R. White and D.J. Scalapino, in D.P. Landau, K.-K. Mon, and H.-B. Schüttler (Eds.), Computer Simulations in Condensed Matter Physics VII, Springer (1994)
18. A.H. Romero, D.W. Brown, and K. Lindenberg, J. Chem. Phys. **109**, 6540 (1998)

5.2 Disordered One-Dimensional Fermi Systems

Peter Schmitteckert

Waghäusler Str. 80
D-68753 Waghäusel, Germany

The interplay between disorder and interaction is one of the most challenging problems in todays solid-state physics. Although much is known for clean interacting systems and for disordered non-interacting systems, little is known about the combined effect of disorder and interaction. The DMRG method allows the study of systems much larger than accessible with exact diagonalization, and with an accuracy not achievable with other numerical techniques.

In one-dimensional systems electron-electron interactions strongly affect physical properties, e.g. the low-energy physics of a clean Fermi system changes from a Fermi to a Luttinger liquid. In addition, it is known for non-interacting electrons in one dimension that disorder always leads to localization [1,2].

In the following section we introduce the model of spinless interacting, disordered fermions in 1D. Then we discuss the application of the DMRG method for disordered systems. In Sect. 4 we present results for a single impurity. Finally, strongly disordered systems are discussed in Sect. 5.

In our implementation of the DMRG method we kept up to 2000 states per block and performed up to 15 finite-lattice sweeps.

1 The Model

In this work we discuss a model of interacting, disordered spinless fermions in one dimension described by the sum of the kinetic energy, a nearest-neighbour interaction and on-site disorder

$$\mathcal{H} = -t\sum_{i=1}^{M}\left(c_i^{+}c_{i+1} + c_{i+1}^{+}c_i\right) + V\sum_{i=1}^{M} n_i n_{i+1} + \sum_{i=1}^{M}\epsilon_i c_i^{+}c_i \,, \tag{1}$$

where M denotes the number of sites, $N = \sum_{i=1}^{M} n_i$ the number of fermions, and $n_i = c_i^{+}c_i$ the fermion density at site i. The ϵ_j are taken uniformly distributed in $[-W/2, W/2]$. The lattice constant is always set equal to 1, and energies are measured in units of t. The twisted boundary conditions are given by

$$c_1 = \mathrm{e}^{\imath\phi}\, c_{M+1} . \tag{2}$$

Having defined our model, we need a measure to distinguish localized and delocalized states. Since the strength of the DMRG method is the calculation of ground-state energies, we study the sensitivity of the ground state with respect to a change of the boundary conditions, i.e. we look at the energy difference ΔE between periodic and anti-periodic boundary conditions,

$$\Delta E = (-1)^N \left[E(0) - E(\pi)\right]. \tag{3}$$

The factor $(-1)^N$ compensates an odd-even effect. With this definition ΔE is always positive. Note that in one dimension $M\Delta E$ is proportional to the charge stiffness $D = E''(\phi=0)$, where the proportionality constant depends on the nature of the system, i.e. the band structure.

2 The Limiting Cases

Clean systems

For free fermions without disorder the ground-state energy is given by sum over the energy levels $E_k(\phi)$,

$$E_k(\phi) = E(k+\phi) = -2t \sum_{k \le k_F} \cos\left(2k+\phi\right), \tag{4}$$

where the energy levels are given by $k_\ell = 2\pi\ell/M$ for M odd and $k_\ell = (2\ell+1)\pi/M$ for M even. Therefore, applying a phase ϕ leads to a shift of the energy values only.

For an interaction in the range $-2 < V/t < 2$ the ground state is a Luttinger liquid. An interesting feature of this ground state is that equation (4) is still valid, only the distribution of the k values changes. At half filling a lengthy calculation within the Bethe ansatz [6] leads to

$$E_M - M\varepsilon_\infty = -\frac{\pi v_F}{6M}\left(1 - 3K\phi^2\right) \qquad -2 < V/t < 2 \tag{5}$$

where E_M denotes the ground-state energy of the M site system, and ε_∞ is the ground-state energy density in the $M \to \infty$ limit. The fermi velcity v_F is given by

$$v_F = t\pi \sin(2\eta)/(\pi - 2\eta) \tag{6}$$

where η measures the strength of the interaction via

$$V = -2t\cos(2\eta). \tag{7}$$

K is the the Luttinger liquid interaction parameter

$$K = \pi/4\eta. \tag{8}$$

The phase sensitivity is now given by

$$M\Delta E = \pi v_F K/2. \tag{9}$$

For $V = -2t$ at half filling there is an instability to a phase separated state, and for $V \ge 2t$ the ground state is a charge-density wave.

Non-interacting disordered system ($V = 0$)

It is a famous result by Anderson [2] that in one dimension even infinitely small disorder leads to localization in non-interacting systems. In the case of spinless fermions in one dimension the conductance g in the localized regime is given by

$$\langle g \rangle \sim \mathrm{e}^{-2L/\xi_0} , \tag{10}$$

where the localization length $\xi_0 \approx 105t^2/W^2$ was calculated numerically by Kappus and Wegner, and Czycholl, Kramer and MacKinnon [4].

Although the conductance is a transport property and the phase sensitivity calculated here is a ground-state property, there exists a connection between these quantities, see Kohn [8].

3 DMRG for Disordered Systems

The DMRG for disordered systems works analogous to the algorithm for reflection symmetric models, see [9]. First one uses the infinite-lattice algorithm to build up the initial blocks. The only modification is that instead of taking the reflection of block A as environment block one performs two density-matrix projections per iteration step. One to project out the important states of block A, and one for block B, see Fig. 1. Once the initial blocks

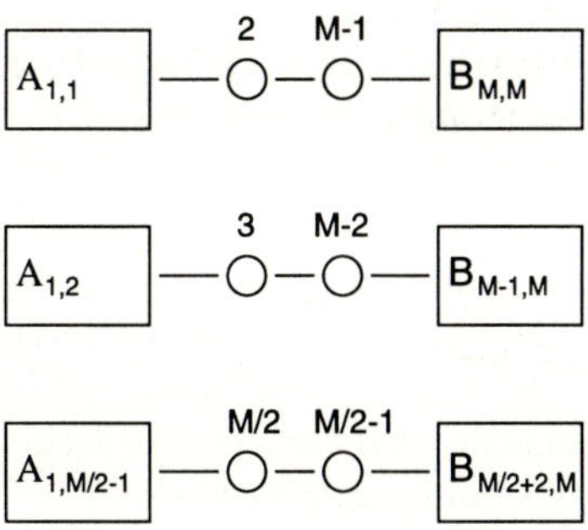

Fig. 1. Initial DMRG sweep for disordered systems. Since the system is not reflection symmetric one has to project on Block A and B separately.

are built one continues with the finite-system algorithm sweeping through the whole lattice. It is important to perform several sweeps, since the first sweeps give a poor result only. This is due to the fact that in the initial building of the blocks the system knows nothing about the disorder of the inner sites. Therefore the system cannot be well adapted to the disorder configuration. Since the first few iterations are inaccurate one does not have to keep a large number of states. In our calculation we typically performed 10 to 15 sweeps, increasing the number of states kept per block by a factor of 1.7 per sweep. Note that after performing the corresponding base changes, we use the ground state of a DMRG step as initial vector for the sparse matrix diagonalization of the next step. Consequently, a DMRG run for large systems can be faster with 15 sweeps than a run with a small number of sweeps, since the number of iterations in the sparse matrix diagonalization can be reduced to 2 or 3.

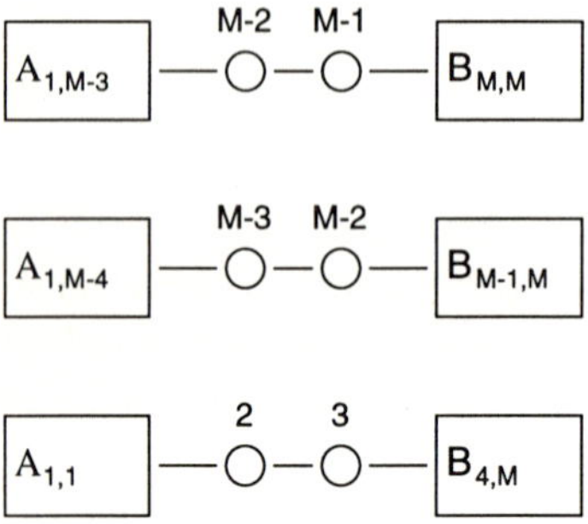

Fig. 2. In the finite-system algorithm one has to sweep through the whole lattice

A dangerous pitfall

In the case of very strong disorder or weak links the DMRG may converge to excited states if not enough states per building block are kept. Fig. 3 shows an example in which the (single target state) DMRG method converges to an excited state if the number of states per block is too small. In this example we have introduced weak links to pronounce the effect.

This is a general problem of the DMRG. One finds eigenstates with a very high precision, but there is no guarantee that one has really found the ground state. Although increasing the number m of states kept per block solves this problem, one does not know whether m is already high enough or not. To overcome this difficulty we performed the DMRG procedure simultaneously for the system with periodic boundary conditions, with anti-periodic boundary conditions and for the clean system. The clean system enforces the existence of states with a finite density of fermions at each site. Without this trick we would not have been able to attack the strong disorder case.

4 A Single Impurity and Weak Disorder

In this section we use the Kane and Fisher scaling to extend $V = 0$ results to the interacting case. Kane and Fisher [7] studied the influence of a single impurity on the conductance of a Luttinger liquid. They found that for attractive interaction an impurity is scaled away, while for repulsive interaction the strength of an impurity is scaled to ∞. We will demonstrate that one can apply their results to a simple derivation of the phase sensitivity as long as one can treat the disorder as a perturbation to the Luttinger liquid state.

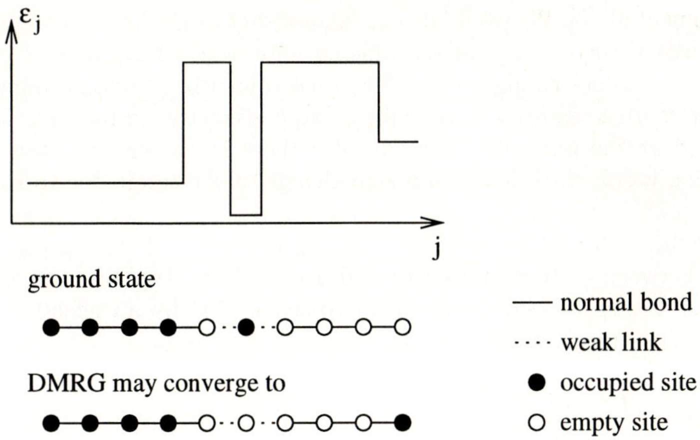

Fig. 3. This example of a system with weak links between site 5,6, and 7 shows a general problem of the DMRG method. Suppose that we start the DMRG iteration from the left and right end of the chain: we may converge to an excited state instead of the ground state if the number of states we keep is too small.

A single weak impurity

For a single weak impurity, $|\epsilon| \ll t$, in a non-interacting system a simple calculation using first-order pertubation theory leads to

$$M\Delta E = \pi t - |\epsilon_0|. \tag{11}$$

We can include interaction by simply replacing the impurity in equation (11) by an effective impurity

$$\epsilon_{\text{eff.}} = \epsilon_0 \left(\frac{M}{M_0}\right)^{1-K} . \tag{12}$$

This leads to the phase sensitivity of an interacting system with a single weak impurity

$$M\Delta E = \pi t - |\epsilon_0| \left(\frac{M}{M_0}\right)^{1-K} . \tag{13}$$

In these expressions M_0 is a short distance cut off, numerically ≈ 2.

A single strong impurity

A strong impurity in a non-interactiong system can be treated with a transfer-matrix calculation,

$$M\Delta E = \frac{4t^2}{|\epsilon_0|}, \tag{14}$$

see Cheung et al. [3]. We would like to remark that in [3] the absolute value of the impurity strength is not taken, which is important for averaging the result for a given disorder configuration. The reason for the additional minus sign for an attractive impurity is that for a large attractive impurity a fermion is trapped at the impurity. Consequently there is one electron less in the conduction band which leads to a sign change of the result due to the odd-even effects.

To include the interaction for a strong impurity we can use a duality relation between a strong impurity and a weak link. Applying the result of Kane and Fisher, we substitute ϵ_0 in equation (14) by an effective strong impurity. As a result we obtain

$$M\Delta E = \frac{4t^2}{|\epsilon_0|}\left(\frac{M}{M_0}\right)^{1-1/K} . \tag{15}$$

Comparison to DMRG results

Figure 4 shows a good agreement between the DMRG and the analytical results for a 60 site system. Deviations from the analytical result are found when a strong impurity becomes so weak that a first-order description is no longer valid. In [10] one can find more results demonstrating that the simple

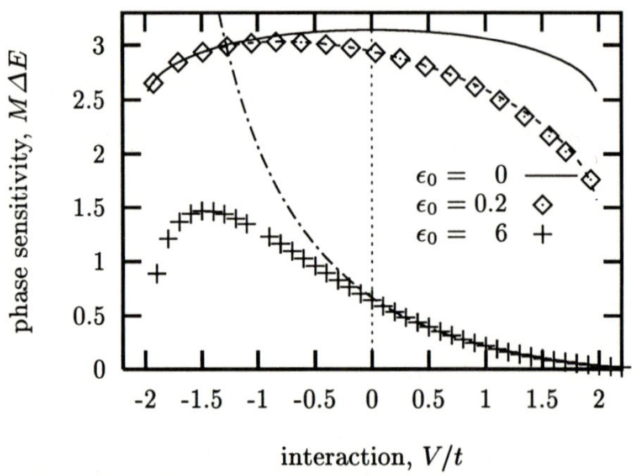

Fig. 4. Comparison of DMRG results for the phase sensitivity with analytical results. The full line corresponds to the Bethe-ansatz solution (9) of the clean system and the dash-dotted lines are given by (13) and (15).

calculations provided in this section give the correct results. Summarizing the results we found that an attractive interaction makes a barrier more transparent, while replusive interaction increases the defect strength.

Weak random disorder

It is a remarkable result that the calculation of the preceeding sections can be extended to the case of weak random disorder. Again we start with a first-order perturbation calculation to obtain the phase sensitivity of a non-interacting Fermi system:

$$M\Delta E = \pi t - |\sum_{n=1}^{M} \epsilon_n \, e^{2\imath n k_F} |, \tag{16}$$

where we introduce disorder by taking the $\{\epsilon_n\}$ uniformly distributed over the interval $[-W/2, W/2]$. Applying the Kane and Fisher scaling to equation (16) and recalling that a random walk leads to a Gaussian distribution of distances we obtain for the average phase sensitivity and the fluctuations

$$M\Delta E = \tfrac{\pi v_F K}{2} - |\textstyle\sum_{n=1}^{M} \epsilon_n \, e^{2\imath k_F n} | \left(\tfrac{M}{M_0}\right)^{1-K} \tag{17}$$

$$\langle M\Delta E\rangle = \tfrac{\pi v_F K}{2} - \tfrac{W\sqrt{M_0}}{\sqrt{6\pi}} \left(\tfrac{M}{M_0}\right)^{(3-2K)/2} \tag{18}$$

$$\sigma^2_{M\Delta E} = \tfrac{W^2 M_0}{12} \left(1 - \tfrac{2}{\pi}\right) \left(\tfrac{M}{M_0}\right)^{3-2K}, \tag{19}$$

where the brakets $\langle\rangle$ denote the impurity average. Details can be found in [10].

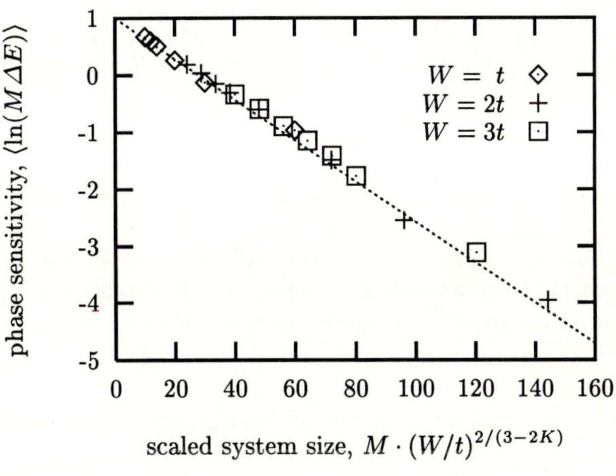

Fig. 5. Average logarithmic phase sensitivity as a function of the scaled system size, for interaction $V/t = 1.2$ and disorder $W/t = 1, 2, 3$.

In Fig. 5 we plot DMRG results for $V/t = 1.2$ and $W/t = 1, 2, 3$ of the phase sensitivity averaged on a logathmic scale versus the scaled system size. In the case of largest systems considered ($M = 60$) we averaged over several hundred realizations, wheras for the smaller systems ($M \leq 20$) we used ensembles of more than 1000 configurations. As one can see there is a nice collapse of the data onto a straight line, even for $M \gg \xi$, where perturbation theory breaks down. The average phase sensitivity, shown in Fig. 5, is for large systems approximately given by

$$\langle \ln(M\Delta E)\rangle = -M/\xi + 1, \tag{20}$$

with the localization length $\xi \approx 28W^{-2/(3-2K)}$.

The width of the distribution of $\ln(M\Delta E)$ shown in Fig. 6 is for small systems proportional to $M^{(3-2K)/2)}$. Note that $\sigma^2_{\ln(M\Delta E)}$ and $\sigma^2_{M\Delta E}$ are directly related to each other, provided $\sigma_{M\Delta E} \ll \langle M\Delta E\rangle$. A crossover is apparent for $M \approx \xi$, i.e. when fluctuations of $M\Delta E$ are comparable to its average. For large systems we find the fluctuations to be proportional to $M^{(2/3)}$, independent of the interaction strength.

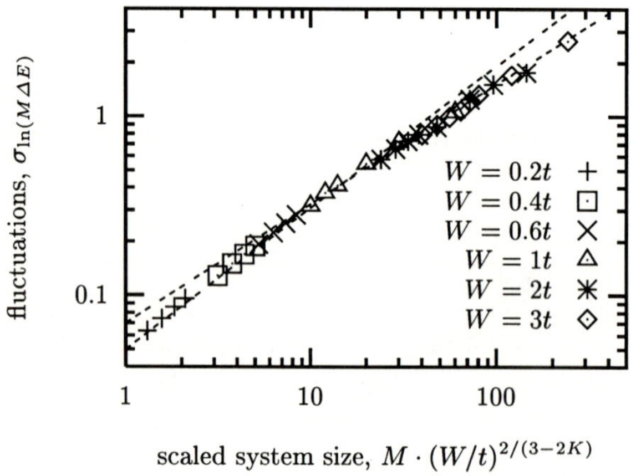

Fig. 6. Fluctuations of $\ln(M\Delta E)$ as a funtion of scaled system size, $V = 1.2$. The full line is the analytic result according to equation (19). For large systems, $M > \xi$, $\sigma_{\ln(M\Delta E)}$ is proportional to $M^{2/3}$ independent of the interaction.

However, the most interesting result of this calculation is that for K $> 3/2$, i.e. $V < -1$, the strength of the defects vanishes so fast that disorder becomes an irrelevant perturbation: there is no localization. Although this transition was predicted by Giamarchi and Schulz [5] using renormalization-group calculations, it was not clear whether the result holds for finite values of the disorder or not. It is interesting that the delocalized phase exists for disorder strength comparable to the band width of the kinetic energy, as one can see in the phase diagram in Fig. 7.

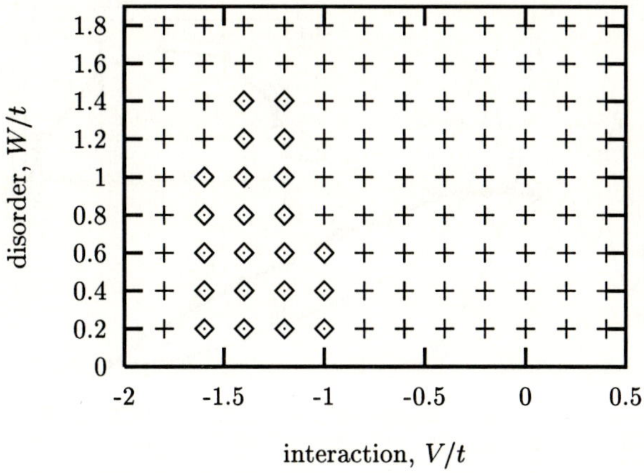

Fig. 7. Phase diagram. The symbol $\diamond$ (+) denotes the region where the variance of the logarithmic phase sensitivity decreases (increases) as a function of the system size. We considered up to 50 sites. The $\diamond$ region corresponds to a delocalized ground state.

5 Strong Disorder

Finally, we present DMRG results for the case of strong disorder. It is clear that we cannot apply our scaling arguments, since we are not expanding around a Luttinger liquid. Looking at the model of 3 sites with $\epsilon_1 = \epsilon_2 = -\epsilon_3 = \epsilon$ one finds an enhancement of the phase senitivity for repulsive interaction, $\Delta E(V = 2\epsilon) = 1/\epsilon$ compared to $\Delta E(V = 0) = 1/\epsilon^2$ and $\Delta E(V \gg \epsilon) = 4/V^2$. Although this example is instructive, the analytical treatment of strong disorder for an interacting system is a difficult task. As we will see below there is a major difficulty: a strong sample dependence of the results.

In Fig. 8 we present DMRG results for the phase density for systems with $W = 9$, $M = 20$, and $N = 10$. The curves (a), (b), (c), and (d) correspond to four different samples. Like these four sample all samples show peaks of enhancement for repulsive interaction. This enhancement can be more than four orders of magnitude compared to the non-interacting value for the same sample. By contrast, the averaged line, where we have averaged over 5000 samples, is a smooth curve, with a small enhancement for small repulsive interaction. Since there are no peaks in the averaged curve we can assume that the peaks of the individual samples are randomly distributed.

In Fig. 9 we plot the local fermion density $n_i(V)$ for sampe (d) of Fig. 8. Looking at the Fourier transformed density $n_k(V)$ in Fig. 10 we find an increase of the $2k_F$ component with V. Comparing these plots we find that at the value of V where the enhancements take place a reorganization occurs. Therefore one can call this phenomenon "delocalization of reorganization".

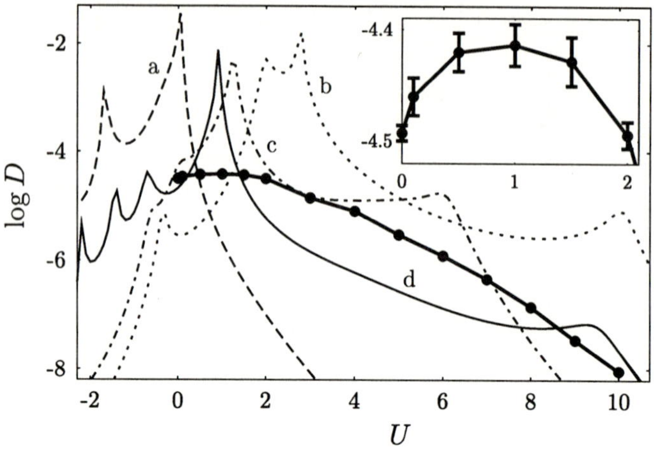

Fig. 8. The logarithmic phase sensitivity $\log D = \log(M\Delta E)$ for four different samples with $N = 10$, $M = 20$, $W = 9$, and $U = V/t$. Thick line: Average $\log(D)$ over 5000 samples.

In [11] we show that this feature is also present in samples away from half filling, but not in all samples.

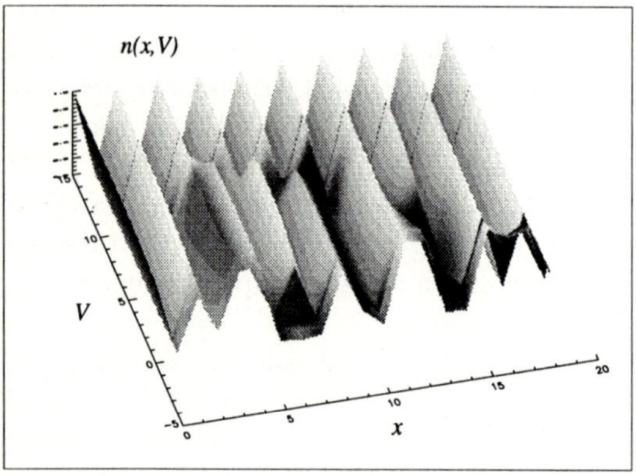

Fig. 9. Local fermion density $n(x, V)$ for a typical sample (d of Fig. 8) for $N = 10$ particles on $M = 20$ sites at $W = 9t$.

6 Summary

We have shown that the DMRG allows the study of interacting, disordered one-dimensional Fermi systems, but: care has to be taken. The scaling arguments provide a test for a DMRG implementation. In addition we have shown that the results of Giamarchi and Schulz can be obtained with Kane and Fisher scaling, and that the scaling results are valid beyond the perturbative regime. In the regime of strong disorder we find a "delocalization

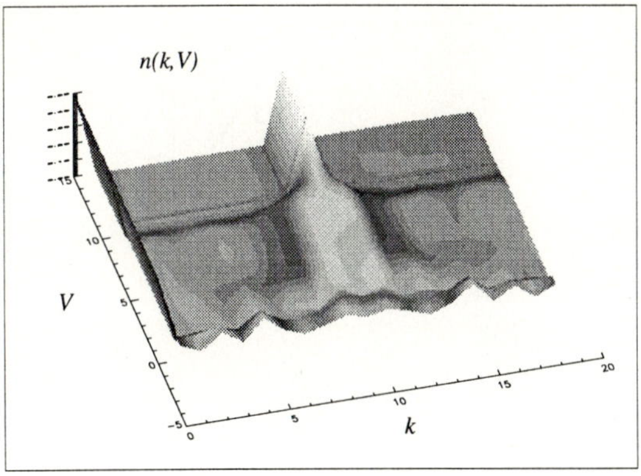

Fig. 10. Fourier transform of the fermion density $n(k, V)$ for a typical sample (d of Fig. 8) for $N = 10$ particles on $M = 20$ sites at $W = 9t$.

of reorganization" effect for repulsive interaction which has a strong sample dependence.

We would like to acknowledge the cooperation with Ulrich Eckern, Peter Schwab, Cosima Schuster, Thomas Schulze, Rodolfo Jalabert, Jean-Louis Pichard, Nic Shannon, and Volker Meden.

References

1. A.A. Abrikosov, Fundamentals of the theory of metals, Elsevier (1988)
2. P.W. Anderson, E. Abrahams and T.V. Ramakrishnan, Phys. Rev. **109** 1492 (1958)
3. H.F. Cheung, Y. Gefen, E.K. Riedel and W.H. Shih, Phys. Rev. B **37**, 6050 (1988); H.F. Cheung, E.K. Riedel and Y. Gefen, Phys. Rev. Lett. **62**, 587 (1989)
4. G. Czycholl, B. Kramer and A. MacKinnon, Z. Phys. B **43**, 5 (1981); M. Kappus and F.J. Wegner, Z. Phys. B **45**, 15 (1981)
5. T. Giamarchi and H.J. Schulz, Phys. Rev. B **37**, 325 (1988)
6. C.J. Hamer, G.R.W. Quispel and M.T. Batchelor, J. Phys. A Gen. **20**, 5677 (1987); F. Woynarovich and P. Eckle, J. Phys. A: Math. Gen. **20**, L97 (1987)
7. C.L. Kane and M.P.A. Fisher, Phys. Rev. Lett. **68**, 1220 (1992)
8. W. Kohn, Phys. Rev. **133**, 171 (1964)
9. P. Schmitteckert, Interplay between interaction and disorder in one-dimensional fermi systems, PhD Thesis, Universität Augsburg (1996)
10. P. Schmitteckert, T. Schulze, C. Cosima, P. Schwab and U. Eckern, Phys. Rev. Lett. **80**, 560 (1998); P. Schmitteckert and U.Eckern, Phys. Rev. B **53**, 15397 (1996)
11. P. Schmitteckert, R. Jalabert, D. Weinmann and J.L. Pichard, Phys. Rev. Lett. **81**, 2308 (1998)

Springer and the environment

Lecture Notes in Physics

For information about Vols. 1–488 please contact your bookseller or Springer-Verlag

Vol. 489: G. M. Simnett, C. E. Alissandrakis, L. Vlahos (Eds.), Solar and Heliospheric Plasma Physics. Proceedings, 1996. VIII, 278 pages. 1997.

Vol. 490: P. Kutler, J. Flores, J.-J. Chattot (Eds.), Fifteenth International Conference on Numerical Methods in Fluid Dynamics. Proceedings, 1996. XIV, 655 pages. 1997.

Vol. 491: O. Boratav, A. Eden, A. Erzan (Eds.), Turbulence Modeling and Vortex Dynamics. Proceedings, 1996. XII, 245 pages. 1997.

Vol. 492: M. Rubí, C. Pérez-Vicente (Eds.), Complex Behaviour of Glassy Systems. Proceedings, 1996. IX, 467 pages. 1997.

Vol. 493: P. L. Garrido, J. Marro (Eds.), Fourth Granada Lectures in Computational Physics. XIV, 316 pages. 1997.

Vol. 494: J. W. Clark, M. L. Ristig (Eds.), Theory of Spin Lattices and Lattice Gauge Models. Proceedings, 1996. XI, 194 pages. 1997.

Vol. 495: Y. Kosmann-Schwarzbach, B. Grammaticos, K.M. Tamizhmani (Eds.), Integrability of Nonlinear Systems. VII, 380 pages. 1997.

Vol. 496: F. Lenz, H. Grießhammer, D. Stoll (Eds.), Lectures on QCD. VII, 483 pages. 1997.

Vol. 497: J. P. Greve, R. Blomme, H. Hensberge (Eds.), Stellar Atmospheres: Theory and Observations. VIII, 352 pages. 1997

Vol. 498: Z. Horváth, L. Palla (Eds.), Conformal Field Theories and Integrable Models. Proceedings, 1996. X, 251 pages. 1997.

Vol. 499: K. Jungmann, J. Kowalski, I. Reinhard, F. Träger (Eds.), Atomic Physics Methods in Modern Research, IX, 448 pages. 1997.

Vol. 500: D. Joubert (Ed.), Density Functionals: Theory and Applications, XVI, 194 pages. 1998.

Vol. 501: J. Kertész, I. Kondor (Eds.), Advances in Computer Simulation. VIII, 166 pages. 1998.

Vol. 502: H. Aratyn, T. D. Imbo, W.-Y. Keung, U. Sukhatme (Eds.), Supersymmetry and Integrable Models. Proceedings, 1997. XI, 379 pages. 1998.

Vol. 503: J. Parisi, S. C. Müller, W. Zimmermann (Eds.), A Perspective Look at Nonlinear Media. From Physics to Biology and Social Sciences. VIII, 372 pages. 1998.

Vol. 504: A. Bohm, H.-D. Doebner, P. Kielanowski (Eds.), Irreversibility and Causality. Semigroups and Rigged Hilbert Spaces. XIX, 385 pages. 1998.

Vol. 505: D. Benest, C. Froeschlé (Eds.), Impacts on Earth. XVII, 223 pages. 1998.

Vol. 506: D. Breitschwerdt, M. J. Freyberg, J. Trümper (Eds.), The Local Bubble and Beyond. Proceedings, 1997. XXVIII, 603 pages. 1998.

Vol. 507: J. C. Vial, K. Bocchialini, P. Boumier (Eds.), Space Solar Physics. Proceedings, 1997. XIII, 296 pages. 1998.

Vol. 508: H. Meyer-Ortmanns, A. Klümper (Eds.), Field Theoretical Tools for Polymer and Particle Physics. XVI, 258 pages. 1998.

Vol. 509: J. Wess, V. P. Akulov (Eds.), Supersymmetry and Quantum Field Theory. Proceedings, 1997. XV, 405 pages. 1998.

Vol. 510: J. Navarro, A. Polls (Eds.), Microscopic Quantum Many-Body Theories and Their Applications. Proceedings, 1997. XIII, 379 pages. 1998.

Vol. 511: S. Benkadda, G. M. Zaslavsky (Eds.), Chaos, Kinetics and Nonlinear Dynamics in Fluids and Plasmas. Proceedings, 1997. VIII, 438 pages. 1998.

Vol. 512: H. Gausterer, C. Lang (Eds.), Computing Particle Properties. Proceedings, 1997. VII, 335 pages. 1998.

Vol. 513: A. Bernstein, D. Drechsel, T. Walcher (Eds.), Chiral Dynamics: Theory and Experiment. Proceedings, 1997. IX, 394 pages. 1998.

Vol. 514: F. W. Hehl, C. Kiefer, R. J. K. Metzler, Black Holes: Theory and Observation. Proceedings, 1997. XV, 519 pages. 1998.

Vol. 515: C.-H. Bruneau (Ed.), Sixteenth International Conference on Numerical Methods in Fluid Dynamics. Proceedings. XV, 568 pages. 1998.

Vol. 516: J. Cleymans, H. B. Geyer, F. G. Scholtz (Eds.), Hadrons in Dense Matter and Hadrosynthesis. Proceedings, 1998. XII, 253 pages. 1999.

Vol. 517: Ph. Blanchard, A. Jadczyk (Eds.), Quantum Future. Proceedings, 1997. X, 244 pages. 1999.

Vol. 518: P. G. L. Leach, S. E. Bouquet, J.-L. Rouet, E. Fijalkow (Eds.), Dynamical Systems, Plasmas and Gravitation. Proceedings, 1997. XII, 397 pages. 1999.

Vol. 519: R. Kutner, A. Pękalski, K. Sznajd-Weron (Eds.), Anomalous Diffusion. From Basics to Applications. Proceedings, 1998. XVIII, 378 pages. 1999.

Vol. 520: J. A. van Paradijs, J. A. M. Bleeker (Eds.), X-Ray Spectroscopy in Astrophysics. EADN School X. Proceedings, 1997. XV, 530 pages. 1999.

Vol. 521: L. Mathelitsch, W. Plessas (Eds.), Broken Symmetries. Proceedings, 1998. VII, 299 pages. 1999.

Vol. 522: J. W. Clark, T. Lindenau, M. L. Ristig (Eds.), Scientific Applications of Neural Nets. Proceedings, 1998. XIII, 288 pages. 1999.

Vol. 523: B. Wolf, O. Stahl, A. W. Fullerton (Eds.), Variable and Non-spherical Stellar Winds in Luminous Hot Stars. Proceedings, 1998. XX, 424 pages. 1999.

Vol. 524: J. Wess, E. A. Ivanov (Eds.), Supersymmetries and Quantum Symmetries. Proceedings, 1997. XX, 442 pages. 1999.

Vol. 525: A. Ceresole, C. Kounnas, D. Lüst, S. Theisen (Eds.), Quantum Aspects of Gauge Theories, Supersymmetry and Unification. Proceedings, 1998. X, 511 pages. 1999.

Vol. 526: H.-P. Breuer, F. Petruccione (Eds.), Open Systems and Measurement in Relativistic Quantum Theory. Proceedings, 1998. VIII, 240 pages. 1999.

Vol. 528: I. Peschel, X. Wang, M. Kaulke, K. Hallberg (Eds.), Density-Matrix Renormalization. Proceedings, 1998. XVI, 355 pages. 1999.

Monographs
For information about Vols. 1–11
please contact your bookseller or Springer-Verlag

Vol. m 12: G. Esposito, Quantum Gravity, Quantum Cosmology and Lorentzian Geometries. Second Corrected and Enlarged Edition. XVIII, 349 pages. 1994.

Vol. m 13: M. Klein, A. Knauf, Classical Planar Scattering by Coulombic Potentials. V, 142 pages. 1992.

Vol. m 14: A. Lerda, Anyons. XI, 138 pages. 1992.

Vol. m 15: N. Peters, B. Rogg (Eds.), Reduced Kinetic Mechanisms for Applications in Combustion Systems. X, 360 pages. 1993.

Vol. m 16: P. Christe, M. Henkel, Introduction to Conformal Invariance and Its Applications to Critical Phenomena. XV, 260 pages. 1993.

Vol. m 17: M. Schoen, Computer Simulation of Condensed Phases in Complex Geometries. X, 136 pages. 1993.

Vol. m 18: H. Carmichael, An Open Systems Approach to Quantum Optics. X, 179 pages. 1993.

Vol. m 19: S. D. Bogan, M. K. Hinders, Interface Effects in Elastic Wave Scattering. XII, 182 pages. 1994.

Vol. m 20: E. Abdalla, M. C. B. Abdalla, D. Dalmazi, A. Zadra, 2D-Gravity in Non-Critical Strings. IX, 319 pages. 1994.

Vol. m 21: G. P. Berman, E. N. Bulgakov, D. D. Holm, Crossover-Time in Quantum Boson and Spin Systems. XI, 268 pages. 1994.

Vol. m 22: M.-O. Hongler, Chaotic and Stochastic Behaviour in Automatic Production Lines. V, 85 pages. 1994.

Vol. m 23: V. S. Viswanath, G. Müller, The Recursion Method. X, 259 pages. 1994.

Vol. m 24: A. Ern, V. Giovangigli, Multicomponent Transport Algorithms. XIV, 427 pages. 1994.

Vol. m 25: A. V. Bogdanov, G. V. Dubrovskiy, M. P. Krutikov, D. V. Kulginov, V. M. Strelchenya, Interaction of Gases with Surfaces. XIV, 132 pages. 1995.

Vol. m 26: M. Dineykhan, G. V. Efimov, G. Ganbold, S. N. Nedelko, Oscillator Representation in Quantum Physics. IX, 279 pages. 1995.

Vol. m 27: J. T. Ottesen, Infinite Dimensional Groups and Algebras in Quantum Physics. IX, 218 pages. 1995.

Vol. m 28: O. Piguet, S. P. Sorella, Algebraic Renormalization. IX, 134 pages. 1995.

Vol. m 29: C. Bendjaballah, Introduction to Photon Communication. VII, 193 pages. 1995.

Vol. m 30: A. J. Greer, W. J. Kossler, Low Magnetic Fields in Anisotropic Superconductors. VII, 161 pages. 1995.

Vol. m 31 (Corr. Second Printing): P. Busch, M. Grabowski, P.J. Lahti, Operational Quantum Physics. XII, 230 pages. 1997.

Vol. m 32: L. de Broglie, Diverses questions de mécanique et de thermodynamique classiques et relativistes. XII, 198 pages. 1995.

Vol. m 33: R. Alkofer, H. Reinhardt, Chiral Quark Dynamics. VIII, 115 pages. 1995.

Vol. m 34: R. Jost, Das Märchen vom Elfenbeinernen Turm. VIII, 286 pages. 1995.

Vol. m 35: E. Elizalde, Ten Physical Applications of Spectral Zeta Functions. XIV, 224 pages. 1995.

Vol. m 36: G. Dunne, Self-Dual Chern-Simons Theories. X, 217 pages. 1995.

Vol. m 37: S. Childress, A.D. Gilbert, Stretch, Twist, Fold: The Fast Dynamo. XI, 406 pages. 1995.

Vol. m 38: J. González, M. A. Martín-Delgado, G. Sierra, A. H. Vozmediano, Quantum Electron Liquids and High-Tc Superconductivity. X, 299 pages. 1995.

Vol. m 39: L. Pittner, Algebraic Foundations of Non-Com-mutative Differential Geometry and Quantum Groups. XII, 469 pages. 1996.

Vol. m 40: H.-J. Borchers, Translation Group and Particle Representations in Quantum Field Theory. VII, 131 pages. 1996.

Vol. m 41: B. K. Chakrabarti, A. Dutta, P. Sen, Quantum Ising Phases and Transitions in Transverse Ising Models. X, 204 pages. 1996.

Vol. m 42: P. Bouwknegt, J. McCarthy, K. Pilch, The W3 Algebra. Modules, Semi-infinite Cohomology and BV Algebras. XI, 204 pages. 1996.

Vol. m 43: M. Schottenloher, A Mathematical Introduction to Conformal Field Theory. VIII, 142 pages. 1997.

Vol. m 44: A. Bach, Indistinguishable Classical Particles. VIII, 157 pages. 1997.

Vol. m 45: M. Ferrari, V. T. Granik, A. Imam, J. C. Nadeau (Eds.), Advances in Doublet Mechanics. XVI, 214 pages. 1997.

Vol. m 46: M. Camenzind, Les noyaux actifs de galaxies. XVIII, 218 pages. 1997.

Vol. m 47: L. M. Zubov, Nonlinear Theory of Dislocations and Disclinations in Elastic Body. VI, 205 pages. 1997.

Vol. m 48: P. Kopietz, Bosonization of Interacting Fermions in Arbitrary Dimensions. XII, 259 pages. 1997.

Vol. m 49: M. Zak, J. B. Zbilut, R. E. Meyers, From Instability to Intelligence. Complexity and Predictability in Nonlinear Dynamics. XIV, 552 pages. 1997.

Vol. m 50: J. Ambjørn, M. Carfora, A. Marzuoli, The Geometry of Dynamical Triangulations. VI, 197 pages. 1997.

Vol. m 51: G. Landi, An Introduction to Noncommutative Spaces and Their Geometries. XI, 200 pages. 1997.

Vol. m 52: M. Hénon, Generating Families in the Restricted Three-Body Problem. XI, 278 pages. 1997.

Vol. m 53: M. Gad-el-Hak, A. Pollard, J.-P. Bonnet (Eds.), Flow Control. Fundamentals and Practices. XII, 527 pages. 1998.

Vol. m 54: Y. Suzuki, K. Varga, Stochastic Variational Approach to Quantum-Mechanical Few-Body Problems. XIV, 324 pages. 1998.

Vol. m 55: F. Busse, S. C. Müller, Evolution of Spontaneous Structures in Dissipative Continuous Systems. X, 559 pages. 1998.

Vol. m 56: R. Haussmann, Self-consistent Quantum Field Theory and Bosonization for Strongly Correlated Electron Systems. VIII, 173 pages. 1999.